AF699174

WITHDRAWN

CHANGE AND PROGRESS IN MODERN SCIENCE

THE UNIVERSITY OF WESTERN ONTARIO
SERIES IN PHILOSOPHY OF SCIENCE

A SERIES OF BOOKS
IN PHILOSOPHY OF SCIENCE, METHODOLOGY,
EPISTEMOLOGY, LOGIC, HISTORY OF SCIENCE,
AND RELATED FIELDS

Managing Editor
ROBERT E. BUTTS
Dept. of Philosophy, University of Western Ontario, Canada

Editorial Board
JEFFREY BUB, *University of Western Ontario*
L. JONATHAN COHEN, *Queen's College, Oxford*
WILLIAM DEMOPOULOS, *University of Western Ontario*
WILLIAM HARPER, *University of Western Ontario*
JAAKKO HINTIKKA, *Florida State University, Tallahassee*
CLIFFORD A. HOOKER, *University of Newcastle*
HENRY E. KYBURG, JR., *University of Rochester*
AUSONIO MARRAS, *University of Western Ontario*
JURGEN MITTELSTRASS, *Universität Konstanz*
JOHN M. NICHOLAS, *University of Western Ontario*
GLENN A. PEARCE, *University of Western Ontario*
BAS C. VAN FRAASSEN, *Princeton University*

VOLUME 27

CHANGE AND PROGRESS IN MODERN SCIENCE

Papers related to and arising from the Fourth International Conference on History and Philosophy of Science, Blacksburg, Virginia, November 1982

Edited by

JOSEPH C. PITT

Department of Philosophy,
Virginia Polytechnic Institute and State University,
Blacksburg, Virginia, U.S.A.

D. REIDEL PUBLISHING COMPANY

A MEMBER OF THE KLUWER ACADEMIC PUBLISHERS GROUP

DORDRECHT / BOSTON / LANCASTER

Q
124.6
.I64
1982

Library of Congress Cataloging in Publication Data

Main entry under title:

CIP

Change and progress in modern science.

(The University of Western Ontario series in philosophy of science; v. 27)
"Sponsored by the International Union of History and Philosophy of Science and Virginia Polytechnic Institute and State University (Virginia Tech)"–Pref.
Includes index.
1. Science–History–Congresses. 2. Science–Philosophy–Congresses. I. Pitt, Joseph C. II. International Conference on the History and Philosophy of Science (4th: 1982: Blacksburg, Virginia) III. International Union of the History and Philosophy of Science. IV. Virginia Polytechnic Institute and State University. V. Series.
Q124.6.C48 1984 509 84-24795
ISBN 90-277-1898-9

Published by D. Reidel Publishing Company,
P.O. Box 17, 3300 AA Dordrecht, Holland.

Sold and distributed in the U.S.A. and Canada
by Kluwer Academic Publishers,
190 Old Derby Street, Hingham, MA 02043, U.S.A.

In all other countries, sold and distributed
by Kluwer Academic Publishers Group,
P.O. Box 322, 3300 AH Dordrecht, Holland.

All Rights Reserved
© 1985 by D. Reidel Publishing Company, Dordrecht, Holland
No part of the material protected by this copyright notice may be reproduced or utilized in any form or by any means, electronic or mechanical including photocopying, recording or by any information storage and retrieval system, without written permission from the copyright owner

Printed in The Netherlands

TABLE OF CONTENTS

PART II. CONCEPTUAL CONSIDERATIONS

PREFACE

The papers presented here derive from the 4th International Conference on History and Philosophy of Science held in Blacksburg, Virginia, U.S.A., November 2-6, 1982. The Conference was sponsored by the International Union of History and Philosophy of Science and Virginia Polytechnic Institute and State University (Virginia Tech). Particular thanks go to L. Jonathan Cohen, Secretary of the Union, as well as to Dean Henry Bauer of the College of Arts & Sciences, Wilfred Jewkes and the Center for Programs in the Humanities, Arthur Donovan and the Center for the Study of Science in Society and the Department of Philosophy and Religion at Virginia Tech. Not only did they come through with the necessary funds, but they were all always ready with a helping hand when things got confusing. Two additional groups of individuals require a special note of thanks. First, considerable appreciation is due the members of the Joint Commission of the International Union of History and Philosophy of Science: Maurice Crosland, Risto Hilpinen and Vladimir Kirsanov. They were more than generous in their advice and co-operation. The Local Organizing Committee (Kenneth Alpern, Roger Ariew, Arthur Donovan, Larry Laudan, Ann La Berge, Duncan Porter, Eleonore Stump and Dennis Welch) not only demonstrated efficiency and insured a pleasant stay for all participants, but also went out of their way on numerous occasions to make everyone feel at home. The success of the meetings is due in large part to their willingness to extend the open hand of hospitality and friendship.

It is unfortunate that all the papers at the conference are not included here. But we all understand about prior commitments, over-commitments and the omni-present unexpected. The authors of the papers here deserve a very special "thank-you" for being so incredibly prompt and cooperative. The delay in the appearance of these proceedings is not their fault by any means. The modern age has extended its reach into the dusty halls of academe and the outcome of the battle between the Virginia Tèch Computer and this editor is not yet clear! The delays, such as they are, are mine.

Some final notes of thanks need to be expressed. In the organization of the conference Tracy Shelor, Sharon

Fenton, and Suzie Van Krey were indispensible. Without their assistance, I would have floundered considerably more often. In the preparation of the volume, Roger Ariew acted unstintingly as Consulting Computer Engineer and Becky Cox demonstrated the most incredible sense of patience while most efficiently and cheerfully letting me change my mind over and over again. For final details and for making it all "look right" thanks go to Harlan Miller and Beth Temple. For generously supplying the oysters we all wish to thank Miss Mary Jane Ellis. And finally a general very special thanks to Donna Pitt.

But in the end it is the Conference participants and contributors to the volume who made it possible to undertake and complete the project and whose whole-hearted cooperation and enthusiasm made this effort as exciting as one as could possibly desire.

Newport, Virginia
May 1984

Joseph C. Pitt

INTRODUCTION

JOSEPH C. PITT

REFLECTIONS ON CHANGE

Change is the theme of this volume, the many faces of change that make science the social and intellectual force it is and the topic of so much critical scrutiny. As the papers here show, there are many kinds of changes that can be addressed when one discusses almost any aspect of science. But before detailing some of these, yet another kind of change needs to be addressed, i.e. the transition that marks the step from adolescence to maturity; in this case it is the maturation of a field of scholarly activity we pause to acknowledge. The time has come to concede that History and Philosophy of Science is a discipline and to desist from the perpetual *apologia* that accompanies work in the field. (It goes without saying then that these comments are not themselves to be construed as an *apologia,* they are rather an objection to such.) This is not to say that History of Science and Philosophy of Science taken separately are no longer viable. It is, rather, to acknowledge that with respect to the interdisciplinary area of History and Philosophy of Science (1) there is an established body of literature, (2) there is a large and interactive group of scholars who approach their work with a perspective that differs in important ways from their ancestoral disciplines, (3) there are departments offering degrees in the subject, and (4) there is a growing number of scholars, publications and activity in the area. And so, it is time to get about our business.

Since introductions ought to be short, the papers introduced being the main issue, I will be brief. These papers present many different aspects of change in science. Some are concerned with global problems of justifying choices between theories or hypotheses and progress in science, others with changes in the social acceptance of scientific practices, still others with the growth and development of individuals. Taken as a whole they allow us to chart some important changes from Descartes and Leibniz to contemporaries such as Piaget. There also emerges a rather spirited appraisal of the character of science that

J. C. Pitt (ed.), Change and Progress in Modern Science, 3–5.
© 1985 by D. Reidel Publishing Company.

helps place the context for change. The picture that develops is of science as a social activity influencing and influenced by other social activities as well as rational considerations, whose acceptable practices are undergoing constant revision, whose criteria are in flux and whose practitioners exhibit both brilliance and mundacity.

But lest it appear that too much is being claimed here by way of organizational structure for this collection, it ought to be observed that the implied order in these materials is primarily the result of the editor's reflection. What order there is is not merely due, however, to some idiosyncratic perception; there is a *prima facie* appealing cut between Historical Dimensions (Part I) and Conceptual Considerations (Part II). But it ought to be clear from the start that this categorization is not hard and fast. All of the papers found in Part I are concerned, at least in part, with some major conceptual issue. Likewise the papers in Part II lean heavily on historical considerations both to illuminate and to punctuate the problems being discussed.

It is also worth pointing out that there are at least two different sets of questions addressed by our authors. First, there is the problem of determining what it is that changes. Many serious discussions of scientific change are misleading to the extent that they give the impression that there is only some one thing (such as a theory) that changes. And so, it is one of the virtues of the material contained here that we are exposed to a range of items that change as the process of science proceeds. Not only do theories change, but so do practices, social groupings, problems (both empirical and conceptual, *pace* Laudan), and concepts such as evidence, confirmation and adequacy. So, it follows that the second set of questions concerning the rationality and universality of science and scientific concepts becomes increasingly more difficult and complicated to resolve as we discover the complexity of the subject.

Where then does this leave us? Given the wealth of material and the complexity of the issues, it should come as no surprise that we are left with more questions than answers. In particular, we are left with a challenge. Specifically, given the results of these investigations, is it possible to develop any adequate account of scientific change which (1) does justice to the historical complexity of the growth of science, (2) is sympathetic to the variety of involuted philosophical issues such historical complexity presents, and (3) is not itself so complex and dynamic as to be amorphous or superfluous?

Since part of what is entailed by change in science is change in our conception of knowledge and our epistemological criteria, surely no univocal account of scientific knowledge will suffice where the historical legacy of science is considered important to our understanding of its current significance. Likewise, since science is a social process in which are embedded and exemplified many different *kinds* of activities, no single-minded concern with one individual or one problem can provide the insight we need to account for the many features of science in a systematic fashion. Thus the challenge seems clear: we need a theory of scientific change which is dynamic (in that it is concerned with the changes that occur within science), multifaceted (in that it is concerned with both the internal levels of complexity within science and with the levels of complexity between science at its various levels and the rest of the social world), systematic (for what good is a theoretical account which fails to bring order out of chaos, even if the order is complex?) and historically sensitive.

The papers that follow do more than point out the complexity of the situation, they also suggest ways in which the goals noted above may be approached. And while no obvious solution emerges, the stage hopefully has been set for further, equally profitable work.

PART I

HISTORICAL DIMENSIONS

ALAN GABBEY

THE MECHANICAL PHILOSOPHY AND ITS PROBLEMS: MECHANICAL EXPLANATIONS, IMPENETRABILITY, AND PERPETUAL MOTION*

Writing to his sister Anne Viscountess Conway in May 1658, John Finch, a former student of Henry More at Christ's College Cambridge, made the following observation: "As to Descartes my Dear I would with all my heart that I could thinke his Philosophy as true as coherent; but coherency is no argument, for he must be a man of mean parts, that forgets himself so far as to make one deduction contradict another..." (Malloch 1917, p. 13). Most of us would agree that Finch, like the young More, was somewhat sanguine about the degree of coherency in Descartes' metaphysics and natural philosophy, but the intriguing point about his remark is the assumption that internal coherence in a philosophical system is a matter of vigilance in its construction, of being continually alert to impending logical solecism. Finch was assuming that it is possible for a philosophical system to be wholly coherent. Whether or not he was right is an engaging meta-philosophical problem that cannot detain us here, yet our *historical* experience suggests that he was being unduly naïve. Every intellectual system in the history of thought has been saddled with its share of inconsistency and incoherence, of antinomies, of problematic conceptual situations, of simple contradiction. *A fortiori*, the same holds for that unmanageably complex shoal of ideas, hypotheses, procedures, theories and systems, that shelters under the umbrella of "The Mechanical Philosophy". Of the numerous conceptual burdens the Mechanical Philosophy had to bear, the two or three I shall examine were not examples of forgetfulness or loose thinking in the protagonists, or obstacles that we might have expected such a progressive philosophy of nature to remove as a normal part of its brief. They illustrate rather conceptual difficulties that seem inherent in the mechanical program, or more precisely in versions of that program.

1. MECHANICAL EXPLANATIONS

To begin, I should like to comment on *explanations* in the mechanical philosophy, one major aim of which was to dis-

J. C. Pitt (ed.), Change and Progress in Modern Science, 9–84.
© 1985 by D. Reidel Publishing Company.

place Peripatetic explanations of natural phenomena. What I have to say here repeats to some extent familiar features of mechanism,[1] yet I think it important to present them within a framework that promises a fresh approach to the analysis of mechanical explanations. The framework I have in mind is that provided by the notion of *structural explanation*, recently explored by McMullin (1978b). Noting the distinction between structural, nomothetic, and genetic explanations, McMullin characterizes the first as obtaining when "the properties or behavior of a complex entity are explained by alluding to the structure of that entity", where the term "structure" "refers to a set of constituent entities or processes and the relationships between them". Structural explanations are causal, "since the structure invoked to explain can also be called the *cause* of the feature being explained" (McMullin 1978b, p. 139). Furthermore, McMullin argues that structural explanations are more satisfactory than nomothetic explanations, in that they make potentially fruitful ontological claims about the causal nexus with which the *explananda* are associated, and in that they involve retroductive inference, which (suitably modified) entails the possibility of explanatory resourcefulness in domains other than those delimited by the *explananda* for which the structural explanations were originally devised (McMullin 1978b, pp. 145-147).

Clearly, explanations in the Mechanical Philosophy are typical examples of structural explanations. The salient features of structural explanation, as outlined by McMullin, hold for the mechanical explanations of a Galileo, a Descartes, a Gassendi, a Boyle, or a Newton. Natural phenomena are given causal explanations by retroductively linking them to bodies with position and motion structured in an appropriate way, and (except in the case of the simplest mechanical effects) it is the *structure*, together perhaps with nomothetic causal relationships, that causes the phenomena, rather than any individual body, motion or disposition.[2] Since the number of possible structures is unlimited, even for a small number of component bodies, as Boyle noted (Boyle 1979, pp. 105-110), mechanical explanations promise a potentially large range of new explanatory tasks, and offer the hope of deductive testing and new predictive domains.

The structural explanations of the mechanical philosophers were particularly successful for those phenomena that could be reproduced, or represented in some respect, by constructing and setting in operation mechanical exemplars on which the proposed explanatory structures were to be modelled. Assuming that Nature is homogeneous in basic composition and uniform in action, the mechanist can confidently devise viable explanations of heat in terms of rapid or percussive motion because we can all produce heat ourselves using such motion, and we can avoid being cold by maintaining ourselves in constant motion. Playing tennis, or ducks and drakes, or watching a blind person using a stick to get about, provide clues (as they did for Descartes) about how certain observed geometrical properties of light might be explained. A promising model for the explanation of the tides is provided by the behavior of water in a non-uniformly moving barge. Galileo is a convenient example of someone who was clear about the value of "duplicating" the *explanandum* by means of a model[3]. In the *Dialogo*, he responded in the following terms to Simplicio's scepticism that the barge model leads to *the* "true cause" of tidal motion:

> ... *Salviati* ... you do not have any stronger indication that the true cause of the tides is one of those incomprehensibles [hidden causes] than the mere fact that among all things so far adduced as *verae causae* there is not one which we can duplicate for ourselves by means of appropriate artificial devices. For neither by the light of the moon or sun, nor by temperate heat, nor by differences of depth can we ever make the water contained in a motionless vessel run to and fro, or rise or fall in but a single place. But if, by simply setting the vessel in motion, I can represent for you without any artifice at all precisely those changes which are perceived in the waters of the sea, why should you reject this cause and take refuge in miracles? (Galileo 1967 (1632), p. 421)

Galileo's dismissal of the lunar influence was somewhat premature, but the issue here is explanatory legitimacy, not total explanatory success. Galileo's approach has elements in common with the new empirical spirit of certain of his contemporaries, such as the sceptic Francisco Sanchez, author of the tag "no one can know what he has not created" (Gregory 1961, pp. 70-77), or more strikingly, the young Descartes, author of the following apparently un-"Cartesian" remark from the *Cogitationes Privatae*: "Man's knowledge [*cognitio*] of natural things is only by analogy [*similitudino*] with those that fall under the senses. And indeed we consider as having philosophized more truly he who has been able to assimilate more happily things sought to those known [*cognitis*] through the senses" (AT(NP), X, pp. 218-219). Later, Mersenne could claim in his *Harmonie Universelle* (1636-37) that "we know the real reason only of things that we can make with our hands, or minds" (Lenoble 1943, p. 384).

Less successful, however, were the mechanists' explanations of those phenomena (notably sensations) for which they could provide explanatory structures – nothing was easier – but which could not be represented in any respect by means of an isomorphic macroscopic mechanism. For example, there is very little, if anything, in our experience that suggests a possible mechanical operation that would *itself, qua* isomorphic explanatory mechanism, produce in us the sensations of this or that color, this or that smell, this or that taste, or, more significantly, would reproduce an action of the will on the body. Not surprisingly, the explanations the mechanists did provide for phenomena of this kind were either extensions to the micro-world of the traditional Peripatetic qualities, or they were simply tautologous, a defect they shared with the corresponding Peripatetic explanations.[4]

In general terms, the explanatory structures in the latter case were specified in such a way that they bore a correspondence, or exhibited a congruity, which *fitted* them to be causes of the *explananda*. Between the *explanans* and the *explanandum* there was posited a relation of *adequation* (or *inadequation*, where appropriate).[5] In other and simpler words, the phenomena to be explained were caused by entities whose structures were such that they caused the phenomena. Previously, opium sent you to sleep because it had a dormitive quality: now it sent you to sleep because it had a particular corpuscular micro-structure that acted on your physiological structures in such a way that it sent you to sleep.

I shall cite just one example of this line of reasoning, which in this respect is typical of the mechanical tradition. In the chapter "Of Odours" in his *Physiologia Epicuro-Gassendo-Charltoniana*, Walter Charlton writes (p. 236):

> The Manner of the Odours moving or Affecting the Sensory can never be explained, but by assuming a certain *Commensuration*, or Correspondency betwixt the Particles amassing the Odour, and the Contexture of the Olfactory Nerves, or Mammillary Processes of the brain delated through the spongy bone. For (1) it is Canonical, that no Immaterial can Operate upon a Material, *Physically*; the inexplicable activity of the Rational Soul upon the body by the mediation of the spirits, and that of Angelical essences excepted. (2) Though an Odour, diffused through the aer, chance to touch upon the hands, cheeks, lips, tongue, &c. yet doth it therein produce no sensation of itself; because the Particles of it hold no proportion to either the pores, or particles of which those parts are composed: but arriving at the organ of smelling, it cannot but instantly excite the Faculty therein resident to an actual sensation, or apprehension of it; in regard of that correspondency in Figure and Contexture, which the particles of it hold to the pores and particles of the Odoratory Nerves.... (Charlton 1659, p. 236)

Furthermore, "the *Diversity* of Odours depends on the Diversity of Impressions made on the sensory, respondent to the various Figures and Contexture of their Particles" (p. 237), rough and "discordant" contextures causing nasty smells, smooth and "concordant" contextures causing pleasant smells.

Several questions arise here. First, we may ask to what extent this deficiency is a serious failing in the mechanical philosophy. In a sense, the explanatory difficulties in the case of sensations would seem to be of the mechanists' own making. They tried to explain everything, which was too much by a long chalk, and their critique of traditional modes of explanation failed to alert them to the dangers of trying to replace Peripateticism *en bloc*. The physics of forms and qualities was tailor-made to explain everything: the mechanical philosophy was not. Explanatory effectiveness was to lie not in comprehensiveness, but in a selective delimitation of the problem domain, and in following through the implications of a developing awareness that plausible explanatory structures on their own were not enough, but required in addition an ontological content that was sufficiently specific to permit mathematical description and/or experimental control.

So we puzzle over the meaning and sincerity of Descartes' extraordinary claim at the end of Part IV (Art. 199) of *Principia Philosophiae* that "there is no phenomenon in nature that is not comprised in what has been explained in this treatise" (AT(NP), VIII (1), 323). We approach with a sense of *déjà-vu* Gassendi's or Charlton's explanatory elaborations on the origin of qualities in general, and on the origins of every possible kind of particular quality: light, colors, sounds, odors, tastes, rarity and density, perspicuity and opacity, subtility and hebetude, smoothness and asperity, gravity and levity, heat and cold, fluidity, stability, humidity and dryness, softness and hardness, flexibility, tractility, ductility. A very familiar Peripatetic list of *explananda,* suggesting an equally familiar metaphorical image: an old and capacious bottle within which some vigorous new wine was circulating. The ontology and explanatory ideals of the mechanists were new, but very often the "problem domain" was largely Peripatetic. We are reminded of a parallel case in astronomy: Copernicus as the last great Ptolemaic astronomer, the format of whose *De Revolutionibus* corresponds, chapter for chapter, to that of the *Almagest*.

On the other hand, the mechanical philosophers were well aware (though Descartes was less explicit on the subject than others) that the causal structure of their respective philosophies contained incorrigibly hypothetical elements. And in the case of sensations in particular, they were acutely aware of the enormous difficulties in finding a solution to the problem of the mind's encounters with corporeal reality. "Where is that Oedipus", asks Charlton, "that can discover any *Analogy* betwixt the Retina Tunica, Optick Nerve, Brain, or Soul therein resident, and any one Colour? and yet no man can deny that there is some certain Analogy betwixt the *Species* and *Sensory*: since otherwise there could be no *Patability* on the one part, nor *Agency* on the other." (Charlton 1654, p. 197)

Among the major responses to the riddle, the Cartesian response merits a few comments. For Descartes, the interactions between mind and body are empirical facts, but at the same time mind and body are really distinct substances with mutually exclusive attributes and modes, which means that the empirical fact of interaction – the substantial union – cannot be intelligibly understood in terms of the ontological and causal categories constituting Descartes' own mechanical philosophy. Descartes resolves these tensions by recalling the Scholastics' illegitimate projection on to physical qualities of the substantial forms that for Descartes properly describe human souls. The soul, he claims, writing to Regius in January 1642, is the *real* substantial form of man, and is the *only* substantial form (AT(NP), III, p. 503, p. 505). The Scholastics were right to see man as body informed by soul, but were wrong to suppose that an analogous interpretation of (say) gravity would serve any purpose in the realm of physical explanation. As Gilson puts it, "since man has invented substantial forms in extending his personal experience to the world of extension, this experience must be that of a substantial form. The copy may be false, but one could not even comprehend its existence if the model were not true" (Gilson 1930, p. 247-248). So Descartes' mechanical world is not devoid of substantial forms at all: there are as many of them as there are persons. Does that mean, however, that the negotiations between mind and *res extensa* are quite beyond mechanical description and explanation?

The surprising answer is no. I offer two arguments. The first derives not from any piece of hard textual evidence, but from something suggested by McMullin's fruitful notion of structural explanation with which I began. It is a standard component of Descartes' account of sensation that there is nothing in the object that resembles anything in the ideas we form of the object's qualities. Among the many texts that make this claim we may note the beginning of *Discours Sixième* of *La Dioptrique* (1637). There Descartes warns us that although there is a resemblance between some arrangement of visible external objects and the retinal and brain patterns it causes while being inspected by the mind, we must not conclude that it is this resemblance that causes the mind to see what it sees, "as if there were other eyes in our brain with which we might be able to perceive it [the "painting" within the brain]". "Rather", he continues, "it is the motions composing it [the brain "painting"] that, acting immediately on our soul, in so far as it is united to our body, are ordained by Nature to make the soul have such sensations [*sentiments*]", and after describing in general terms how motions in the sensory nerves cause the corresponding sensations, he notes that it all happens "without there having to be any resemblance between the ideas the mind [*âme*] conceives and the motions that cause these ideas." (AT(NP), VI, pp. 130-31).

Now in the case of color, *La Dioptrique* and *Les Météores* present the speculation that it is caused by some mechanical disposition analogous to the rotation of tiny spherical particles, and that particular colors correspond respectively to particular ratios between the speeds of rotation and of linear motion of the model particles. So in analyzing the rainbow in *Discours Huitième* of *Les Météores*, Descartes hypothesizes that "those [particles of subtle matter] that have a tendency to rotate much more strongly [than to move in a straight line] cause the colour red, and those that have a tendency to rotate only a little more strongly cause yellow ... green appears where they rotate only a little more slowly, and blue where they rotate much more slowly" (AT(NP), VI, pp. 333-34). The rotation hypothesis then enables Descartes to offer a qualitative account of dispersion.

Structures abound in this explanatory scheme. The explanatory model is structured in virtue of the scale of ratios of rotational to linear speeds of its constituent particles. Accordingly, the actual mechanical process that causes color sensations is also structured, in virtue of the hypothetical one-to-one correspondence between the scale of real physical (and mathematizable) differentiating characteristics of the process and the scale of ratios in the model. There is the qualitative structure in the scale of rainbow colors as sensed in the mind, the steps in the scale bearing a one-to-one causal correspondence to points on the scale of differentiating characteristics of the actual mechanical process, and thereby bearing a one-to-one analogical correspondence to steps in the scale of speed ratios in the model.

So although a *particular* color bears no resemblance to a *particular* mechanical process causing it, it does not follow that color *as a general property* bears *no* resemblance to some corresponding mechanical explanatory scheme.[6] There is a resemblance in the latter case, a resemblance that consists in specified comparisons between colors being analogically linked to specifiable ratios *between* ratios of rotational to linear speeds, the corresponding causal structure deriving its order and intelligibility from Nature, or more correctly, from divine ordinance. The situation is very similar to that obtaining in Descartes' early

Compendium Musicae (written 1618, first published 1650), in which a basic theoretical assumption is that the mathematical relationships holding between the physical variables in sound production form a causal isomorphism with the musical relationships heard by the listener. As Descartes remarks in the section on consonants, "tones are to each other as [the lengths of] the strings".[7] Viewed in this way, it is as though Descartes' account of color is nudging him in the direction of a concept of mind that will at least leave open the possibility of speaking of the mind's *mechanical* behavior, when in a passive state.

My second argument does depend on a piece of textual evidence, according to which the mind's *actions* on corporeal substance can also be viewed mechanically, and in a directly mathematical way. The text in question is a passage from the "excerpta Anatomica",[8] which were copied by Leibniz during his 1670s Paris visit from the Descartes papers then in Clerselier's possession. In the passage Descartes is trying to work out a mathematical analysis of the ways in which various kinds of force (principally gravity) can move a body through media whose resistances bear varying proportions to the speed of motion. Descartes writes:

> If any body is activated or impelled into motion always with an equal force, that is a force imparted to it bz the mind [*mens*] (for no other such force can exist), and moves *in vacuo*, it always takes three times as long to travel from the beginning of its motion to the middle as it does from the middle to the end, and so on. But because no such vacuum can be granted, there being space [*spatium*] of some sort, it always resists in some way, and does so in such a way that the resistance always increases in geometric proportion to the speed of the motion, so that eventually it reaches a point where the speed no longer increases appreciably; and one can determine a particular finite speed which it will never equal.
>
> As for those bodies impelled by the force of gravity, since gravity does not always act equally, like the mind [*anima*], but there is some other body already in motion, never can it impel a heavy body as fast as it itself moves,

but even *in vacuo* the impulse will always diminish in geometric proportion. Indeed a decrease due to two or more causes in geometric proportion is equivalent to the decrease that arises when all these are considered together as a single cause also in geometric proportion, and the result will still be the same. And similarly, if another cause restrains [the body] with an arithmetic force, a diminution in geometric proportion will arise. But if some other force impels constantly in geometric proportion simultaneously with that which diminishes geometrically, the geometric force will eventually cease and only the arithmetic will remain and increase the motion which, as was said above, is what the mind [*anima*] does *in vacuo*. Finally, if the impulse increases geometrically and [at the same time] decreases or even increases arithmetically, the speed will increase indefinitely [*in infinitum*] in the composite proportion. This can be explained geometrically by taking the proportional [increases in speed which] the triangle and the curved line [represent], and combining them this way [Fig. 1] for addition, and this way [Fig. 2] for subtraction, so that the speed during the first [interval] of time is to the speed during the second as the area *abc* is to the area *aced* (AT(NP), XI, pp. 629-631).

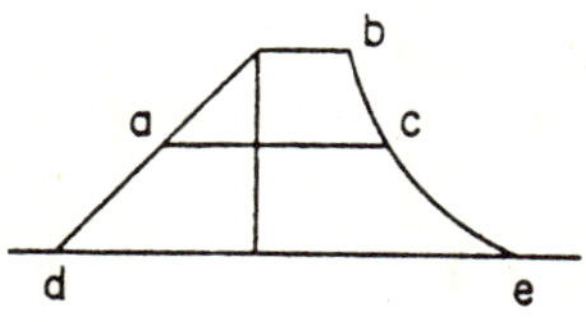

Fig. 1

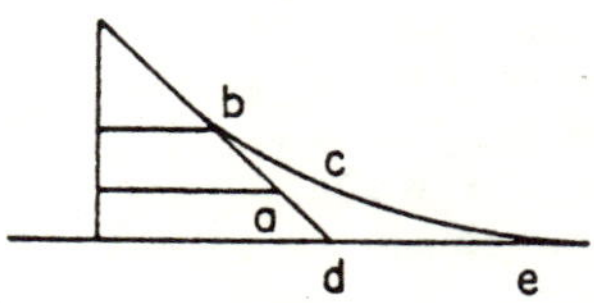

Fig. 2

One might wish for greater overall mathematical clarity and detail in this fragment,[9] but there seems little doubt about what is claimed for the mechanical powers of the human mind. In any purely corporeal interaction, which for Descartes includes the action of gravity, the force transmitted to a body at a given time from the same mechanical agency varies with the speed of the body at that time. The amount of motion a body receives from the gravitational vortex (presumed to have a constant "effective" speed) is greater when it is at rest than when it is already moving; for example, it receives no further motion at all when the "effective" speed of the vortex equals the body's speed of fall (the "terminal velocity").

Not so for the mind, which Descartes claims can add equal increments of speed to a body under its influence whatever the speed of the latter. This fragment from "Excerpta Anatomica" therefore complements the tentative inference I drew from the treatment of the rainbow. Whether in an active or passive state, the mind can behave mechanically in the sense that certain of its operations can be mathematically described because they form part of a causal nexus involving corporeal structures that in themselves are open to a relatively more straightforward mathematical treatment.

But does mathematical description imply *law-like* mechanical behavior? What does the passage from "Excerpta Anatomica" tell us about the mind as causal agent in the corporeal world which is subject to "the laws of nature"? The implications of the passage are troubling, for Descartes. If the will is free, and if the mind can add even one increment of speed, however small, to a body on which it is acting (via the pineal gland), it follows immediately that the total quantity of motion (*i.e.* motive force) in the Universe is *not* conserved, but increases each time a mind wills a bodily action entailing an incremental change of speed in the pineal gland – animal spirits network. And if the mind can add successive *equal* increments of speed, however small, it follows that each of us at will can cause a monotonic increase in the total quantity of motion in the universe.[10] Each time I voluntarily raise my arm I violate the inalienably Cartesian conservation law that appears as the nomothetic centrepiece of *Principia Philosophiae*, Part II.[11]

It is scarcely possible that Descartes was unaware of this awkward consequence of his conception of the mind as an active participant in the corporeal world. That would explain why there does not appear again in his writings after 1635 anything approaching the directness of the fragment from "Excerpta Anatomica". We meet with general claims to the effect that "the little gland which is the principal seat of the soul ... can also be moved in diverse ways by the soul", to borrow a phrase from *Les Passions de l'Ame* (AT(NP), XI, pp. 354-355), or with attempts to play down the possibility that the mind can *add* motion (via the pineal gland) to parts of the body, coupled with appeals to the concept of *determination* as a means of giving mind some purchase in the corporeal realm, as for example in the *Responsio ad Quartas Objectiones* (1641):

> ... even within ourselves it is not the mind (or soul) that moves *immediate* the external members of the body, but only that it can determine the flow of that very subtle fluid we call animal spirits, which, flowing continually from the heart through the brain into the muscles, is the cause of all movements of our limbs, and can often cause several different movements, each of them with equal facility ... (AT(NP),VII, p. 229)

As Descartes never tires of telling us in *La Dioptrique* and elsewhere, "determination", that is, the directional mode of a body's motive force (see my 1980*a*, pp. 248-61), must be distinguished from the force (motion) itself, and one can change without the other necessarily changing. So restricting the mind's actions in the corporeal world to changes in determination, without changes in the speed of the bodies in question, has the effect of regaining the integrity of the conservation law, while retaining an active non-mental role for the mental partner in the substantial union.

At least one of Descartes' contemporaries was not beguiled by this move. The shrewd Henry More did not need prompting from the "Excerpta Anatomica" (of which he was unaware) to include the following query in his letter to Descartes of 23 July 1649 (receiving no reply[12]):

> On Art. 36 [of *Principia Philosophiae*, Part II]. I ask: when the human mind stirs the [animal] spirit by thinking more attentively and for a longer time and, moreover, rouses up the body

> itself, is it not the case that it is then increasing the motion of the universe? (AT(NP), V, p. 385).

Leibniz obviously knew of the incompatibility between the conservation law and the mind's corporeal actions by 1676 at the latest, since it was during his stay in Paris, which ended in 1676, that he took a copy of the "Excerpta Anatomica", adding the significant marginal comment: "Hence (NB) the force of the mind in a vacuum is arithmetical" (see note 10). However, his later recollection of the matter, or rather his interpretation of Descartes' position, as he found it in both the "Excerpta" and other writings, was often more liberal than is warranted by Descartes' texts. For example, in the *Monadology* (1714) we read:

> 80. Descartes recognized that souls cannot give force to bodies because the same quantity of force is always conserved in matter. He believed, however, that the soul could change the direction of bodies [*des corps*, which Loemker mistranslates as "the body"]. But this was because the law of nature was still unknown in his day, according to which matter conserves also the same total direction. If he had noticed this, he would have fallen upon my system of pre-established harmony. (Leibniz 1969, p. 651).

And similar accounts of Descartes' position, though expressed at greater length and in different terms, turn up in the correspondence with Arnauld (30 April 1687: Gerhardt, *Phil.*, II, p. 94), and in the *Theodicy* (Gerhardt, *Phil.*, VI, pp. 135-136). It seems to me a misrepresentation for Leibniz to claim that according to Descartes the mind cannot give force to bodies *because* of the conservation law. Rather, the mind *does* give force (at least to the pineal gland), and *therefore* problems arise when this situation is confronted by the conservation law. A more accurate account of Descartes' position does appear, however, in "Eclaircissement du Nouveau Système de la Communication des Substances, pour servir de réponse à ce qui en est dit [par M. Foucher] dans le Journal [des Sçavans] du 12 Septembre 1695':

> ... You know that Monsieur Descartes believed that the same quantity of motion is conserved in bodies. It has been shown that there he was mistaken; but I have shown that it is always the case that what is conserved is the same motive force [*la force mouvante*], for which he mistook the quantity of motion. However the changes that take place in bodies as a consequence of the soul's modifications embarrassed him, because these modifications seemed to violate the law [of conservation]. So he thought he had found a way out, which is in fact ingenious, by saying that one must distinguish between motion and direction, and that the soul can neither increase nor diminish the motive force, but that it changes the direction, or determination, of the flow of animal spirits, and that that is how voluntary motions take place. It is true that he was very far from explaining how the soul manages to change the flow of these bodies, that seeming as inconceivable as to say that it gives them motion, unless with me one were to turn to the pre-established harmony. But one must realize that there is another law of nature which I have discovered and demonstrated, and which Monsieur Descartes did not know: not only is the same quantity of motive force conserved, but so also is the same quantity of direction, whatever direction in the universe is chosen ... this law, being as beautiful and general as the other one, did not merit violation either, and that is how it is in my system, which conserves force and direction, and in short all natural laws of bodies, notwithstanding the changes that occur in them as a consequence of those in the soul. (Gerhardt, *Phil.*, IV, pp. 497-498)

To me this reading of Descartes' position is more convincing than what is to be found elsewhere in Leibniz.[13] It conveys the tension Descartes must have felt in trying to reconcile his laws of nature with his theory of the mind-body relation, a tension that becomes more pointed when we set the above excerpt from the *Responsio ad Quartas Objectiones* alongside the passage from "Excerpta Anatomica".

And we read with some sympathy the following escape clause in *Principia Philosophiae*, Part II, Art. 40, that is, the article devoted to the Third Law of Nature,[14] which derives in part from the conservation law:

> ... And all the particular causes of the mutations that happen to bodies are comprised in this third law, at least those causes that are themselves corporeal; for I am not examining now whether the minds of angels and humans have a force to move bodies, nor what kind of force it might be. This is something I am saving for the treatise *de homine*. (AT(NP), VIII(1), p. 65)

What a pity that treatise was never written.

In formulating his understanding of Descartes' position in this area Leibniz had to handle not only Descartes' writings, but also those of some of the Cartesians, whose teachings he believed reflected faithfully those of Descartes himself. Indeed he found the disciple Clauberg "clearer than the master", though at the same time neither Clauberg nor the others (including Clerselier) "added anything to the discoveries of their master", publishing "only paraphrases of their leader" (To Jacob Thomasius, 30 April 1669: Loemker, *Phil. Papers*, p. 94) Accordingly, it is instructive to look at Clerselier's attempt to grapple with the same difficulties concerning the mind-body union. His letter to Louis de la Forge of 4 December 1660 (Clerselier, III, 1667, p. 640-46)-described by Clerselier himself as "Observations de Monsieur Clerselier, touchant l'action de l'Ame sur le Corps" – presents a neater version of what Descartes might have meant, or might have wished he could have claimed without embarrassment, than can be found anywhere in Descartes' own writings, though on reading it one wonders what Clerselier made of the quoted passage from "Excerpta Anatomica", or what kind of discussion he and Leibniz had on the curious piece as Leibniz was making his copy.

Shortly after the beginning of the letter Clerselier notes that all properties belong to one or other of the only two substances we know: the spiritual and the corporeal. Among the properties that do not belong to corporeal substance, but only to spiritual substance, is that of being able "to give the first motion to body, or to implant in it a completely new motion which increases the quantity of motion already in the world" (p. 641). However, it is only infinite spiritual substance, *i.e.*, God, "that is capable of implanting the first motion in body ... of creating or making a thing that does not exist be and have existence" (pp. 641-642). It might be objected that since motion is only a mode of body, an infinite power is not required to create it. But Clerselier thinks that to move and animate body requires a power of the same order as that which created body itself from nothingness, so no finite mind would be sufficiently powerful to produce motion, "unless the will of the Creator were to join with [that of such a mind]" (p. 642). Even if matter had been moved and divided for the first time on the word of an angel, "his word would have been only the instrument of that of God, whose power [*vertu*] alone would have worked this wonder, it not being possible that the nothingness of motion [*le neant du mouvement*] obey other than an infinite power [*puissance*]" (*ibid.*).

On the other hand, finite spiritual substance – the human mind – is "capable only of determining the motion [*determiner le mouvement*] that is there already" (*ibid.*). This situation causes Clerselier no difficulty, since

> ... the determination of motion . . . adds nothing real in Nature, and ... tells us nothing more than motion itself, which cannot be without determination. So it is not to be marvelled at that the soul possesses the faculty of determining motion, as our own experience convinces us it has; for that does not prevent God from being the Author of all forms [*forme*] through which matter passes, which are all effects, consequences, and dependencies of the motion He introduced into it, and conserves in it, and so that does not prevent Him from being really the Creator of all things. (pp. 642-643)

The determinative faculty depends on the substantial union, not on the mind alone, and it achieves its effects only when there is an "appropriate disposition of the body, when the soul wills and determines for itself some motion or other, of which disposition it is possible for the soul not to have any knowledge" (p. 643).

As for the crunch question of how the soul's determinative actions actually operate, Clerselier admits that we do not know, no more than we understand how God created all things from nothing. But our experience shows that the soul's determinations of bodily motions are real, so the union of body and soul can have no cause other than "the will of the One who has united and joined them together" (p. 646). As we always knew, at the heart of Cartesian natural philosophy lies a divine mystery

Texts such as the Clerselier letter and the "Excerpta Anatomica" broaden our appreciation of the difficulties the Cartesian mechanical programme had to face. To entertain the possibility of a partial mechanico-mathematical treatment of a substantial form (mind-body), while retaining both divinely-granted free will and divinely-grounded laws of nature, was a major mistake, though one that had to be made. To the extent that mind-body interaction was an empirical datum, we might say that Descartes' mistake was to be too empirical, and to believe thereby that the experientially validated substantial union might be describable in terms that were devised specifically for only one partner in that union.

We also see more clearly something that will be news to philosophers: the indissoluble links between mechanics and theories of mind and the will in the 17th century. Descartes' conservation law gets in the way of the freedom of the will to move the body. His solution to that difficulty involves his notion of determination, which for its intelligible application in the mind-body problem would have required not only some minimal account of how the mind can alter determinations of motion, but also a theory of oblique collisions, which Descartes did not provide, and which, as I have suggested elsewhere (1980*a*, pp. 256-257), would have been well-nigh impossible to construct on the basis of his conservation law and his concept of determination. In any event, the conservation law still gets in the way of the mind-originated motions of the pineal gland, which it seems is moved by the mind *immediatè* prior to the gland determining the motions of the animal spirits. As Leibniz realized, with Huygens' help. Descartes' conservation law is invalid in that it conserves quantity of motion (mass x speed), rather than quantity of force (*vis viva*, mass x speed^2), and Leibniz discovered (again with Huygens' help) that there is at work in the corporeal world another conservation law, which conserves quantity of motion measured in a given direction (see *e.g.*, Costabel 1960). It follows, argued Leibniz, that Descartes' theory of mind-body interaction is wrong on (at least) three counts: *a*) the conservation law that embarrasses Descartes the substantial-union theorist is the wrong one, *b*) the correct conservation law (mv^2) is still in danger from Descartes' substantial union, but it is inconceivable how the mind could *create* motion in a body, *c*) the new conservation law ($m\vec{v}$), which holds in purely mechanical situations, makes nonsense of Descartes' deployment of determination in attempting to explain the supposed action on bodies of an incorporeal substance, since it is also inconceivable how the mind could change the *direction* of motion of a body. Occasionalism is a possible answer, but that would detract from God's wisdom and workmanship, and would invoke an unsatisfactory *Deus ex machina* (Leibniz 1969, p. 269). Much better to embrace the pre-established harmony, the two Leibnizian conservation laws, Leibniz's doctrine of substance and the theory of monads. Then all Cartesian entanglements disappear. As one reflects on Leibniz's handling of the Cartesian legacy, one becomes more and more convinced that his mature doctrines were some kind of natural and even inevitable *terminus ad quem* in the twin histories of the mechanical philosophy and the theory of mind.

The difficulties in reconciling Descartes' conservation law and his theory of the mind's action on body have attracted only sporadic attention from Descartes scholars in the present century.[15] In the older works, the most extended discussion seems to be that of Jean Laporte (see note 15), who concludes that because of the mind's violation of the laws of Descartes' own mechanics, "it must be admitted that the union of mind and body carries with it an element of unintelligibility" (Laporte 1950, p. 248), a view which I share, especially in the light of the evidence in "Excerpta Anatomica". Yet the quoted passage from the "Excerpta" seems to have been overlooked by almost all modern commentators on the Cartesian mind-body problem (the exception being Daniel Garber 1983), which is unfortunate, given its pivotal role in the matter. As I hope has been made clear, the passage shows that in the purely corporeal world there is a causal input originating from minds, which are of course non-mechanical in their essence, yet which have the purely mechanical effect of increasing the total amount of motive force in the universe.

However, it might after all be a mistake to see an incompatibility between Descartes' conservation law and his mind-body doctrine. This is the position adopted by Garber and Remnant (1983 and 1979, respectively). They argue that the conservation law applies only to the corporeal world created by God and sustained in existence by Him with a constant quantity of motion and rest: it does not apply to the corporeal world as acted on by minds in the exercise of the will. And as part of their evidence they cite precisely the passage from Art. 40 of *Principia Philosophiae*, Part II, that I described, when quoting it above, as an "escape clause". Perhaps that passage should be read as a warning against assuming that the world of which the conservation law and the Third Law of Nature are true is co-extensive with that in which mind can produce mechanical results. Clearly, if that were the correct reading, the question of a violation of the conservation law would not arise, and Garber would be justified in claiming that "there is reason to believe that Descartes may never have been committed to the position that his conservation law holds universally, and may have allowed for the possibility that animate bodies lie outside the scope of the laws that govern inanimate bodies" (1983, section 2).

This reading of the situation seems to me to create as many difficulties as it solves. To begin with, it implies that the power of human minds is in a curious respect superior to that of the divine mind. God could change the quantity of motion He created, since His power and will are infinite, but He does not do so, because He is eternal and immutable (*cf.* Descartes to Mersenne, 15 April 1630: AT(NP), I, pp. 145-146). We on the other hand can create as much extra motion as our mortal frames can stand, because of the freedom of our will to act through the mind-body union, and we do in fact create extra motion, because we exercise that will in the daily execution of the most mundane movements of our bodies. This particular difficulty could perhaps be met by citing the following strange remark in Descartes' letter to More of August 1649:

> ... the moving force [of a body] can come from God himself conserving just as much motion [*translatio*] in matter as He put in it at the first moment of creation; or again, it can come from a created substance, such as our mind or any other thing whatever, to which He will have given the force of moving the body ... (AT(NP), V, pp. 403-404)

It might be that Descartes intends the mind to enjoy divine aid in creating new forces in bodies, but then that would imply that he is unwittingly allowing for the mind's participation – as God's agent – in divine abrogations of the divinely-grounded conservation law. And if occasionalism is not just round the corner in such an interpretation of Descartes' intentions, where does that leave the freedom of the will in the context of bodily movement?

Second, if the total amount of motion in the universe at any given time depends on *our* will and behavior, it is not clear what explanatory purpose is served by the law stating the conservation of the same quantity of motion (force) initially created by God.

Third, Garber's and Remnant's distinction between "animate" and "inanimate" bodies is unavailing in this context, since as far as the range of applicability of laws of motion is concerned, "the machine of our body", as Descartes describes it (*e.g.*, AT(NP), XI, pp. 226, 331, 341), that is the machine comprising all parts of our body (including the pineal gland), is no different from any other machine or system of bodies in the external corporeal world. After all, that was the whole point of Descartes' mechanistic physiology. Furthermore, if (say) my body "lies outside the scope of the laws that govern inanimate bodies", then it must follow that any external corporeal system with which I come into continuous operative contact – my bicycle, for example – is also "animate" within the meaning of the act, and is therefore not subject to Descartes' Laws of Nature. Clearly, it is easy to imagine a science-fiction set-up in which the whole of Nature is "animate" in this sense, and therefore not subject to the Laws of Nature, a situation of unmanageable contradiction.

But why substitute a hypothetical contradiction for one that lurks already in the irremediably incoherent mechanics of the substantial union?

2. IMPENETRABILITY AND FORCE

These difficulties with the mind-body interaction were not accompanied by a free run in the matter of body-body interactions. There too the mechanists encountered a variety of conceptual entanglements that were to plague them throughout the period under review. I shall select just one issue that illustrates this point: the treatment of impenetrability, and its relation to force, at the hands of Descartes and Euler.

Traditionally, it was held that there could be no empty space, that there could be no interpenetration of dimensions, because two bodies could not occupy the same place at the same time. Thus body possessed three fundamental properties: three-dimensional extensionality, impenetrability, and resistance (see Grant 1978 and 1981). Now the mechanists who held to a revived atomism took extension, impenetrability, and shape to be the defining attributes of body, with resistance arising in some way from the forces with which God endowed bodies in motion or at rest.

For Descartes, deeply committed to a programmatic quest for simplicity and economy in his natural philosophy, extension alone was sufficient to define body, as every textbook and study keeps reminding us. Less often, if at all, are we reminded that for Descartes a body's impenetrability is a *consequence* of its being extended.[16]

The clearest evidence for this feature of Descartes' account of body is found in his correspondence with Henry More (1648-49). More had argued that the idea of extended substance is of wider reference than the idea of body. Extension is not characteristic of body alone, since God and angels are also extended beings. According to More, body is properly defined as a tangible or impenetrable substance, and not simply as *res extensa* (More to Descartes, 11 December 1648: AT(NP), V, pp. 238-240). Furthermore, tangibility and impenetrability are not proper to extended substance *per se*, since More had no difficulty in "clearly" conceiving extended substances that are in no way tangible or impenetrable (More to Descartes, 5 March 1649: AT(NP), V, p. 301). For example, God can penetrate bodies (11 December 1648: AT(NP), V, pp. 238-240).

In reply to these proto-Newtonian criticisms, Descartes made his familiar claims (*a*) that God, like angels and minds, is not an extended substance, except perhaps in the loose and improper sense that "God is everywhere"; and (*b*) that only body is extended substance, and *vice versa*, and that where there is extension there must be, irrespective of the sensible appearances, corporeal substance (Descartes to More, 5 February 1649: AT(NP), V, pp. 267-269). However the pressure of More's criticisms, especially those in his letter of 5 March 1649, induced Descartes to clarify a consequence of his position that he had previously mentioned only in passing in *Responsio ad Sextas*

Objectiones (AT(NP), VII, p. 442), and which is merely hinted at in the sections preceding the collision rules in *Principia Philosophiae*, Part II. In his letter to More of 15 April 1649 Descartes maintained that one *res extensa* cannot penetrate another, otherwise a part of their extension would become a literal "nothing", and therefore not a substance which could penetrate anything. Undermining More's distinction between impenetrable extended body and penetrable extended substance, Descartes argued that impenetrability is in a secondary sense an essence of *res extensa*:

> If you conceive extension by means of the mutual disposition of parts, then you cannot deny that each single one of them touches other neighbouring parts. And this tangibility is a real property, and is intrinsic in body [*rei intrinseca*], but not so for that tangibility which takes its name from the sense of touch.
>
> Furthermore, one cannot understand one part of extended substance penetrating another part equal to itself, without at the same time recognizing that [*half*] the extension [... *mediam partem eius extensionis*: read *dimidiam*] is removed or annihilated. But whatever is annihilated does not penetrate anything else, so it is demonstrated, according to me, that impenetrability belongs to the essence of extension, but not to that of anything else. (AT(NP), V, pp. 341-342).

Note however that impenetrability is of the essence of *res extensa* in a derived sense, since extension is still the primary essence by which body is to be defined. The letter to More of 5 February 1649 clarified this point:

> ... body is not to be defined by impenetrability, but by extension. This is confirmed by the fact that tangibility and impenetrability relate to parts, and presuppose the idea of division or termination; but we can conceive a continuous body of indeterminate size, or an indefinite body, in which nothing other than extension enters into consideration. (AT(NP), V, p. 269).

Finally, we must not confuse the *tangibilitas* that accompanies this impenetrability with that which we experience through the sense of touch, nor confuse the intrinsic hardness of extended bodies at rest with the hardness we feel when our hands happen to come into contact with certain external objects. The latter kind of hardness is explained by the mutual rest of the particles constituting the body: the former kind derives from the extensionality of the individual particles.

Now the intriguing thing about this coupling of impenetrability with extension (the traditional line) is that it is not linked to considerations of *force*. The non-interpenetration of *res extensae* is not simply *caused* by repulsive forces; it seems sufficient for it to be a logical consequence of the principle of the permanence, *Deo volente*, of extended corporeal substance, and of that untried Cartesian axiom that nothing cannot have properties, and therefore cannot act.

Yet to say that two physical bodies in the real world are mutually impenetrable invites a causal explanation of *why* they stay apart, or at most remain in contact. To find such an account of the required physical causes, we might turn to *Principia Philosophiae*, to those articles in Part II that deal with collisions and with the associated forces that express God's maintenance of bodies in their modes of motion or rest. But there is no mention of the forces that might maintain a body's *shape*, and in those texts there is no explicit mention of impenetrability, and no mention at all of the doctrine of the impossibility of the interpenetration of dimensions being a logical correlate of the bodies' extension.

There seems therefore to be no relation between the logico-geometrical necessity of impenetrability and the physical causation that prevents bodies from penetrating each other. It might be objected that the physical impenetrability of Descartes' bodies refers simply to rigidity, but Descartes' account of the latter idea carries its own well-known difficulties. In conclusion, therefore, there is an unsatisfactory gap between the logico-geometrical and the physical at the heart of Descartes' account of corporeal substance.

Before considering Euler's failure to solve the same problem of the relation between force and impenetrability, it must be noted that other figures in the mechanistic tradition – Newton, Locke, More, Gassendi, Boyle – who specified impenetrability as one of the essential characteristics of bodies (together with extension) did not succeed either in explaining satisfactorily the causality that makes corporeal impenetrability possible. For them impenetrability seems to have been a datum from which all mechanical action flowed.

Euler's views on the subject appeared at length in two texts, the "Recherches sur l'origine des forces" (Euler 1752), and the "Introductio" to the *Theoria motus corporum solidorum seu rigidorum* (Euler 1765). The differences between the two texts lie more in format and organization than in the content, which is virtually identical in the two texts. For convenience therefore I shall follow the French text.[17]

Euler begins the "Recherches" with two basic principles: (*a*) that a body at rest will remain at rest, and (*b*) that a body in rectilinear motion will continue in rectilinear motion. If a body initially at rest begins to move, it does so because some *external* cause has appeared to produce the change, and similarly for changes in a body's rectilinear motion, though in the latter case it is not clear if Euler thinks a moving body has a force maintaining it in motion. It seems that he does not think that it does have such a force. In both cases Euler appeals to the principle of sufficient reason, though he does not explain why straightline motion is the causally free norm.

The persistence of a body in one or other of the two states is for Euler an *essential property* of body, and any cause producing a change of state cannot arise from within the body, but must be external to it. This property is the body's *inertia*: "an essential attribute of all bodies, in that it would be impossible for there to be bodies deprived of this property" (p. 110). However, the term *inertia* seems more appropriate for bodies at rest than for those in motion: continuation of motion is customarily seen as an action, whereas *inertia* signifies opposition to all action. Euler resolves the apparent contradiction by noting that the same principle accounts for the body's persistence in both states, so it is reasonable to use the same term to describe the principle when at work in either case. "So the name must be linked to a meaning that marks an aversion to any change" (p. 111), except of course those changes that consist in uniform rectilinear displacements from place to place. It follows that any supposed properties of body that are incompatible with inertia in the sense described must be proscribed.

Any cause producing a change of state is called *force*, but the term "force of inertia" is ill-conceived:

> 9. ... since the effect of inertia consists in the conservation of the same state, and the effect of forces tends to change the state of bodies, it is evident that these two effects are directly contrary to each other, and that inertia signifies rather something completely opposed to the idea of forces ... (p. 112)

What then is the *origin* of the forces associated with the initiation of or changes in speed observed in all bodies? For Euler the answer lies in the *impenetrability* of bodies.

His argument reduces to the following. Consider two bodies A and B, with A at rest and B striking it; it does not matter whether the bodies are hard or elastic, and they may be assumed to be of comparable size. When B strikes A a sudden change occurs in each of them. Why? Because, if A remained at rest, B could not continue its motion, as required by its inertia, without passing through A. But that is impossible, because of the impenetrability of each body. So if one or other of the bodies were not impenetrable, one would pass through the other, and no change of state would take place.

> 13. So there we have straightaway a very remarkable state of affairs. Impenetrability, which being involved in the two bodies not being able to continue in their state, must contain probably the cause of the change that takes place in the collision, so all possible attention must be accorded to it in this investigation. Now impenetrability is, in the opinion of all philosophers, a property as general and as essential to all bodies as extension, or inertia. However, it seems to me that most of them have not pinned down with sufficient exactness the idea we ought to have of this property. For when one says that all bodies are impenetrable, it must not be thought that a sponge, for example, constitutes an exception, in that it allows water to penetrate it. We know that it is not particles of the sponge that the water penetrates, but the pores, which allow the water to

> enter as the matter lodged in them leaves, although they may be filled with air or some other matter. This example serves to remove several doubts that one might have on the generality of this quality [*qualité*] of bodies.
> 14. Hence it is clear that impenetrability is a property of bodies such that it is because of it that a body, being in a place, in so far as it is in that place, does not allow another body to occupy the same place at the same time. The condition I have inserted into this definition, contained in the words "in so far as it is in that place", protects it from objections that might be made against it. For when water enters the pores of a sponge, the water does not occupy the place that the sponge now occupies, but it moves into the spaces [*espaces*] that another matter had occupied previously and has [now] left on the water entering... (pp. 113-114)

Euler goes on to note that impenetrability is an *essential* property of body, like extension or inertia: "extended bodies, even endowed with inertia, would not even deserve the name of body if they did not have this property" (*ibid.*) A body without impenetrability would not affect our senses, and could not therefore apprise us of its existence. Also, a penetrable body would not differ from an empty space. Firstly, the body's supposed occupancy of its place would not prevent other bodies occupying the same place, which would render the penetrable body wholly superfluous as being other than simply identical to the place occupied. Secondly, the parts of a penetrable body could be made to occupy the same place, however small, and thus the body could be reduced to a mere point, or nothing at all.

Furthermore, Euler argues that there are no "degrees of impenetrability". Either bodies are impenetrable, or they are not. If a body were not wholly impenetrable, there would be a penetrable part, which would accordingly be a nothing.

There remains the question of determining the origin of the force that produces the changes of state that A and B undergo, "for, because of the inertia, there is no doubt but that this change arises from a specific cause [*d'une certaine cause*]; and every cause capable of changing the state of bodies is included under the name force" (p. 115). Is this force therefore necessarily linked to the impenetrability, or is it separable from it?

> ... In the latter case the impenetrability could subsist without this force. Now, on abolishing this force, that is the cause of the change of state of the bodies, its effect ought also to cease, and so each body ought to remain in its state. But that is impossible, since the bodies would then have to penetrate each other. So it follows necessarily that the force in question is absolutely linked to the impenetrability, and that it is in no way separable from it. (*ibid.*)

Euler's argument for an "absolute" link between impenetrability and force does not work. On the one hand, it depends on the *prima facie* plausibility of it being possible to separate impenetrability from force in such a way that impenetrability *alone* can be said to produce the same effect (*i.e.*, change of state) as impenetrability + force. Tautologously, impenetrable bodies cannot penetrate each other, which means accordingly that they must change each other's state if they collide. But if the term "force" covers "every cause capable of changing the state of bodies", then the supposed separation of impenetrability from force is logically impossible, since impenetrability is itself a species of "force", *ex definitione*.

On the other hand, if the link between impenetrability and force is one of logical necessity, then Euler's attempt to set up a relation of physical causality between them becomes superfluous, unless he intends it to be understood that causal relations are or can be also relations of logical necessity. Furthermore, Euler assumes that there is no *tertium quid* that might be ontologically prior to both impenetrability and force, and which might possibly forge the claimed necessary link between them.

At this point in the text, however, there is a turnabout in Euler's thinking. Immediately after the passage just examined, he goes on to claim that impenetrable bodies are "endowed with" the force required to *maintain* the impenetrability:

> 19. Therefore, once we recognize the impenetrability of bodies, we are obliged to admit

> that impenetrability is accompanied by a force sufficient to prevent penetration. And in fact if it is impossible for the bodies to penetrate each other, there must be insurmountable obstacles which oppose the penetration, and if the penetration cannot be avoided without the bodies changing state, there must be in the bodies themselves, in so far as they are impenetrable, *forces sufficient to produce this change of state* [my italics], without which [*i.e.*, the change of state] the impenetrability could not continue. It would therefore be an open contradiction to maintain that the bodies are impenetrable and at the same time to deny them the forces that are absolutely necessary to maintain the impenetrability and to oppose penetration.
>
> 20. In so far as the bodies are impenetrable, therefore, they are also endowed with forces necessary for the maintenance of this property, and without these forces it would be impossible even to form to oneself an idea of impenetrability; so that it is impenetrability, which is the first source of the forces that subsist in the world and produce an infinity of changes in the state of bodies. Now these forces come into play [*se déployer*] only when it is a question of avoiding [*prévenir*] penetration. For as long as the bodies can persevere in their state, either of rest or of motion, without their impenetrability being violated, no change in their state will take place, just as if the forces joined to the impenetrability did not exist. But when the bodies can no longer remain in their state without them penetrating each other, it is then that the forces of impenetrability act, in changing the state of the bodies as much as is necessary to prevent any penetration. (p. 116)

And Euler continues with the warning that we must beware of a definition of force that would see it as being a *continual* effort to change state:

> ... [these forces] do not act continually, but only when the bodies by continuing in their state would be penetrating each other.

> Furthermore, the change of state that they produce is only an indirect effect, since their principal effect consists in the maintenance of the impenetrability, and since they change the state of the bodies only in so far as is required to prevent penetration . . (pp. 116-117).

What can one conclude from this sequence of arguments? My conclusion is that Euler's account of the relation between force and impenetrability is incoherent. Firstly, he argues that the *origin* of force is impenetrability, in the sense that it is because bodies are impenetrable that they exert forces to change each other's state. On this view impenetrability is ontologically prior to force. Yet he also argues that appropriate forces are exercised *to prevent* penetration taking place, which is a reversal of the ontological priority in the first view. The upshot is that "l'origine des forces" remains quite unclear. Even if it be conceded that Euler is justified in seeing an "absolute" link between impenetrability and force, the most that can be claimed is that "impenetrability" and "(exercised) force" are twin terms describing the same corporeal property. But what then might be the origin of that corporeal property? Euler does not tell us. His attempt to ground force ontologically in impenetrability is a telling instance of the apparently ineradicable difficulties attending any Cartesian account of body and its dynamical properties. That Euler is a Cartesian in this context is hard to deny, if certain aspects of the "Recherches" paper are compared with the difficulties in Descartes' account of impenetrability examined above. For Descartes extension (corporeal essence) implies impenetrability, but there is no causal link between any corporeal force and the impenetrability that is a prerequisite for corporeal interaction. Euler shifts the essence of body from extension to impenetrability,[18] and explicitly relates impenetrability to force, but the nature of that relation remains elusive, and in sum Euler's position on the matter is no more illuminating than that of Descartes. More significantly, in both cases the problems are the result of what Gaukroger correctly describes as "a strong Cartesian desire to base mechanics upon something which was intuitively self-evident".[19] For both Descartes and Euler that something was primally geometrical, and therefore unlike the forces that bodies display in their interactions. Euler's 1750 "Recherches" paper does not succeed in its main purpose, but it constitutes an important attempt to clear up a puzzle of long standing in what we might abbreviate as the "non-Leibnizian" mechanical tradition.

3. PERPETUAL MOTION AND THE *MACHINA MUNDI*

My final problem area for the mechanists takes as starting point certain cosmological implications of hardness or impenetrability being essential corporeal characteristics at work at the particulate level. The key example is Isaac Beeckman, one of the earliest authors of a collision theory, who shows his immediate awareness of these implications in a passage of the *Journal* that rounds off his 1618 discussions on collision theory.[20] One of Beeckman's collision rules is that for equal hard (*i.e.*, totally non-elastic) bodies colliding head-on with equal speeds, they "come directly to rest, the motion of each being annihilated" (*Journal*, I, p. 266). He then applies this rule to the general case of head-on collision. Given that speed and "corporeity" [*corporeitas*] conjointly measure force or motion, equal and opposite forces undergo mutual annihilation, irrespective of the speeds and corporeities taken separately. If two unequal bodies collide head-on with unequal speeds, there will be a loss of motion equal to twice that of the body with the lesser force, and the remaining force will be distributed conservatively between bodies so that they move after the collision with the same speed in the same direction (*ibid.*). However, these conclusions lead Beeckman to the following prophetic "quaery":

> [Marginal heading:] Motion in a vacuum never increases, but decreases. Why, therefore, is there not a final universal inactivity [*quies*]?
>
> Taking all this [the collision rules] for granted, [it will be seen that] motion in a vacuum can never be understood as tending towards a faster motion, but because of equal collisions [*aequales occursus*, involving equal and opposite forces] all bodies eventually tend towards rest. From which it follows that only Almighty God was able to maintain [*conservo*] motion by moving [at the Creation] once and for all time the largest bodies with the greatest speed, which successively activate and regenerate the remaining bodies perpetually tending towards rest. (*Journal*, I, pp. 266-267).

Beeckman's principle of "mutual annihilation", coupled with his atomistic interpretation of nature, implies a universe in which in the long run all motion will have disappeared. Finding this melancholy conclusion disturbing, Beeckman does not reject the principle (as Descartes was to do), but retains both it and the idea of persistence of motion by charging God with the initial creation of what seems to be an infinite quantity of motion. Equal motions will eventually disappear through head-on collisions, but the indefinitely large initial quantity of motion ensures that the final *quies* is postponed for an indefinitely long time, thus ensuring a state of activity at any given time in the future. Beeckman's first impulse in his quest for a solution to the dilemma is to appeal to God's omnipotent creative power.

Later, Beeckman speculated on two other possible ways out of the difficulty, the first involving a counterfactual conditional, the other the elasticity of natural bodies. In a *Journal* entry of September 1629 (*Journal*, III, pp. 128-129), the full sense of which is not easy to establish and requires a degree of interpolation, he argues that the prospect of an ultimate universal *quies* would recede if the motive and resisting forces of bodies did not depend on their corporeity, but only on their speeds. On such a hypothesis, there would still be mutual annihilations when bodies (of whatever size) collided head-on with equal speeds. But there would be compensations for these losses, he seems to be saying, because large bodies at rest could be set in motion by even the smallest bodies, and with the *same* speeds as the smaller bodies had before (and of course after) impact, thereby augmenting in a sense the availability of the original motive force. But such a hypothesis is nonsense, for "who would believe that an iron ball of the greatest weight, shot from a cannon, could be checked by a single atom moving equally fast?" Corporeity remains a factor, on all fours with speed, in determining the motive power of moving bodies, so there cannot be any compensations of the sort created by the hypothesis: the principle of mutual annihilation is still cosmologically unsettling.

So far Beeckman had been thinking solely in terms of absolutely hard bodies, that is bodies devoid of elasticity.[21] Accordingly, another way out of the impasse might be to consider the effects of elasticity, as he does in a note of November 1634 which vies with the September 1629 entry in interpretative difficulty. Again the argument seems to depend on the compensation idea:

> [Marginal heading:] How motion may not be destroyed.
> Two bodies colliding with equal force [*vis*] come to rest, unless they rebound, and moreover are compressed together in such a way that they disengage at a later appropriate time and move in a greater degree than was the case at the moment [of collision]. So that for all parts [of the bodies] connected with the rebound, *etc.*, taken together, the original motion is maintained [*conservo*] and renewed, just as a bent steel plate springs back at a given suitable moment. Therefore motion, being created by God once and for all time, is conserved [*conservo*] eternally no less than corporeity itself. (*Journal*, III, p. 369).

The passage is not free from obscurity, but it seems safe to conclude that in his later years Beeckman was beginning to entertain the possibility that elasticity (whatever its mechanism) might provide a means of averting an unpalatable consequence of his collision theory.

Now Descartes knew Beeckman personally, he had read the *Journal* and was acquainted with the guiding ideas of its author, and Beeckman's consequent influence on Descartes' natural philosophy was very considerable, as is by now widely recognized.[22] In particular, there can scarcely be any doubt that the theory of hard body collision in the *Journal* provided Descartes with his starting point in the elaboration of his own rules of collision. Furthermore, although it is (I think) impossible to prove through an appeal to textual evidence, I am satisfied that the cosmological difficulties of Beeckman's collision theory were very much in Descartes' mind when he was working on his Laws of Nature and the seven collision rules in *Principia Philosohiae*, Part II. Like Beeckman, Descartes chose perfectly hard bodies as the subjects of his collision theory, but unlike Beeckman, Descartes proposed from the outset a divinely-grounded law of conservation (or "non-vectorial" motion) with reference both to the universe and to systems of interacting bodies viewed in isolation. There was therefore no principle of "mutual annihilation" in Descartes' physics. God's conserving power did for Descartes the job done for Beeckman by God's infinite power at the Creation.

Leibniz found a different solution to the problem, though he was (presumably) unaware of either Beeckman or his cosmological conundrum. We need not labor the familiar Leibnizian themes that save the Leibnizian universe from a final unwinding: the universal pre-established harmony, monadic autonomy, the conservation of *vis viva*, the rejection of absolutely hard bodies coupled with the elasticity (to varying degrees) of all corporeal particles.[23]

Equally unaware of Beeckman was Isaac Newton, in whom we find a repeat of the Beeckman story, though with a Newtonian ending. Again it would be to labor the now-familiar to detail Newton's struggles in the Quaeries to the *Opticks* (4th ed. 1730) to come to terms with the cosmological implications of interactions between hard bodies endowed only with the passive *vis inertiae* (see McGuire 1968, and McMullin 1978*a*). This kind of mechanism implied the final decay of all motion, for the reparation of which active principles, the non-mechanical agents of God, must be on hand to stay the universal collapse that Beeckman had feared 80 years earlier. Note by the way the striking similarity between Beeckman's note of November 1634, quoted above, and the last sentence of the following excerpt from the *Opticks*:

> ... by reason of the Tenacity of Fluids, and Attrition of their Parts, and the Weakness of Elasticity in Solids, Motion is much more apt to be lost than got, and is always upon the Decay. For Bodies which are either absolutely hard, or so soft as to be void of Elasticity, will not rebound from one another. Impenetrability makes them only stop. If two equal Bodies meet directly *in vacuo*, they will by the Laws of Motion stop where they meet, and lose all their Motion, and remain in rest, unless they be elastick, and receive new Motion from their Spring ... (Newton 1730, p. 398).

Generally, therefore, for Beeckman, Descartes, Newton, Leibniz, and in short all supporters of the mechanical philosophy, it was believed, though on widely differing grounds, that the universe is continually on the move, that it has been continually on the move since the

Creation, and that it will remain on the move until God's appointed time for its dissolution in the final Conflagration. Now the mechanists' universe was a mechanical universe, a machine, "an or *self-moving engine*", as Boyle described it in his *Origin of Forms and Qualities* (Boyle 1979, p. 49). The machine of the universe was therefore a *perpetual motion machine*. At the same time, the mechanical philosophers (at least those who understood mechanics) claimed that mechanical perpetual motion was an impossibility: no purely mechanical perpetual motion machine could be constructed. An examination of the perpetual motion question in the period, and of this *prima facie* paradox, promises therefore to throw some sort of new light on the mechanical philosophy.

For convenience, and by way of clarification, I shall distinguish four types of *perpetuum mobile* (PM, also abbreviating *perpetua mobilia*) which can be identified in the literature.[24] This four-way division is suggested in large measure by distinctions implied or explicitly made by those who wrote on or constructed PM, and in particular, in drafting the descriptions of the second, third, and fourth types of PM, I have had at my elbow the three-way division of PM given by Lana Terzi in his *Magisterium Naturae et Artis* (Lana Terzi 1684-92, I, pp. 483-487). I shall also distinguish between "Old Style" (O.S.) and "New Style" (N.S.) PM, designations that refer respectively to the periods before and after the creation of the first two laws of thermodynamics in the mid-19th century.[25] To avoid cluttering the pages with references, the historical examples mentioned can be found easily in one or other of the following fundamental sources (listed in note 24): Wilkins, Schott (both works), Lana Terzi, Dircks (both works), de Solla Price (1959), White, Needham, and (probably) Ichak.

PM 1 (O.S.). A system or device that maintains indefinitely the motion it already has, in the absence of dissipative influences. Examples: a single body moving "inertially" through empty Newtonian space; a symmetrical rigid body rotating in empty space about an axis of symmetry; Stevin's *clootcrans* in disequilibrium;[26] or to take a N.S. equivalent, an electric current induced in a supercooled metal ring. PM 1 (O.S.) correspond to PM 3 (N.S.), and operate normally in idealized conditions.

PM 2 (O.S.). A device that moves perpetually with the same motion or the same repeated sequence of movements, overcoming dissipative influences alone, or in addition doing useful work of some kind. The device receives no power from sources other than the *inertia* (*impetus*, in pre-Newtonian terminology), or the *weight* and inertia, of its constituent bodies. Examples: Bhāskarā's "mercury wheels" (*c.* 1150), Villard de Honnecourt's "mallet wheel" (1235); the automaton perfected by Comenius in a state of Pansophico-Pythagorean elation by 1639, which consisted of three weights moving on seven rolling cylinders (Nováková, 1979); the wheel devised and presented before Charles I (*c.* 1640) by Edward Somerset, second Marquis of Worcester; J. E. Elie-Bessler Orffyreus's notorious wheels of misfortune (1712-1719); or more ambitiously, closed-cycle water mills, in which the water-wheel operates an Archimedean screw which raises water to the height from which it fell (indeed higher, "in practice"), the water falling again to turn the wheel, which meanwhile operates also a grindstone or some other productive unit. There are hundreds of examples of PM 2 (O.S.) in the history of the subject, and those that could promise power in addition to mere perpetuity were in great demand during the 17th and 18th centuries.[27] PM 2 (O.S.) are examples of PM 1 (N.S.), and correspond to PM (Lana) 1. Lana's *primum genus* of PM is "purely artificial", that is, it is a purely mechanical device involving only the motions of heavy or light bodies.

PM 3 (O.S.). As for PM 2 (O.S.), except that the power sources, which are still internal to the device, are principally what at the time were described as *physical* or *natural* agencies (usually magnetism, but also the actions of chemical mixtures, fiery spirits, *etc.*), rather than inertia or weight, although the latter normally contribute in some way to the performance. Examples: Peter Peregrinus of Maricourt's magnetic wheel (1269), aided by impetus and gravity, repeated in many guises in the succeeding centuries; John Taisnier's magnetic "hemicycle" (before 1579), in which a loadstone at the top of an inclined plane draws up an iron ball which then falls back via a semicircular groove to resume its motion up the plane; the chemical PM of the Paracelsians; the arrangement proposed by John Bernoulli (his 1690, Appendix), in which two miscible liquids of different specific gravities separate at a filter attached to the bottom end of a vertical tube placed in the mixture, the lighter fluid thereby rising up the tube and overflowing through its open top end back into the mixture. There are many examples of PM 3 (O.S.) in the literature. Like PM 2 (O.S.) they are examples of PM 1 (N.S.), and correspond to PM (Lana) 3. Lana's *tertium genus* of PM is physico-mechanical (see description of PM (Lana) 2 below), but the perpetuity is itself artificial in the sense that the constructor decides on and has control over the disposition of the natural or physical power sources as integral components of the device: in the case of PM (Lana) 2, which are also physico-mechanical, the natural agents of power (*e.g.*, atmospheric or celestial influences) lie outside the constructor's control.

PM 4 (O.S.). A device that performs the same kinds of motions or tasks as PM 2 or PM 3 (O.S.), but avails itself of an endless supply of power from natural or physical sources *external* to itself, such as the sun, moon, or other celestial bodies, winds, rivers and natural fountains, seasonal or daily atmospheric changes. Examples: Peter Peregrinus of Maricourt's *terrella* (1269), a spherical loadstone mounted on a rotatable axis orientated along the N-S celestial axis, the idea being that the loadstone rotates sympathetically with the diurnally rotating celestial sphere, which is the source of the magnet's "natural virtue"; clocks and other engines operated by the tides; Jean de Hautefeuille's perpetual pendulum clock (his 1678), whose operation is based on the lengthening and shortening of a specially prepared rod of fir, which are caused by changes in atmospheric humidity as night follows day, the shortening being used to raise a weight. There seem to be fewer examples of PM 4 (O.S.) in the history of PM than there are of PM 2 and PM 3. Viewed in a *cosmological* context, PM 4 (O.S.) are examples of PM 2 (N.S.), and they correspond to PM (Lana) 2. Lana's *secundum genus* of PM is "physico-mechanical", that is, it is a "mixture" of the natural (physical) and the artificial (mechanical): natural in that it employs an external natural source of power to do useful work and to offset the internally-created deficiencies of pure mechanism: artificial in that art is indispensable if natural powers are to be available in manageable concrete form.

Of this quartet of types of PM ("O.S." henceforth taken as understood), which of them were thought possible, which impossible, during the "Old Style" period? To take PM 1 first, their possibility as other than abstractions would have been admitted only by those who subscribed to a non-plenumist account of the material world, to some form of a persistence-of-motion principle, and (for the case of a single body in rectilinear motion) to an infinite universe.

PM 1 form a special class in this context: although they are certainly *mobilia*, they do not happily bear the appellation "machine" or even "device".

PM 2 were held to be impossible, both in theory and in practice, by those who were alive to certain implications of Peripatetic natural philosophy, and by those who were *au fait* with the principles of mechanics. For the former there was the distinction between the celestial and terrestrial regions, and between the differing kinds of motion corresponding to these regions. There were the corresponding principles that nothing violent, and therefore nothing produced by Art, that is, nothing mechanical, can be perpetual; that all natural motions beneath the moon have rest as their *terminus ad quem*, and that there are on earth no *vires infatigabiles*. Also, there was the realisation, shared by Peripatetics and non-Peripatetics alike, that the composite bodies in the sub-lunar world cannot be everlasting and cannot therefore constitute the fabric of any machine. Such notions informed the anti-PM writings of Leonardo da Vinci and Cardano (Duhem I, 1905-6, pp. 52-60), and the (translated) title alone of Domenico Mellini's tract against PM conveys the gist of the story: "A Discourse in which it is proved, against the opinion of some, that there cannot be found by artificial means, nor given to a body composed of corruptible matter, a continuous and perpetual motion" (Mellini 1583). A century later, Lana supported his dismissal of the possibility of PM 1 (Lana) - equivalent to PM 2 - by noting that it would be contrary to the nature of the heavy (or light) bodies composing an alleged PM of that kind. They tend naturally towards rest, and have no internal principle of perpetual motion, no internal principle of continual power or activity; *impetus* and *momentum* do not last forever.

For those who knew something of mechanics, whether tacitly as practical men, or designatively as theoretical men, there was the inescapable fact that in the domain of mechanical artifacts and their operations, "nothing will come of nothing". This truth took different forms, which were variations on what was to become the principle of virtual displacements (virtual work), first enunciated by John Bernoulli in a letter to Varignon of 1715. From its early implementation in the statical work of Jordanus de Nemore (*fl. c.* 1220), through to the 17th century, when it received a fairly clear formulation from Descartes (To Huygens, 5 October 1637: AT(NP), I, pp. 431-447), the principle of work, expressed either in terms of displacements or the operationally equivalent velocities, was shared by all serious writers on mechanics, and was the theoretical basis of arguments rejecting PM 2 on mechanical grounds. In his unpublished "Traicté de Mechanique et specialement de la conduitte et Elevation des Eaux" (early 1640s?), Roberval stated the work principle, and made what seems a veiled allusion to PM 2 inventors: "... one must not hope to save both time and power [*puissance*], nor to produce a large effect with little force, unless over a long time; and this is where ignorant people usually go astray, and cause others to make fun of them, and of science too, on which other ignorant people often wrongly lay the blame" (Duhem, II, p. 210). The first important published rejection of PM 2 on these grounds was Philippe de La Hire's short article in the *Journal des Sçavans* for 1 August 1678:

> Of those who claim to have discovered perpetual motion, there is not one of them who does not agree that two weights, disposed so that they can move in their natural direction in times or along paths that are reciprocally as their weights, will remain in equilibrium. Nevertheless, there is no perpetual motion from which one cannot draw a conclusion quite opposed to this principle . . (de La Hire 1678, p. 165; also in Dircks 1968 (1861 + 1870), I, p. 102 (bad translation), p. 514 (French)).

To these arguments we may add Leibniz's well-known use of the work principle and the impossibility of PM 2 as *principia* in his demolition of the Cartesian measure of force (Costabel 1960; Iltis 1971).

Rejection of PM 2 was therefore *de rigueur* for all those who were not cranks or at best innocent utopians. However, the same does not hold for PM 3 and PM 4. As far as I can judge from the evidence, *no one* in the "Old Style" period - whether crank, utopian, Peripatetic natural philosopher, adept in mechanics, or adherent to whatever degree of the mechanical philosophy - rejected outright the possibility of some version of a PM 3 or PM 4. Before exploring the reasons for this dichotomy between PM 2 and PM 3 and 4, and its implications for our understanding of the mechanical philosophy, I must present some of the evidence, which is culled from the writings of both major and minor writers in the period. The presentation is approximately chronological.

Isaac Beeckman firmly rejected the notion that in practical mechanics one could dispense with the principle that gains in force cannot be achieved without corresponding losses in time (the equivalent of the work principle). In his *Journal* there are several discussions (in Dutch) on the futility of attempting to construct PM 2, and in each case the argument turns on the "force-time" principle. For example, one of the marginal headings (May-June 1627) reads: "To move anything whatever with any force whatever in any time whatever is to discover perpetual motion".[28] Yet the *Journal* contains accounts of possible PM 3 and 4, including the PM 3 that Cornelius van Drebbel presented to James I of England shortly after entering the King's service as engineer in 1605. In this device a "fiery spirit" was harnessed to activate an ensemble of globes that imitated the motions of the sun, moon, and heavens.[29] Beeckman took such devices seriously; he was certainly interested in them, and he was in favor of attempts to set up PM 4 by utilising natural powers such as tidal movements (*Journal*, II, p. 353).

As we have seen, Descartes stated the work principle explicitly, and so he ridiculed pretended PM 2 he learned were being produced in Holland (To Huygens, 18 February 1643: AT(NP), III, p. 806; cf. Huygens to Descartes, 26 May 1642: *ibid.*, p. 790). Yet in his youth he himself had speculated on the possibility of achieving PM using magnets to activate an automaton, or employing a body (if such could be found) whose weight varied with the phases of the moon (*Cogitationes Privatae*: AT(NP), X, pp. 230-231). More significantly, in *Regula* XIII appears the following passage:

> ... we commit a sin of omission each time we disregard some condition required for the determination of a question, which is contained expressly in the question or is to be taken as understood in some way. For example, suppose we are in search of perpetual motion [*motus perpetuus*], not the natural kind, such as the perpetual motion of the stars or of fountains, but a perpetual motion produced by human industry. Now several have believed that it could be done, reckoning that the earth rotates perpetually on its axis and that the loadstone retains all the properties of the earth. Sup-

> pose then that someone thinks he is going to discover a perpetual motion in this way if he arranges the stone so that it moves in a circle [*in orbem*], or at least so that it communicates to iron its motion together with its other powers [*virtus*]. Even should he succeed, he would not have produced a perpetual motion by art [*arte*], but would have used only a motion that is natural [*naturalis*], just as if he had placed a wheel in the flow of a river so that it turned for ever [*semper moveretur*]. Therefore he would be omitting a condition required for the determination of the question, etc." (AT(NP), X, pp. 436-437).

Leaving aside a possible confusion between Maricourt's magnetic wheel (which incorporated a loadstone-bearing arm pivoted at the centre of a circular iron-toothed arrangement) and his *terrella* (see above), it is clear from this passage that for Descartes it was against the rules in the quest for PM to employ *natural* powers, which include at least the powers of magnetism and flowing water, with fountains and celestial perpetuities as exemplifications of such powers in operation. The quest for PM properly speaking was the quest for PM 2, which Descartes believed was futile. Devising PM 3 and PM 4 was not playing the game according to the rules: but note that Descartes was not dismissing them out of hand, as he did PM 2.

For Huygens, the impossibility of PM 2 derived directly from "Torricelli's Principle", which Huygens used to good effect many times in his mechanics (see my 1980*b*, pp. 168, 189-192): if the principle were not true, it would be possible to build a PM 2. But it did not follow that PM 3 or PM 4 were impossible. In January 1660 Johannes Joachim Becher, the creator of phlogiston, visited Huygens at The Hague to tell him about two PM that he had built at Mainz, and which moreover had been going for six months. Writing to Pierre Carcavy on 26 February the same year, Huygens mentioned that Becher had sent him plans of his machines, though without the explanatory text that would reveal the secret. All Huygens could deduce from the plans was that

> ... one of the machines (for he has two different inventions for perpetual motion) is purely mechanical and the other physico-mechanical.

> As for the latter, the thing does not seem to me absolutely impossible, but as for the other I confess it passes belief. (*Oeuvres*, III, p. 28).

Towards the end of his life, Huygens wrote to Leibniz that whoever thought perpetual motion was possible *mechanicè* did not know what he was talking about, but as for perpetual motion "*physico-mechanice*, it still seems there is some hope, such as by using the loadstone" (Huygens to Leibniz, 12 January 1693: *Oeuvres*, X, p. 384). This mention of the loadstone, the perennial cynosure of the PM visionaries, reminds one of a passage from an earlier letter to Oldenburg of 26 June 1669:

> I have seen in a letter from Monsieur du Hamel mention of another of Monsieur Hooke's clock inventions, which moved with the help of a loadstone: if he uses no other motor [*moteur*], it would be a means of perpetual motion and would be an admirable invention. I beg you to send me any news of the matter, in so far as it is possible to learn anything of it. (*Oeuvres*, VI, p. 460).

In the late 1660s Isaac Newton toyed with a few ideas for PM using magnetic and gravitational forces. In the manuscript notes entitled "Quaestiones quaedam philosophicae", begun after 1664, Newton speculated that PM might be possible if one could harness the continual stream of subtle particles that he then conjectured might be the cause of gravitational force. "Whither y[e] rays of gravity may bee stopped by refecting [*sic*] or refracting y[m], if so a perpetuall motion may bee made one of these two ways", and he follows with sketches of two wheels, one turning horizontally under the impact of the gravitational flux descending obliquely on what seem to be wedges distributed uniformly round the circumference, the other turning vertically under the impact of the downward flux striking one half of the wheel (fitted presumably with paddles), the other half being shielded from the flux by a block of some sort. Magnetic force might also be due to streams of subtle particles, so we find Newton offering hints and suggestions for magnetic PM, for example: "Whither a loadstone will not turne round a red hot iron fashioned like wind-mill-sailes, as ye wind doth y^{m}", or: "Whither a magneticall pendulum is perpendicular to ye Horizon or not, & whither iron is heaviest w^{n} impregnated, or when ye north pole or south pole is upmost. Coroll. A perpetuall motion." (Westfall 1980, pp. 90-92, Figs. 3.1 and 3.2). It is worth noting that because of Newton's mechanistic interpretation of magnetic and gravitational forces, his gravitational wheels, and the adumbrated second magnetic device given above, are PM 4, and to this extent they share similarities with Maricourt's *terrella*. The half-dozen devices he has in mind in this text are in fact all PM 3 or PM 4: it is inconceivable that Newton could have entertained the possibility of PM 2, although to my knowledge there are no texts that are explicit on the matter.[30]

As I have already mentioned, Lana rejected the possibility of PM 2, that is PM (Lana) 1. However, he claimed that PM (Lana) 2, corresponding to PM 4, were both theoretically and practically possible, "verum etiam de facto"; and he believed that PM (Lana) 3, corresponding to PM 3, were also feasible, and could in fact be constructed in many different ways. Lana was clear about what was possible and what was not: "If anyone expects to obtain purely artificial perpetual motion solely by the motions of heavy and light bodies, by the disposition and arrangement of some mechanism, they will labor in vain. But it will be otherwise if they use some other natural motion as an auxiliary" (Lana Terzi 1684-92, I, Tract. III, Book VIII, Prop. IX, p. 487).

The physicist and horologist from Orléans, the Abbé Jean de Hautefeuille (1647-1724), best known perhaps for his 1675 priority dispute with Huygens on the invention of the spring balance, provides us with several illuminating pages on the conceptual issues attaching to the PM question. In his earliest publication on the matter, his *Pendule Perpetuelle* (1678), he presents (among other things) the perpetual pendulum clock briefly described above as an example of PM 3. After describing the instrument, Hautefeuille broaches the question of whether or not his invention will be a PM, "should it succeed". Some say no, on the grounds that it employs a natural agent, not solely weights or springs. Others say yes, on the grounds that "weight and spring are natural agents like all the others", so for them Hautefeuille's perpetual pendulum, which employs a natural spring to raise a weight, represents "a happy marriage between Art and Nature".

Naturally, Hautefeuille shares the latter view. He has examined the purely mechanical approach, in which only weights or springs of some sort are used (including pneumatic or hydraulic devices), or in which the machine "has in it the principle of its own motion, and continues it on its own without requiring a new external cause that might renew this principle." He has examined also the purely physical approach, represented for example by a machine "that imitates the movement of the heart in animals, the movements of the tides, and some other similar movements" (p. 7). As for the first approach, however, he sees in it a contradiction in terms:

> ... How is a machine meant to have within it the principle of its own motion, and not receive it from Nature? For perpetual motion to be possible mechanically, by art, it would have to be the case that a machine set itself in motion on its own, without that being natural; that a body implant more motion in another body than it has itself; that a body not lose its motion in communicating it to another; that a heavy body rise on its own, or not resist being lifted; that it be heavy at one time, and light at another; all of which are impossible, and imply a manifest contradiction. (p. 8)

No wonder therefore that the greatest mathematicians have believed mechanical PM to be impossible: Descartes, Huygens, de La Hire (in his *Journal des Sçavans* article which appeared just three days previously). So it is a ridiculous waste of time and energy to search for "something that is impossible and contrary to the laws of Nature" (p. 8).

Only the physical approach offers hope of success, and Hautefeuille lists some candidates: the alternate mixing and separation of oil of vitriol and oil of tartar, which would imitate the circulation of the blood in animals; sea-water in a hollow torus imitating tidal movements (from Drebbel); a rotating globe of sulphur (an electrostatic device?); a rotating spherical loadstone imitating the rotating magnetic Earth; the heliotropic substance given to Athanasius Kircher by an Arab merchant. But as far as Hautefeuille is aware, none of these candidates has yet made the grade; indeed some of them, far from moving perpetually, have not even begun to move. To achieve real perpetual motion, therefore, on must "join Art to Nature", which is what Hautefeuille considers he has done with his hygrometric pendulum.

Thirty-four years later Hautefeuille published *La Machine Arpantante* (1712), which took the form of a letter to Fontenelle, appropriately Secrétaire Perpétuel of the Académie des Sciences. In the context of a plea that the Académie improve the situation of inventors by establishing a "Compagnie des Nouvelles Découvertes", Hautefeuille mentions his own inventions, and makes the following important theoretical point about PM:

> I have found perpetual motion, not the mechanical kind using the weight of bodies, whose impossibility I have shown, and of which several mathematicians have provided demonstrations, but physical perpetual motion, based on the magnet and its properties. I used to think the latter impossible, and that the attraction between magnet and iron was analogous to that pushing bodies to the centre of the earth. However, on reflecting on the matter afresh, I saw that there is a difference between the two effects. The magnet attracts the iron by one end [*côté*] and repels it by the other, whereas bodies are pushed to the centre of the earth on all sides [*côtez*]. The power of the magnet can be diverted, changing its direction, but this cannot be done with the central force [of gravity]. The magnet can be surrounded by a machine so that part of the machine is outside the magnet's sphere of activity, which is not practicably possible on the terrestrial globe. Magnetic matter is in continual motion, and we can say that it is an invisible river, and that it is only a question of setting it going, and arranging things in an appropriate way, to receive its influence (pp. 5-6).

Finally, a recapitulation of his views on PM, together with references to his earlier work, was included in Hautefeuille's *Le Mouvement Perpétuel Magnétiqu* (1719), in which he describes how to restore continually the shortfall at the end-points of pendular swing by the use of fixed attractive poles carefully arranged in conjunction with a magnetically-loaded pendulum.[31]

This sequence of individual case-histories shows that the period of "the birth of modern science", of "the mechanization of the world-picture", saw a continuation of attitudes and aspirations that one feels must have attended the earliest awakening ideas of humanly-originated perpetual motion. Beginning in 12th-century India, passing to the Islamic world of the 13th century, and from thence to Medieval Christendom, the idea of PM almost certainly derived from centuries-long contemplation of the eternal motions of the heavens, and of the great natural cycles in the sub-lunar world, or from cognate philosophico-religious notions such as the endless cycles of *kalpas* of Hindu cosmology ("for Hindus as well as Taoists, the universe itself was a perpetual motion machine" (Needham 1965, p. 540). No doubt there was also the desire to imitate and extrapolate to their apotheosis such cyclical machines as water-clocks, of which Su Tzu-Jung's great astronomical clock-tower (1090) at Khaifêng might have been the most inspiriting example for visitors from further west (see Needham, Price, White, in note 24). Yet it is extremely unlikely that the PM idea, whatever its immediate origin, was not related in some way to the celestial and meteorological perpetuities that have been among the most awe-inspiring and most universally known natural phenomena in the history of our species. It would have been natural to wonder if analogous perpetuities on a more human scale could be reproduced by human industry and ingenuity. Could there be devised a machine that would enable Art to pass the supreme test of its emulation of Nature? Hence the appearance of devices that combined the "sheer love of machinery' with the perennial longing to create with one's own hands a working model of the eternally moving macrocosm (de Solla Price 1980, p. 76).

The task was impossible for those who put their trust in simple mechanical contrivance, but their broken hopes did not mean the dissipation of the Promethean vision. Rather, that vision needed to be re-shaped, an insight we owe probably to Peter Peregrinus of Maricourt. *Ars aemula naturae* still epitomized the aspiration, but now it could be satisfied only by coupling the simplicities of mechanical artifice, ruled by the intelligible and incorrigible laws of mechanics, with the complexities of natural process, ruled by the equally incorrigible yet only darkly perceived and understood laws of physics. Viewed in this light, the protagonists in our case-histories shared the same stage with Peregrinus, playing similar roles, and using props from the same wardrobe.

The framework of their efforts to marry Art to Nature was clearly Peripatetic. To borrow Schmitt's comment on a 16th-century figure dealing with a quite different example of the Art-Nature partnership, we could say that the general viewpoint shared by our protagonists "is acceptable within the canons of Peripatetic philosophy and [in addition] utilizes the new way of thinking in which man becomes an active participant in the natural process" (Schmitt 1976, p. 548). Yet that general viewpoint contains elements of a more technical nature that are equally Peripatetic in inspiration, though it may not be usual to interpret them in that way.

The arguments against PM 2, whether those of "moderate Peripatetics" like Lana (Heilbron 1979, p. 111) or those of mechanicians like Descartes, Huygens, or de La Hire, were either direct expressions of or logically related to the Peripatetic view that heavy bodies by their nature strive to reach the centre of the earth, and accordingly, that they do not act contrary to their nature by rising of their own accord. In this context, and purely for convenience, we may label this view "Torricelli's Principle", which in Torricelli's own formulation (1644) states that two heavy bodies joined together cannot move on their own unless their common centre of gravity descends (Duhem 1905-1906, II, pp. 1-2). Linking the Principle to Galileo's insights in his studies on the pendulum (Galilei 1638, pp. 96, 162-164), Huygens developed a generalized formulation in his *Horologium Oscillatorium* (1673), Pt. IV, Hyp. I: "If any number of weights begin to move by the force of their own gravity, the centre of their composite gravity cannot rise higher than where it was at the beginning of the motion" (*Oeuvres*, XVIII, pp. 246-250). Now clearly, if a weight, however small, could rise *sua sponte* through however small a distance, the work principle would simply disintegrate. Torricelli's Principle is a necessary condition for the establishment of the work principle, though it is not sufficient for the specification of its mathematical form. So Torricelli's 1644 formulation, its Huygenian generalization, and the work principle, are all directly dependent on (though not fully determined by) a fundamental component of Peripatetic natural philosophy. Earlier, when explaining why PM 2 were thought impossible, I distinguished between Peripatetic natural philosophy and "the principles of mechanics". We can now see that the distinction is not as substantive as my attempt at expository clarity might have implied. In short, PM 2 were impossible for basically Aristotelian reasons.

There is another equally informative way of explaining the impossibility of PM 2. They were impossible because the ubiquitous field of gravitational force is immune to human manipulation (*pace* Newton with his anti-gravity screen), so there cannot be a machine using only gravitational force acting in *different* directions at *different* times; and such vectorial variability is a prerequisite for any cyclical sequence of non-kinematical movements, and therefore for any PM. By contrast, in the case of magnetic and other "natural" forces, whose effects are in general neither ubiquitous nor immune to human intervention, constructional ingenuity alone will ensure the directional variations through time presumed to be appropriate for kinetic perpetuity. That is what Hautefeuille was getting at in the passage quoted above from *La Machine Arpantante*. And this basic contrast between gravitational and other forces helps to explain why "the science of weights" has been of central importance in the history of *mechanics* right from the beginning, and why on the other hand the science of mechanics has had a much harder struggle to gain a position of central importance in the history of *physics*.[32]

The picture so far is at odds with the mechanist ideologies advanced by the principal protagonists whom we have had sharing the stage with Peter Peregrinus. Descartes will serve as the paradigm case. For him, mechanics is just "a small part of the true physics" (to Plempius for Fromondus, 3 October 1637: AT(NP), I, p. 421), all physical causes are *ipso facto* mechanical causes, and there is no distinction between the artificial and the natural, since both are mechanical in composition and operation:

> ... I recognise no difference between artifacts [*arte facta*] and natural bodies, except that the operations of artifacts are carried out for the most part via components [*instrumentum*] so large that they are easily apparent to the senses, for that is necessary so that they can be made by men. But on the contrary natural effects depend almost always on certain organs

> so small that they escape all our senses. And indeed there are no considerations [*ratio*] in Mechanics that do not apply also to Physics, of which Mechanics is a part or species. It is no less natural for a clock with this or that movement [*ex his vel illis rotis composito*] to tell the time, than it is for a tree grown from this or that seed to produce such and such a fruit ... (*Principia Philosophiae*, Pt. IV, Art. 203: AT(NP), VIII (i), p. 326. Cf. Descartes to More, 5 February 1649: AT(NP), V, pp. 277-278).

Descartes certainly recognises the inferiority of human to natural productions (to Reneri for Pollot, [April or May 1638]: AT(NP), II, pp. 40-41), but as far as the rule of mechanical law and the universality of mechanical structure are concerned, nature is on all fours with the works of man.

Yet the work principle, which prohibits PM 2, seems inapplicable in the case of PM 3 and PM 4, whose possibility must therefore be in some way compatible with that principle; and that compatibility cannot derive from any Cartesian conservation principle that applies both to machines in general and PM 2 in particular, and also to the *machina mundi*, or to large systems within it (the meteorological sphere containing the earth, for example). Descartes' principle of conservation of motive force (size x speed) is not related to his formulation of the work principle (also expressed in terms of "force": AT(NP), I, pp. 435-436).

It might be objected that his denial of PM 2 simply means his intuitive recognition of the validity of the work principle as an expression of some kind of conservation principle, and that on the other hand his acquiescent attitude towards PM 3 and PM 4 concords, say, with his claim that God is the creative source of all activity in the universe, from which it would follow that the conservation of the large (though not unlimited) amount of motive force created in the beginning ensures that Nature will always be able to supply PM 3 and PM 4 with power. However, the dissolution of the Nature-Artifact distinction, entailed by his mechanist ideology, means that *any* system of bodies, whether "artificial" or "natural", is a machine in precisely the same sense that a PM 2 would be, given that weight is a purely mechanical effect. Hence the work principle ought to apply to any such system of bodies, whether constructed or natural. I say it *ought* to apply, since the point holds only if Descartes' work principle is equivalent to or can be derived from an appropriate, universally applicable "force" conservation principle. But this is not the case for Descartes, since *his* conservation principle is the wrong one for this purpose. So the work principle, as stated by Descartes (and I imagine by everyone before Leibniz), is not a *principium mechanicae* (or even *physicae*), but is an empirical generalisation, a law-like statement about the operation of simple machines under the ubiquitous and inviolate influence of gravity. That is one of the reasons the work principle makes no appearance in *Principia Philosophiae*.

It might also be objected that for Descartes any given system of bodies, precisely because it is purely mechanical, cannot move perpetually on its own in any event, simply because of friction, so PM 3 and PM 4 are still possible, since they avail themselves of the almost endless supply in Nature of divinely-created force. Quite so, but then a difficulty arises. On this view, and given that *vacuum non dari*, it seems intuitively (a rigorous demonstration is impossible), according to Descartes' Laws of Nature and his seven collision rules, that there should be a final universal *homogeneity* of distribution of motive force throughout the universe. Beeckman and Newton feared a universal final state of rest: Descartes ought to have feared a final state of uniformly distributed motive force, in which no diversity is present. Descartes had no such fears, of course, though he provides no clue as to how his principles ensure physical variety in the universe at any future time (cf. More to Descartes, 5 March 1649, and reply, 15 April: AT(NP), V, pp. 315, 347-348). God's plans for the universe supply the answer, but they do not form part of Cartesian natural philosophy. These considerations on PM show the tension between the explanatory simplicities of the mechanical philosophy and the empirical laws of *nature*, between the Nature-Art dichotomy revealed by attitudes concerning PM, and the Nature-Artifact identity entailed by the canons of the mechanical philosophy

I see little reason to suppose that similar considerations do not hold, *mutatis mutandis*, for most of the other protagonists in this story. Hautefeuille's argument against PM 2 is plain enough, but if PM 2 are 'contrary to the laws of Nature", he does not tell us why the same laws of nature do not rule out PM 3 and PM 4, which he advocates as being feasible, since they represent the marriage of Art to Nature. For Huygens, his generalisation of Torricelli's 1644 principle, the most important single hypothesis in his mechanics, ensures the impossibility of PM 2, but Huygens provides no clue as to how it squares with his belief that magnetic PM 3 are not beyond possibility. As in the case of Descartes' work principle, Huygens' "Hypothesis I" in Pt. IV of *Horologium Oscillatorium* is an empirical generalisation, not a *principium mechanicae*. Generally, the world described by the proponents of the mechanical philosophy was not congruent with the world of which their mechanical theorems were true, since there was one important mechanical rule that did not apply to the *machina mundi*: that prohibiting the perpetual motion of *machinae artificiales*.

Two figures do not fit neatly into the overall picture: the young John Bernoulli, and Leibniz. The two-liquid arrangement that the 23-year-old Bernoulli described in the Appendix to his first published work, *Dissertatio Chymico-Physica de Effervescentia & Fermentatione* (1690), and which I cited above as an example of a PM 3, was in fact intended by Bernoulli to be a "Mobile Perpetuum pure artificiale". Rather than accept the traditional distinction between mechanical PM 2 and natural PM 3 and 4, Bernoulli refreshingly turns the tables by claiming the possibility of PM 2 *precisely because* (*a*) perpetual motions exist in Nature, and (*b*) Nature is mechanical in operation:

> ... Does not Nature herself, which is said always to operate according to mechanical laws, show that a perpetuum mobile is possible? To mention just one example, what is the unceasing [*perennis*] flow of seas and rivers, if not a perpetual motion? Are all things not accomplished mechanically? Consequently, it must be acknowledged that whatever does not go beyond the limits of mechanical laws is not impossible. So what prevents us from following Nature's lead in this matter, however imperfectly we may imitate her? ... (*Opera Omnia*, I, pp. 41-42; Dircks 1861 + 1870, I, p. 532.)

Bernoulli correlates *ars aemula naturae* and *machina mundi*, the laws of mechanics governing all. However, his unqualified enthusiasm in 1690 - assuming we can take the above text at face value - was soon to become tempered by more realistic considerations, probably as a result of Leibniz's influence.[33] In their correspondence of the mid-1690s we find Bernoulli mooting the possibility of mechanical PM (to Leibniz, 17/27 December 1695: Gerhardt, *Math.*, III, p. 224) in answer to its use by Leibniz in his *reductio ad absurdum* "mv^2" argument. In his reply Leibniz remarks that "whoever defends perpetual mechanical motion as being possible will also think it possible for water to rise up mountains on its own" (to John Bernoulli, 28 December 1695: *ibid.*, pp. 227-228). Bernoulli takes this to be an oblique reference to his 1690 *Dissertatio* and maintains that there is nothing absurd about PM in which a common centre of gravity rises either *sua sponte* or (as in the case of his own PM) because of some internal motion of the particles. But note the following clarification: "However, I believe, as you do, that perpetual motion is impossible to obtain through manipulating a machine of solid bodies, operated by artifice alone and by human industry" (to Leibniz, 18/28 January 1696: *ibid.*, p. 237). So Bernoulli had at last joined the tradition I have been describing, and I need scarcely underline the irony in the author of the above passage from the 1690 *Dissertatio* being also the author of the first formal statement, twenty-five years later, of the principle of virtual displacements.

In his next letter, Leibniz assures Bernoulli that he did not have the 1690 filter device in mind at all (otherwise he would not have written so candidly), and indeed goes on to express admiration for Bernoulli's invention, agreeing with Bernoulli's view that

> ... internal motion arising from principles other than gravitational ought to be taken into consideration, as for example if there were in a liquid an enduring or periodic fermentation, and you have in mind only physical perennial motion. For as you very rightly say, if the lighter liquid is placed on the heavier, and then thoroughly mixed with it, their common centre of gravity will rise; from whence accordingly it is established that some extraneous or physical force is required for the mixture. And perhaps then another physical force will be required to manage the filtration or to surpass whatever is responsible for [*causa*] maintaining

> the mixture, which maintaining cause might possibly be for example the simple tenuity of the parts. So I did not intend to scoff at your ingenious study. (to Bernoulli, 28 January 1696: *ibid.*, p. 242.)

There is no reason to doubt Leibniz's sincerity in this letter. After all, some 25 years earlier he himself had devised a PM 3, which consisted basically of a cylindrical bellows, one end of wood the other of lead, and immersed in water wooden end upwards, perpetually rising and falling as upward and downward thrusts accompany the bellows' alternate expansions and contractions.[34] Instances of this sort from Leibniz's early work seem to be in simple accord with the later position typified by the following remark from the plan for a projected *Elementa physicae* (1682-1684):

> ... the power of the effect and of the cause are equal to each other, for if the effect were greater, we should have mechanical perpetual motion, while if it were less, we should not have [perpetual] [*perpetuus*, which Loemker translates as "continuous"] physical motion ... (Leibniz 1969, p. 279).

It might therefore appear that Leibniz was straightforwardly another representative of the tradition that contrasted impossible PM 2 with possible PM 3 and 4 in the ways I have explained. That simple conclusion would be mistaken. In his middle years (roughly the last two decades of the 17th century), Leibniz's account of PM shows that he had something new and profoundly original to say on the mechanical philosophy and on its relations to the Nature-Art distinction.

Leibniz's coupling of the equality of cause and effect and the impossibility of PM 2, and their central role in the replacement of Descartes' conservation of quantity of motion by the conservation of *vis viva*, are familiar ground - though it may still be asked which principle takes epistemological precedence, or why the effect-cause equality is not a tautology, another example of "adequation". My concern here is rather the precise correspondence Leibniz sees between the effect-cause principle and the reality, indeed necessity, of *physical* perpetual motion. If there were an imbalance between the powers of cause and effect in favor of the effect, PM 2 would be possible, and if the imbalance were at the expense of the effect, there would result (one presumes, though Leibniz is not explicit on the matter) a continual depletion of corporeal activity in the universe, a situation effectively that of the final *quies* that worried Beeckman and Newton. The powers of (full) cause and (total) effect must be in perfect balance, and therefore physical perpetual motion must obtain in the universe God has chosen to create.

Given its provenance, this kind of physical perpetual motion cannot be identical to that exhibited by PM 3 and PM 4. In his "Illustratio Tentaminis de Motuum Coelestium Causis" (1689), Leibniz does argue at one point that perpetually moving automata "cannot be had unless by a perpetual miracle, for if ... we suppose a corporeal cause supplying forces in abundance, it would be not a mechanical but a physical perpetual motion ..." (Gerhardt, *Math.*, II, pp. 275-276). But all this means is that if forces in abundance compensate for dissipative losses, the automata are performing Leibnizian physical perpetual motion (because of the consequent cause-effect equality), and to that extent every PM 3 or PM 4 would be for Leibniz an example of physical perpetual motion in his sense. However, the converse does not hold: a Leibnizian physical PM is not necessarily describable as a PM 3 or PM 4.

To see why this is so, we turn to the *Essay de Dynamique* of 1692. Definition 3 concerns mechanical perpetual motion, which Leibniz here describes in evocatively Peripatetic terms:

> Mechanical perpetual motion (which is being sought in vain) is a motion in which the bodies are in a violent state [*un état violent*], and acting to leave it make no progress, but at the end of a certain time the whole thing is once again in a state not only as violent as that which obtained initially, but even moreso, since in addition to the re-establishment of the initial state the machine must also be capable of producing some effect or be of some mechanical use, without any external cause contributing to the whole operation. (Costabel 1960, p. 98).

Then in the accompanying illustrative Scholium we find an implied definition of physical perpetual motion:

> It is true that if accidental impediments are removed the falling bodies can rise again on their own exactly to their initial height. And that is necessarily so, otherwise the same force would not be conserved, and if the force diminishes the full effect will not be equal in power [*equivalent*] to the cause, but will be less than it. So it can be said that there does exist physical perpetual motion, for example a perfectly free pendulum; but this pendulum will never go beyond its initial height, and it will not even reach it if it operates to produce the smallest effect along its path, or if it overcomes the least impediment, otherwise it would be a mechanical perpetual motion. (*Ibid.*).

These lines recall the *Theoria motus abstracti* (1671) and its "Problema generale" on the generation of bodies, figures, and motions from given lines, motions, and bodies. There Leibniz draws a three-way distinction in the methods of generation: "the Geometrical, that is, imaginary but exact; the Mechanical, that is, real but not exact; the Physical, that is, real and exact... the Physical [method of generation] comprises those ways in which nature can create things, that is, in which bodies produce themselves by themselves" (Gerhardt, *Phil.*, IV, pp. 234-235; cf. Costabel 1960, pp. 44-45).

A Leibnizian physical PM is therefore a real system within which there is at all times a precise equivalence between causes and the effects they produce, and accordingly, that at a given moment the *sum* of the *vires vivae* and the powers of the effects produced up to that moment is the same as at any other moment. At a more basic level, and given Leibnizian universal elasticity, this means that the total amount of *vires vivae* of the particles constituting the system is conserved. So the perpetuity of motion in a Leibnizian physical PM is ensured primarily by the conservation of these powers within the system, it is not the result of a "natural" power continually compensating for the deficiencies of the system's internal mechanisms (though such an arrangement might easily be the means whereby the cause-effect equality is ensured). PM 3 and 4 are therefore subsets of the set of Leibnizian physical PM, which we now see include also PM 1.

It follows also that Leibniz's freely-swinging pendulum is in a significant sense a PM 4. During the period of his initiation into a more Leibnizian way of thinking on dynamical matters, John Bernoulli had argued, on reading *Specimen Dynamicum* (1695, Pt. I), that the Cartesian measure of force could be retained if "the effects" of the actions of two moving bodies were taken to be the respective distances gone on entering some resisting medium (to Leibniz, 8/18 June 1695: Gerhardt, *Math.*, III, pp. 188-190). In reply, Leibniz explained that to measure motive force correctly, the "cause-effect" axiom must be applied in a situation in which the motive forces of the test bodies are wholly absorbed by the medium during the production of "the effect", *and are wholly returned to them* during the inverse exercising of "the force of the effect" to produce the original motive forces. But in the case of Bernoulli's resisting medium, the forces are not "returned" at all. On the other hand, whatever the precise mechanical operation of gravity, "whose cause I do not deny comes from surrounding [bodies]", Leibniz observes that the gravitational medium both absorbs and returns fully the (maximum) motive force of (say) the pendulum bob used in *Specimen Dynamicum* to derive his measure of *vis viva* (to John Bernoulli, 24 June 1695: *ibid.*, pp. 193-194). The freely-swinging pendulum moves perpetually because at all times full cause = total effect, that is, at any given moment the *vis viva* + the power of the effect achieved is a constant. But it is also a PM 4, since the maintaining cause of its special kind of cyclical motion is an ever-present "natural power" external to the pendulum.

The importance of this point lies in the fact that although Leibniz shares with other mechanists the view that gravity is ultimately a mechanical effect, it was he who found a way of using the traditional formulation of an empirical law (the work principle) to derive a new measure of motive force and a new law of its conservation whose universality had the effect of making (in principle) all "natural powers" mathematically manageable mechanical powers, and at the same time of making the mechanical effects of gravity (also elasticity) into a "natural power". In short, Leibniz transformed the distinction between PM 3 and 4, and PM 2, into the distinction between systems in which *potentia* is conserved and those in which it is increased *ab nihilo*.

A formalized presentation of most of these new theoretical moves appears in Part II of the little-studied *Dynamica de Potentia et Legibus Naturae Corporae* (1689). In Definition I of Section I ("De Causa et Effectu Activis"), Leibniz introduces the notion of "an active state of affairs [*thema*]" or "a state of affairs endowed with power", from which a change [*mutatio*] may or may not result. Definition 2 then reads:

> If there are two states of affairs [*themata*] possessing power [*potentia*], and if from the action alone of one there follows the other, or at least it follows without it being a required supposition that its power already co-exists with the former, then the first state of affairs is the cause, and the later state of affairs the active effect or the absolute effect. But if from the cause there cannot follow another additional effect existing at the same time, the cause will be full [*plena*] and the effect total [*integer*]. (Gerhardt, *Math.*, II, p. 435).

An "Axiom and Definition 3" then naturally follow:

> The total effect is equivalent [*aequivaleo*] to the full cause, and so there is no perpetual mechanical motion, or if you like a cause cannot produce an active effect which can do more than the cause itself, but neither can it produce a total effect which can do less than the cause itself. For if part of the power is absorbed by impediments, it is nevertheless not destroyed, but is transferred to the impediments, which are to be counted in the total effect. And the measure of the quantity of power in a state of affairs is its quantity in another determined active state of affairs, which exists in the first state of affairs itself possessing power, or in its full cause, or total effect.
>
> The proposition that the total effect is equivalent to the full cause derives from a higher Metaphysics, and does not ply mere words, but deals with universal truths of things. Nature observes this law with the greatest constancy, and its truth can be

> understood from the fact that if it is removed there remains no way of estimating powers, or of determining from causes anything concerning the size of effects. (*Ibid.*, p. 437).

Of the 42 propositions that follow to complete Section I, Propositions 5-8 are of especial interest in the present context. In Proposition 5 Leibniz shows that "The total effect can reproduce the full cause or its equivalent [*gemellus*]", in Proposition 6 that "As far as the size of the power is concerned, it does not matter whether some total effect is produced mediately or immediately", and in Proposition 7:

> There is always the same power in any system of bodies not in communication with others.
>
> For since the bodies are not in communication with others (ex hyp.), any later state of the bodies will be the total effect of an earlier state (by Defn. 2), and so (by the Axiom and Prop. 6) will be equal in power. Thus the quantity of power is always the same.
>
> Hence if there is one body, it will always retain the same power, and if there are several colliding among themselves, there will always be the same power in their sum. (*Ibid.*, pp. 439-440).

From which it follows, Proposition 8, that "There is always the same power [but not the same quantity of motion] in the Universe", since the Universe is a system of bodies not in communication with any other bodies (*Ibid.*, pp. 440-441).

To complete the picture, we have Leibniz revealing in "De ipsa natura" (1698) that for him "the true, but little noted, distinction between *nature* and *art*" is that "universal nature is, so to speak, the artifice of God [*artificium Dei*], and such a work, indeed, that any natural machine whatever consists further [read "truly"] of infinite organs [*organis constet prorsus infinitis*], ... and so entails an infinite wisdom and power on the part of its author and governor" (Loemker, *Phil. Papers*, p. 499; cf. Leibniz-John Bernoulli letters of 1 March - 3 June 1699: Gerhardt, *Math.*, III, pp. 577, 580, 582-584, 590). Artificial machines, by contrast, consist - if at all - of only a finite number of artificial *organa* (which, it will be recalled, had the classical sense of *instruments* or *engines*). As Leibniz expressed it in the later metaphysical framework of *La Monadologie* (1714),

> ... a machine made by human art is not a machine in each of its parts; for example, the tooth of a brass wheel has parts or fragments which are not artificial so far as we are concerned, and which do not have the character of a machine, in that they fit the use for which the wheel was intended. But the machines of nature, living bodies, are still machines in their smallest parts, into infinity. It is this that makes the difference between nature and art, that is, between the divine art and ours. (Leibniz 1969, p. 649).

In the first section of this study I suggested that Leibniz represented "a natural and even inevitable *terminus ad quem* in the twin histories of the mechanical philosophy and the theory of mind". The evidence of this concluding section seems to me to support this view of Leibniz's position in the former history. The tension that emerged in the ordinary mechanical "tradition" between the Nature-Art dichotomy, revealed by attitudes to PM, and the Nature-Artifact identity prescribed by mechanism, disappears with Leibniz. The dichotomy is still there in Leibniz's thought, but it has become transformed through the fact that the natural law which Nature observes "with the greatest constancy", and which forbids PM 2, is the same law that ensures physical perpetual motion, that is conservation of *potentia*, in any system of bodies considered in isolation from others. The identity is there too in the sense that Leibniz is a mechanist, holding that all phenomena in nature are explicable in mechanical terms, yet again there is a profound transformation in that the *artificium Dei* consists in an infinity of natural machines within natural machines *ad infinitum*, each machine pursuing its history in accordance with Leibniz's new laws of dynamics, which govern artificial machines as well, and in accordance with the pre-established harmony, thereby avoiding the final homogeneous distribution of activity threatened by Cartesian mechanism. The tensions have gone, and one is left with the conviction that to claim that Leibniz is *the* pivotal figure in the history of the mechanical philosophy is to propose a well-founded hypothesis.

The Queen's University of Belfast

NOTES

*In addition to its initial presentation at Virginia Polytechnic Institute and State University, Blacksburg, Virginia, shorter versions of this paper were given, during November 1982, in the Department of Philosophy, Princeton University, and as the Woodward Lecture, Department of History of Science, Yale University. I wish to record my thanks to the participants at the Blacksburg Congress, to Daniel Garber and his Princeton colleagues, and to Martin Klein and his colleagues at Yale, for their comments and criticisms. I take this opportunity to record a special tribute to the late Derek de Solla Price, who was among Professor Klein's colleagues on that occasion, for his advice and encouragement, and for the continuing inspiration to be derived from his writings.

I am particularly indebted to the Research and Scholarships Committee, Academic Council, The Queen's University of Belfast, for the financial support that enabled me to carry out a major part of the research involved in the preparation of the present paper.

Thanks are also due to my departmental colleague Robert E. Hall for several fruitful discussions on the themes of Section 3, and for enabling me to read Mellini's 1585 *Discorso* by proxy; to Maren Dunsby, Department of German, Queen's University of Belfast, who facilitated a painless comprehension of some material in German; and to my colleague Adrian Mallon for helpful information on the *perpetuum mobile* question in the Academie des Sciences.

Except where otherwise indicated, all translations are my own.

[1] Though the tautologous nature of some of its explanations is mentioned in the literature less often than one might expect. See however Westfall (1971, p. 41)
[2] In discussion McMullin made the pertinent observation that the problem of *force* looms large in any kind of mechanical explanation, and that it is integral to a wholly satisfactory exploration of the mechanical philosophy in terms of structural explanations. I agree entirely, so although force will figure importantly in Section 2, this section of the paper is to be read only as an attempt to draw some partial conclusions from an application of McMullin's account of structural explanation to the mechanical philosophy.
[3] For an extended analysis of this feature of Galileo's methodology, employing the notions of causal structures and causal proportionality, see the important articles by Donald Mertz (1980 and 1982).
[4] Hence the irony in (*e.g.*) the following passage from Boyle's *The Origin of Forms and Qualities according to the Corpuscular Philosophy* (1666):

> ... if (for instance) it be demanded how snow comes to dazzle the eyes, they [the Schools] will answer that it is by a *quality* of whiteness that is in it, which makes all very white bodies produce the same effect; and if you ask what this whiteness is, they will tell you no more in substance than that it is a *real entity*, which denominates the parcel of matter to which it is joined *white*; and if you further enquire what this real entity which they call a *quality* is, you will find, as we shall see anon, that they either speak of it much after the same rate that they do of their substantial forms ... or at least they will not explicate it more intelligibly. And accordingly, if you further ask them how white bodies in general do rather produce this effect of dazzling the eyes than green or blue ones, instead of being told that the former sort of bodies reflect outwards - and so to the eye - far more of the incident light than the latter, you shall perchance be told that it is their respective natures so to act: by which way of dispatching difficulties they make it very *easy*

> to solve all the phenomena of nature in general, but make men think it *impossible* to explicate almost any of them in particular. (Boyle 1979, p. 16)

Of course the irony is tempered by the fact that Boyle himself was by no means innocent of the difficulties of finding proper mechanical explanations for many natural phenomena.

[5] The Scholastic term *adaequatio*, understood broadly, combines the notions of being suitable, fully representative, commensurate, in a certain proportion, or (more strictly) of being equal in some sense. Yet where appropriate the term can be understood as denying the possibility of *degrees* of suitability, *etc*. See Goclenius (1613, pp. 61-63) and Lalande (1972, p. 26).

[6] See Nancy L. Maull, 1980, in which she demonstrates the sharp distinction in Descartes theory of perception between mere sensations of secondary qualities notably color, and judgements about sensations, most importantly (she persuasively argues) judgements of a spatial kind that derive from the mind's innate "natural geometry". There is no resemblance between color sensations and the corporeal motions or pattens that produce them. However, "figures, unlike colours, can be the subject matter of geometry", so Descartes' natural geometry is his "way of explicating a sense in which the content of our sensations may be 'similar' to the physical objects that they represent" (p. 36). Maull's arguments do not necessarily conflict with the point I am trying to make, and in any event her main purpose is to call attention to the "foundational' role of Descartes' natural geometry in his overall scientific programme. Furthermore, it is obvious that the validity of my case depends (at least partly) on what I mean by "resemblance". Until I find the courage to tackle that intimidating exercise in definition, I must assume that there is at large some skeletal sense of the term that at the very least makes my claims intelligible.

[7] " ... sonus se habet ad sonum, ut nervus ad nervum", the *nervi* in question being lute-strings: AT(NP), X, p. 97.

[8] Although the "Avertissement" and running head in AT's presentation of the excerpts imply that they can be described collectively as "anatomical" (AT(NP), XI, pp. 545, 549-634), the passages occupying pp. 621-634 are strictly speaking not anatomical, but deal with a wide vari-

ety of topics, as indicated by their subheading "Problemata" and their re-allocation in the Index under "Varia" (p. 710). The "Excerpta Anatomica" (including the "Problemata") were composed by Descartes at different times, the passage under discussion dating probably from about 1635.

[9] A mathematical analysis of the passage is attempted in Ch. 6 ("Terminal Velocity") of my doctoral thesis, "Descartes' Dynamical Thought: A critical study of some problems, principles and concepts" (Queen's University of Belfast, 1964, unpublished, pp. 289-355, 347-55).

[10] Leibniz was sufficiently struck by this implication of the passage to write on his copy, beside the second reference to the action of the mind, "Ergo (NB) vis animae in vacuo arithmetica" (AT(NP), XI, p. 630, note *b*).

[11] See Art. 36, which is headed: "That God is the primary cause of motion and conserves always the same quantity of motion in the universe" (AT(NP), VIII(1), p. 61).

[12] That is, More received no reply to that part of the letter in which the query appears, because at the time Descartes was too busy preparing for his Swedish venture to devote his full attention to More's letter. However, Descartes did find time to draft a reply (AT(NP), V, pp. 401-405) to the opening pages of More's letter before leaving Egmond for Stockholm. For a history of the Descartes-More-Hartlib correspondence see my "Avertissement" to More's letter to Clerselier, [July-August 1655?], in AT(NP), V, pp. 628-42.

[13] In their respective articles (1979) and (1983), Peter Remnant and Daniel Garber argue that Leibniz was mistaken to attribute to Descartes the view that the mind changes *only* the direction of motion of those bodies it influences, their grounds being that there is no evidence for this view in Descartes' writings. Their claim is justified in so far as it relates to the "only", *i.e.*, to the question of the mind's ability to change also the bodies' speed, but there is sufficient evidence that Descartes did hold to a change-of-direction account of the mind's actions on body. The key idea here is "determination" [*determinatio, détermination*], the directional mode of a body's force, a concept that Descartes uses (in its verbal form) several times to describe the mind's action. Above we have had him writing that the mind "can determine [*determinare*] the flow of that very subtle fluid", and in *La Description du Corps Humain* (1648) he claims that even voluntary motions cannot arise

unless the organs in question are "well disposed", whatever our will, "though it is the soul that determines them [*i.e., the motions*] [... qui les determine]" (AT(NP), XI, p. 225). And on the next page appears the phrase "all motions that are not determined by our thought" ("tous les mouvements qui ne se determinent point par nostre pensée). Now it is true that for a given body changes in *détermination* can occur through changes in speed alone, the direction of motion remaining fixed [see my (1980*a*), p. 253)], and in the case of the mind's governance of the pineal gland such an understanding of the concept would seem appropriate. Yet Descartes' intention in using the concept in the context of mind-body interaction must be to emphasize changes in *direction*, since it was mind-originated changes in speed that he was trying (unconvincingly) to avoid, as I have argued above.

[14] "This third law of nature is that when a moving body collides with another, if it has less force to continue along a straight line than the other has to resist it, then it is turned in the other direction, and while retaining its motion loses only the determination of the motion. If however it has more force, then it moves the other body with it, and forfeits as much of its motion as it gives to the other body ... " (AT(NP), VIII(1), p. 65).

[15] *Cf.* Smith (1902, pp. 82-83); Hamelin (1911, pp. 372-73); Laporte (1950, pp. 245-48); Descartes (1966, p. 92, note 1); and more recently, Remnant (1979), and Garber (1983).

[16] What follows resumes note 27 of my (1980*a*, pp. 299-300). See also my (1977, p. 401).

[17] Thereby complementing Stephen Gaukroger's choice of the *Theoria* account for similar and other more extended purposes in his (1982). Gaukroger does not spotlight the incoherence in Euler's account of the force-impenetrability link that I argue for in the following pages, but deals with a related difficulty:

> If the only forces that acted were due to impenetrability, ... and if these acted solely to avoid penetration, then there would be no reason why the bodies should not simply stop, since this would be quite sufficient to avoid penetration, as well as being very economical. Inertia must therefore be dynamically effective, and if it is, then recourse to extra forces

> deriving from impenetrability is otiose at best. Indeed, it is wholly obscure how impenetrability, which is unquantifiable, could give rise to a quantifiable force: moreover, we would have to imagine each body having a potentially infinite reservoir of repulsive force so that it would have sufficient force to repulse any body in impact. So far as I can see, any coherent conception of a dynamically effective impenetrability would have to be along the lines of Boscovich's and Kant's accounts, where impenetrability is a function of repulsive force, and not *vice versa,* and this is a conception Euler could not have accepted. (Gaukroger 1982, p. 148)

On Euler's concept of inertia see further Harman (1983), especially pp. 238-39.

[18] See further Gaukroger (1982, pp. 138-140).

[19] *Op. cit.*, p. 141. Elsewhere in the same study Gaukroger writes:

> To simplify somewhat, for Euler as for d'Alembert it was a question of squeezing Newtonian mechanics into a Cartesian shape, a shape which, it was hoped, was to render it more certain and more fruitful. The certainty of the new mechanics was to derive from its structure, but more importantly from the nature of the foundations of that structure. These foundations were of a rigorously Cartesian type: they could not be doubted without fear of contradiction. This, I suggest, was the essence of Euler's foundational project. Newtonian dynamics, suitably reformulated, had to be shown not just to be true but to be necessarily true. (p. 134)

[20] For a more extended (yet less accurate) presentation of these cosmological speculations of Beeckman, see my (1973, pp. 380-385).

[21] The editor of the *Journal*, Cornelis de Waard, refers several times to Beeckman's collision rules as applying to *soft* bodies (*cf.*, *Journal*, I, p. 265, n. 6). In this de Waard was mistaken, as can be seen from *Journal*, I, p. 25; II, p. 45.

[22] The most recent account of the Beeckman-Descartes relationship appears in the valuable van Berkel (1983, *passim* and especially pp. 292-301), which is (surprisingly) the first monograph on Beeckman's life and work.

[23] On Leibniz's rejection of absolute hardness and admission of universal elasticity, see (*e.g.*) *Specimen Dynamicum* (1695) (Loemker, *Phil. Papers* pp. 446-447), the letters to John Bernoulli of 2 and 25 July 1697, and 22 July, 30 September 1698 (Gerhardt, *Math.*, III, pp. 438-439, 447-48, 515, 544-545), and to Thomas Burnet [1699] (Gerhardt, *Phil.*, III, p. 260).

[24] There is an extensive literature on PM, not all of it informed by historical understanding. The following items make up what seems to me the most valuable "short list". Wilkins (1648, pp. 211-246); Schott (1657, Pt. II, Class II ("De Machinis motum perpetuum aemulantibus"), pp. 329-382), also his (1661, pp. 621-660 ("Analecta Mathematica, sive Theoreses Mechanicae Novae . . nec non De Motus Artificialis Perpetui Possibilitate")); Lana Terzi (1684-92, vol. I, Tractatus I I ("De multiplici motus localis genere"), Bk. VIII ("De motus perpetuatione"), pp. 483-526.) Schott, Lana Terzi and Wilkins give excellent surveys of PM of their period. Diderot & D'Alembert (1751-1780, XII ("Perpétuel")); de Lalande, in Montucla (1799-1802, III, pp. 813-20 (depends heavily on D'Alembert)). Dircks (1968 (1861 + 1870)), though disorganized and devoid of historical analysis, is by far the best compilation of devices, machines, projects, plans and patents, texts and articles; also H. Dircks (1865) which contains a useful bibliography. Ichak (1914) which I have not seen (and seems unobtainable), but is recommended by others; Riddell (1917); Thorndike (1923-1958), III-VIII, as per indexes; Michel (1927); Hiebert (1962, *passim*): White (1964) (1962, pp. 129-134); Needham (1965, pp. 538-546); Costabel (1972); Elkana (1974, *passim*, especially pp. 28-30); de Solla Price (1959) and (1975 pp. 36-38).

[25] A PM 1 (N.S.) is a mechanism with 100 + % efficiency, and which thereby violates the First Law of Thermodynamics (principle of conservation of energy). A PM 2 (N.S.) satisfies the First Law, but violates the Second Law of Thermodynamics (the principle of entropy), in that it transforms completely a given amount of heat into an equivalent amount of mechanical energy, or more generally, that it operates in such a way that there is no decrease in the total amount of available energy within the system. A PM 3

(N.S.) simply moves forever, without resistance, and without useful output.

[26] Stevin's *clootcrans* is one of the most famous theorem-proving dodges in the history of science, as its presumably symbolic appearance on the spine and cover of the *Dictionary of Scientific Biography* seems to recognize. Stevin imagined a wreath of frictionless spheres draped over a smooth wedge placed so that its largest face is horizontal. Then he argued that the whole arrangement must be in equilibrium, and therefore, after subtraction of the catenary portion of the wreath beneath the wedge, that the unequal portions lying on the inclines about the vertex must also be in equilibrium, from which he deduced his equivalent of the "W $\sin\theta$" relation for the inclined plane. If the arrangement were not in equilibrium, Stevin claimed, the spheres would begin to move and, because of symmetry, continue to move, which would constitute an instance of perpetual motion, which Stevin took to be an absurdity. See *De Beghinselen der Weeghconst* (1586), Bk. I, Pt. 2, Th. XI, Prop. XIX, in S. Stevin, (1955-1966, I, pp. 174-179).

Curiously, it has not often been noted that Stevin's demonstration is invalid, and that its invalidity derives from the fact that *within the terms he himself would have accepted*, which were those of pre-Newtonian mechanics, such a perpetual motion would not be absurd at all. (See Dijksterhuis's editorial comment in Stevin (1955-1966, I, p. 179, note 1), and Dijksterhuis, (1961, p. 327.)) The consequent of Stevin's counterfactual supposition is an example of constant motion of an idealized closed cycle of bodies under a constant force, which was a straightforward situation in pre-Newtonian physics. His demonstration can be restored to apodictic health, however, by moving to a post-Newtonian framework, in which the resultant constant force arising from the presumed disequilibrium would produce a constant acceleration of the spheres, yielding a steadily increasing amount of kinetic energy over a limitless period of time, which is an absurdity. Dijksterhuis's analysis of Stevin's mistake (1961, *loc. cit.*) seems to me to miss the point. See rather Costabel (1972).

[27] So it comes as no surprise to find, as another instance of Francis Bacon's prophetic insight, that in the engine-rooms of Salomon's House "we have divers curious clocks ... and some perpetual motions" (*The New Atlantis*, 1626: *Works*, III, p. 164). These perpetual motions would

presumably have included also (perhaps preponderantly) examples of PM 3 and PM 4 (O.S.).

[28] (*Journal*, III, p. 3). On the same theme see further the large number of references to PM (most of the material being in Dutch) in the excellent Index in vol. IV of the *Journal*.

[29] See index references to Drebbel in vol. IV of the *Journal*. On 15 March 1631 Beeckman copied into his *Journal* (ff. 294v - 295v) a letter of 1613 from Drebbel to James I of England, in which Drebbel described some of his inventions, one of which was a perpetual clock. The text of Beeckman's copy, whose source was unknown to De Waard, is given in (*Journal*, III, pp. 439-442). Note that this letter is not the one (also to James I of England) in which Drebbel described the PM 3 mentioned in the text: that letter appeared in Drebbel's *Wondervondt van de eeuwighe beweging* (Alkmaar 1607).

I classify Drebbel's assemblage of celestial globes as a PM 3 because the active source of its motion was a "fierie spirit", which Drebbel extracted "out of the minerall matter, joining the same with his proper aire", according to Thomas Tymme, who described the device in his *A Dialogue Philosophicall* (London, 1612): (Dircks 1861 + 1870, II, pp. 10-16). However, PM 4 might be a more accurate classification, if (as seems likely) Drebbel believed that the operations of the fiery spirit were themselves directed by celestial influences, a belief of Paracelsian inspiration.

[30] It seems that Newton did not reply to W. J. 'sGravesande's letter of 7 August 1721, in which an impressed 'sGravesande describes the construction and operation of the wheel Orffyreus (see paragraph on PM 2 in text) made for Landgrave Charles of Hesse-Cassel (Newton 1977, pp. 143-147, also in translation in Dircks (1861 + 1870, I, pp. 39-42)), 'sGravesande maintained that since our knowledge of the laws of nature is imperfect, we cannot be certain that PM are impossible, and in fact he produced a demonstration of the invalidity of claims of the impossibility of PM 2. See further Hall's notes (*ibid.*, pp. 146-147), and Dircks (1861 + 1870, I, pp. 38-39, 187-190, 196-197; II, 112-115). The story goes that Orffyreus smashed up his wheel in a fit of rage on discovering that 'sGravesande was so bold as to examine for himself the wonder to find out how it worked. Newton did not reply either to J. C. F. von Hatzfeld, who wrote in a sour mood to Newton in 1723 or 1724 to inform him how disagreeable it was to have been refused admittance to Newton's house on the grounds

that he did not have a recommendation, when he had come only to show him his PM. If Newton refused this second invitation, Hatzfeld wrote, it would show that he was as unwilling of being convinced of his errors as most other men (Newton 1977, pp. 253-254).

[31] More case-histories could be added to this list, but considerations of space must prevail. I will however note one manuscript collection of curious items on PM mostly from the late-18th century: Archives Nationales (Paris), Ms O[1] 1293, dossier entitled "Bâtiments, Inventions: mecanique, horlogerie, mouvement perpētuel 1691 - 1786" (nos. 47-98). Most of the documents are from PM inventors to the Crown begging recognition and support. One of them is the physician Le Roux, who pestered members of the Académie des Sciences and the scientific advisors of the Comte d'Angiviller, Directeur Général des Bâtiments under Louis XVI, with his PM projects. Exasperated by the lack of enthusiasm for his ideas, he finally burst into print with a 4-page *Traité abregé d'un mouvement perpētuel, en partie méchanique, en partie élémentaire, aux savants et aux amateurs distingués* (Paris, 1784), which announced, without undue profuseness of explanatory detail, an arcane contraption that would work day and night "without the aid of men or horses", and would cost nothing (except for the expense involved in replacing the materials). (A copy of the *Traité* is included in the dossier in question in O[1] 1293.) Note Le Roux's unexpected use of the term "élémentaire" to mean "physique": like the more distinguished figures in our case-histories, Le Roux distinguished between purely mechanical PM, whose impossibility he recognised, and mechanico-physical PM, of which his invention was of course a successful example.

It seems to me odd, and for the moment historically enigmatic, that it was not until 1775 that the Académie des Sciences took a formal decision not to consider any more PM (PM 2, judging by the text) submitted to it for examination: *Histoire de l'Académie Royale des Sciences. Annee M. DCC. LXXV* . . (Paris, 1780), pp. 92-101. (See also Montucla (1799-1802, III, p. 820; and Dircks (1861 + 1870, I, pp. 190-191, 521-522)). This is not to say that the Academie had necessarily been sympathetic to PM submissions before that date. For example, my colleague Adrian Mallon points out that Philippe de La Hire would not have been permitted to describe himself as an academician in the title of his 1678 article in *Journal des Sçavans* unless (*a*) the piece had been read to the Académie, and (*b*) it had been approved for publication. Still, instances of this sort do not constitute formal pronouncements from the Académie on the impossibility of PM.

[32] For a good account of the whole question of the scope and meaning of "physics" during the 17th and 18th centuries, and of the problem of magnetism within that domain, see Heilbron (1979, esp. Ch. 1, pp. 9-97).

[33] The influence of his elder brother Jakob cannot be excluded. In his *Dissertatio de Gravitate Aetheris* (1683), Jakob argued that it was a "vain hope" to expect to construct a PM using capillary attraction (pp. 244-245).

[34] Hannover, Niedersächsische Landesbibliothek, Leibniz-Handschriften: "Technika", ff. 202 *recto-verso*; *c*. 1671. This text requires more careful study than I was able to accord it during the preparation of this paper; in particular, it is not clear how exactly the expansions and contractions of the bellows are supposed to take place. Maier was not impressed by Leibniz's PM: in her (1955, p. 269) she compares his thinking here unfavorably with what she claims was the greater degree of "scientific maturity" on the question of PM to be found among the medievals.

REFERENCES

Bacon, F., (1857-74), *The Works of Francis Bacon*, eds. J. Spedding, R. L. Ellis, D. D. Heath, 14 vols., London. (Referred to as *Works*.)

Beeckman, I. (1939-1953), *Journal tenu par Isaac Beeckman de 1604 a 1634*, ed. C. de Waard, 4 vols. The Hague: Nijhoff. (Referred to as *Journal*.)

Berkel, K. van, (1983), *Isaac Beeckman (1588-1637) en de mechanisering van het wereldbeeld ... with a summary in English* (Thesis for Doctor of Letters in Rijksuniversiteit, Utrecht) Amsterdam: Rodopi.

Bernoulli, Jakob (1683), *Dissertatio de Gravitate Aetheris*. Amsterdam: H. Wetsten.

Bernoulli, John (1690), *Dissertatio Chymico-Physica de Effervescentia & Fermentatione ... cum descriptio alicuius perpetui mobilis ...* Basel.

______________ (1742), *Johannis Bernoullii ... Opera Omnia ...*, ed. G. Cramer, 4 vols. Lausanne and Geneva: M. M. Bousquet. (Referred to as *Opera Omnia*.)

Boyle, R. (1979), *Selected Philosophical Papers*, ed. with introduction by M. A. Stewart. Manchester, Manchester University Press; New York: Barnes & Noble.

Charlton, W. (1654), *Physiologia Epicuro--Gassendo---Charltoniana: or A Fabrick of Science Natural, Upon the Hypothesis of Atoms, Founded by Epicurus, Repaired by Petrus Gassendus, Augmented by Walter Charlton* ... London: T. Heath. Facsimile ed. with indexes and new Introduction by Robert H. Kargon, 1966. New York and London: Johnson Reprint Corp. (Sources of Science, No. 31).

Clerselier, C. (1667), "A Monsieur de la Forge, Medecin à Saumur, Observations de Monsieur Clerselier, touchant l'action de l'Ame sur le Corps. A Paris le 4 Decembre 1660", *Lettres de M. Descartes* ... ed. Claude Clerselier, 3 vols. Paris: C. Angot, 1657-1667: vol. III (1667), Letter CXXV, 640-46. (Not in AT or AT(NP))

Costabel, P. (1960), *Leibniz et la dynamique: les textes de 1692*, Paris: Hermann.

____________ (1972), Article on "Perpétuel (Mouvement)" *Encyclopedia Universalis*, 20 vols. Paris: vol. 12.

Descartes, R. (1964-74), *Oeuvres de Descartes, publiées par Charles Adam & Paul Tannery, Nouvelle Présentation, en co-édition avec le Centre National de la Recherche Scientifique*, eds. P. Costabel, J. Beaude, B. Rochot, 11 vols. Paris: Vrin. (Referred to in text as AT(NP))

____________, (1966), *Les Passions de l'Ame*, Introduction & Notes by Geneviève Rodis-Lewis, 2nd ed. Paris: Vrin.

Diderot, D. & D'Alembert, J. (1751-1780), *Encyclopédie, ou Dictionnaire raisonné des sciences, des arts et des metiers*, ed. Diderot & D'Alembert, 35 vols. Paris and Amsterdam.

Dijksterhuis, E. J., (1961), *The Mechanization of the World Picture*, trans. C. Dikshoorn. Oxford: Oxford University Press.

Dircks, H. (1861 + 1870), *Perpetuum Mobile, or, A History of the Search for Self-Motive Power, from the 13th to the 19th Century ...*, 2 vols. (First and Second Series), London. Facsimile reprint: 2 vols. Israel, Amsterdam, 1968.

_________ (1865), *The Life, Times, and Scientific Labours of the Second Marquis of Worcester. To which is added, a reprint of his Century of Inventions, 1663, with a Commentary thereon*. London: B. Quaritch.

Duhem, P. (1905-06), *Les origines de la statique*, 2 vols. Paris: Hermann.

Elkana, Y. (1974), *The Discovery of the Conservation of Energy*, with a Foreword by I. B. Cohen. London: Hutchinson.

Euler, L. (1752), "Recherches sur l'origine des forces", *Mémoires de l'Academie des Sciences de Berlin*, 6 (1750, pub. 1752), 419-447. In *Leonhardi Euleri Opera Omnia*, series II, vol. 5. (Commentationes mechanicae: Principia mechanica, ed. J. O. Fleckenstein), Lausanne, (1957), 109-13-.

_________ (1765), *Theoria motus corporum solidorum seu rigidorum ex primis nostrae cognitionis principiis stabilita, et ad omnes motus, qui in huiusmodi corpora cadere possunt, accommodata*, A. F. Röse, Rostock & Greifswald, 1765. In *Leonhardi Euleri Opera Omnia*, series II (Opera mechanica et astronomica), vols. 3-4 (ed. Charles Blanc). Berne. 1948-1950.

Gabbey, A. (1973), Review of Wilson L. Scott (1970), *Studies in History and Philosophy of Science*, 4, 373-385.

_________ (1977), "Anne Conway et Henry More: Lettres sur Descartes (1650-51)", *Archives de Philosophie*, 40, 379-404.

_________ (1980*a*), "Force and Inertia in the Seventeenth Century: Descartes and Newton', in S. Gaukroger (1980), 230-320.

_________ (1980*b*), "Huygens and Mechanics", in *Studies on Christiaan Huygens*, eds. H. J. M. Bos *et al.* Lisse: Swets & Zeitlinger, 166-199.

Galilei, G. (1632), *Dialogue Concerning the Two Chief World Systems - Ptolemaic & Copernican*, trans. Stillman Drake, Foreword by A. Einstein, 2nd ed. Berkeley: University of California Press, 1967.

_________ (1638), *Two New Sciences Including Centers of Gravity & Force of Percussion*, trans., Introduction and Notes by Stillman Drake. Madison, Wisconsin: University of Wisconsin Press, 1974.

Garber, D. (1983), "Mind, Body, and the Laws of Nature in Descartes and Leibniz", forthcoming in *Midwest Studies in Philosophy*, 8.

Gaukroger, S. (1980), Editor, *Descartes: Philosophy, Mathematics and Physics*. Sussex: Harvester Press; New Jersey: Barnes & Noble.

____________ (1982), "The Metaphysics of Impenetrability: Euler's Conception of Force", *The British Journal for the History of Science, 15*, 132-154.

Gilson, E. (1930), *Etudes sur le rôle de la pensée médiévale dans la formation du système cartésien*, 3rd ed. (1967). Paris: Vrin.

Goclenius, R. (1613), *Lexicon Philosophicum, quo tanquam clave philosophiae fores aperiunter ...*, Frankfurt, 1613. Facsimile ed., Olms, Hildesheim, 1964 (together with *Lexicon Philosophicum Graecum ...*, Marburg, 1615).

Grant, E. (1978), "The Principle of the Impenetrability of Bodies in the History of Concepts of Separate Space from the Middle Ages to the Seventeenth Century", *Isis, 69*, 551-571.

_________ (1981), *Much Ado about Nothing: Theories of space and vacuum from the Middle Ages to the Scientific Revolution*. Cambridge and New York: Cambridge University Press.

Gregory, T. (1961), *Scetticismo ed empirismo: studio su Gassendi*. Bari.

Hamelin, O. (1911), *Le système de Descartes*, Paris: Alcan.

Harman, P. (1983), "Force and Inertia: Euler and Kant's *Metaphysical Foundations of Natural Science*", in *Nature Mathematized: Historical and Philosophical Case Studies in Classical Modern Natural Philosophy*, ed. William R. Shea. Dordrecht: Reidel, 229-249.

Hautefeuille, J. (1678), *Pendule Perpetuelle, avec un nouveau balancier: et la maniere d'élever l'Eau par le moyen de la Poudre à Canon, et autres nouvelles inventions, contenuës dans une Lettre de Monsieur de Hautefeuille, écrite à un de ses Amis, A Paris ce quatriéme d'Aoust 1678*, n.p.

______________ (1681), *L'Art de respirer sous l'eau et le moyen d'entretenir pendant un tems considerable la Flamme enfermée dans un petit lieu*, n.p.

______________ (1712), *La Machine Arpantante, avec laquelle on mesurera facilement & exactement les Terres et les Chemins*, n.p.

______________ (1719), *Le Mouvement Perpétuel Magnétique, Avec L'ap[p]lication à une Horloge a Pendule, laquelle ira continuellement sans Poids & sans Ressorts*, n.p.

Heilbron, J. (1979), *Electricity in the 17th and 18th Centuries: A Study of Early Modern Physics*, Berkeley: University of California Press.

Hiebert, E. (1962), *Historical Roots of the Principle of Conservation of Energy*, Madison Wisconsin: State Historical Society of Wisconsin.

Huygens, C. (1888-1950), *Oeuvres complètes*, 22 vols. The Hague: Nijhoff. (Referred to in text as *Oeuvres*.)

Ichak, F. (1914), *Das Perpetuum mobile*. Leipzig.

Iltis, C. (1971), "Leibniz and the *Vis Viva* Controversy". *Isis*, *62*, 21-35.

La Hire, P. de (1678), "Demo[n]stration de l'Impossibilité du Mouvement Perpetuel envoyée comme il s'ensuit à l'Auteur du Journal par M. de la Hire de l'Académie Royale des Sciences, à l'occasion de plusieurs de ces Mouvemens qui ont paru depuis peu", *Journal des Sçavans, du Lundi I août 1678*, Paris 1724 ed., 165-166. (Original ed., 304-305)

Lalande, A. (1972), *Vocabulaire technique et critique de la philosophie*, 11th ed. (1st ed. 1902-23). Paris: Presses Universitaires de France.

Lana Terzi, F. (1684-92), *Magisterium Naturae et Artis, Opus Physico-Mathematicum* ..., 3 vols. Brescia: M. Ricciardo.

Laporte, J. (1950), *Le rationalisme de Descartes*, 2nd ed. Paris: Presses Universitaires de France.

Leibniz, G. (1849-63), *Leibnizens Mathematische Schriften*, ed. C. I. Gerhardt, 7 vols. Berlin and Halle. Facsimile reprint, Olms: Hildesheim. (Referred to in text as Gerhardt, *Math*.)

__________ (1875-1890), *Die Philosophischen Schriften*, ed. C. I. Gerhardt, 7 vols. Berlin. Facsimile reprint, Olms: Hildesheim. (Referred to in text as Gerhardt, *Phil*.)

__________ (1903), *Opuscules et fragments inedits de* Leibniz: Extraits de *manuscrits de la Bibliotheque royale de Hanovre*, ed. Louis Couturat. Paris: Felix Alcan. (Referred to in text as Couturat, *Opuscules*)

__________ (1969), *Philosophical Papers and Letters*, ed., trans. and with Introduction by Leroy E. Loemker, 2nd ed., D. Reidel, Dordrecht/Boston. (Synthese Historical Library, vol. 2) (Referred to as Loemker, *Phil. Papers*.).

__________ (1717), *The Leibniz-Clarke Correspondence* ..., ed. with Introduction and notes by H. G. Alexander, Manchester: Manchester University Press, 1956.

Lenoble, R. (1943), *Mersenne, ou la naissance du mécanisme*. Paris: Vrin.

Le Roux, C.-J., (1784), *Traité abregé d'un mouvement perpétuel, en partie méchanique, en partie élémentaire*. Paris: Didot.

McGuire, J. E. (1968), "Force, Active Principles, and Newton's Invisible Realm", *Ambix*, *15*, 154-208.

__________ (1972), "Boyle's Conception of Nature", *Journal of the History of Ideas*, *33*, 523-542.

McMullin, E. (1978*a*), *Newton on Matter and Activity*. Indiana: Notre Dame University Press.

__________ (1978*b*), "Structural Explanation". *American Philosophical Quarterly*, *15*, 139-147.

Maier, A. (1955), *Metaphysische Hintergründe der Spätscholastischen Naturphilosophie*. Rome: Edizioni di Storia e Letteratura.

Malloch, A. (1917), *Finch and Baines*: *A Seventeenth-Century Friendship*. Cambridge: Cambridge University Press.

Maull, N. (1980), "Cartesian Optics and the Geometrization of Nature", in S. Gaukroger (1980), 23-40.

Mellini, D. (1583), *Discorso nel quale si prova contra l'oppenione di alcuni non si potere artifizialmente ritrovare, ne dare ad un corpo composto di materia corrottibile un movimento che sia continovo & perpetuo*. Florence.

Mertz, D. W. (1980), "On Galileo's Method of Causal Proportionality", *Studies in History and Philosophy of Science*, *11*, 229-42.

__________ (1982), "The Concept of Structure in Galileo: its Role in the Methods of Proportionality and *Ex Suppositione* as applied to the Tides", *Studies in History and Philosophy of Science*, *13*, 111-31.

Michel, J. (1927), *Mouvements perpétuels*: *leur histoire et leurs particularités depuis les premières tentatives du XII siècle jusqu'aux engins des inventeurs modernes*, Desforges, Girardot et Cie. Paris. (The cover bears a different title: *Mouvements perpétuels*: *98 fallacieux mécanismes qu'imaginerent maints chimeriques inventeurs du moyen age a nos jours*.)

Montucla, J.-E. (1799-1802), *Histoire des Mathématiques ... Nouvelle édition*, 4 vols., H. Agasse, Paris, An VII - An X. (Vols. III-IV were completed and published after Montucla's death (1799) by Joseph-Jerome de Lalande.)

Needham, J. (1965), *Science and Civilisation in China*, Vol. 4, Pt. II (Physics and Physical Technology: Mechanical Engineering). Cambridge: Cambridge University Press.

Newton, I. (1977), *The Correspondence of Isaac Newton*, ed. Rupert A. Hall and Laura Trilling, vol. VII (1718-1727). Cambridge: Cambridge University Press.

__________ (1730), *Opticks, or A Treatise of the Reflections, Refractions, Inflections & Colours of Light*, based on 4th ed. 1730, Foreward by A. Einstein, Introduction by E. Whittaker, Preface by I. B. Cohen, Analytical Table of Contents by D. H. D. Roller. New York: Dover, 1979.

Novakova, J. (1979), "Comenius's first Report on Perpetual Motion", *Acta Comeniana*, *4*, 389-406.

Price, D. J. de Solla (1959), "On the Origin of Clockwork, Perpetual Motion Devices and the Compass", *Contributions from the Museum of History and Technology, United States National Museum Bulletin* (Smithsonian Institution), No. 218, Papers 1-11: Paper 6, 82-112.

__________ (1975), *Science since Babylon*, enlarged edition. Yale University Press.

__________ (1980), "Philosophical Mechanism and Mechanical Philosophy: some notes towards a philosophy of scientific instruments", *Annali dell' Instituto et Museo di Storia della Scienza di Firenze*, *5*, Pt. I, 75-85.

Remnant, P. (1979), "Descartes: Body and Soul", *Canadian Journal of Philosophy*, *9*, 377-386.

Riddell, H. (1917), "The Search for Perpetual Motion", *Reports and Proceedings of the Belfast Natural History and Philosophical Society*, Session 1915-16, 62-89.

Schmitt, C. (1976), "John Case on Art and Nature", *Annals of Science*, *33*, 543-559.

Schott, G. (1657), *Mechanica Hydraulico-Pneumatica, qua praeterquàm quod aquei elementi natura* ... FrankfurtamMain: H. Pigrin.

__________ (1661), *Cursus Mathematicus, sive absoluta omnium mathematicarum disciplinarum encyclopaedia, in libros XXVIII digesta* ... Würzburg: J. Hertz.

Scott, W. (1970), *The Conflict between Atomism and Conservation Theory 1644-1860*. London/New York: Macdonald/Elsevier.

Smith, N. (1902), *Studies in the Cartesian Philosophy*, London: Macmillan.

Stevin, S. (1955-66), *The Principal Works of Simon Stevin*, ed. E. J. Dijksterhuis, 5 vols. Amsterdam: Swets & Zeitlinger.

Thorndike, L. (1923-1958), *A History of Magic and Experimental Science*, 8 vols. New York: Columbia University Press.

Westfall, R. (1971), *The Construction of Modern Science: Mechanisms and Mechanics*. New York: John Wiley & Sons.

___________ (1980), *Never at Rest: A Biography of Isaac Newton*. Cambridge: Cambridge University Press.

White, L. (1962), *Medieval Technology and Social Change*. Oxford: Oxford University Press.

Wilkins, J. (1648), *Mathematical Magick: Or, the Wonders that may be performed by Mechanical Geometry*, London, 1648. Facsimile ed. of *The Mathematical and Philosophical Works of the Right Rev. John Wilkins* (London, 1802). London: Frank Cass & Co., 1970, 85-246.

KATHLEEN OKRUHLIK

GHOSTS IN THE WORLD MACHINE: A TAXONOMY OF LEIBNIZIAN FORCES

1. INTRODUCTION

The problem to be examined in this essay is arguably the most central and certainly the most discussed of all the difficulties which placed mechanism under strain in the seventeenth and eighteenth centuries. It is the problem of how to account for activity in nature once matter has been stripped of all internal agency and rendered "indifferent to motion". The ghosts of my title are simply the sources of agency which move the otherwise inert world machine.

It is perhaps difficult to justify this choice of topic since it has been studied so extensively by others. The closest and most careful work to date, however, has concerned the status of force and activity in Newton's physics[1], and I plan to take advantage of this work by applying it to Leibniz's dynamical theory of matter.

There are at least two considerations which favor such a proposal. First, no one in the seventeenth century was as aware as Leibniz of the conceptual and empirical difficulties attending the mechanistic theories of matter. No one was more self-conscious or more articulate than he in trying to specify these difficulties. Accordingly, it is less necessary in the case of Leibniz than in other cases to read between the lines and to guess at his intentions. Leibniz's motivations and his arguments (both good and bad) are spelled out in considerable detail.

The second advantage of examining the *general* problem of agency in nature through the *particular* eyes of Leibniz is closely related to the first. There are a number of fairly obvious and not-so-obvious ways in which one might attempt to find a place for agency in the world machine; and because Leibniz *was* so concerned with the problem, he discussed most of these over the course of his long career. As a result, tracing Leibniz's research gives us an opportunity to do a sort of general taxonomy of ghosts in the 17th century.

J. C. Pitt (ed.), Change and Progress in Modern Science, 85–105.
© 1985 by D. Reidel Publishing Company.

2. CARTESIAN STAGE-SETTING

Descartes' matter theory provides the best background against which to examine Leibniz's efforts to locate agency in the world machine. The merits of the Cartesian theory are obvious. When Descartes divorced mind and matter, he laid the foundations for a determinate mathematical treatment of matter. The essence of matter on this account is, of course, extension, which is purely passive and geometric in character. All other properties of matter are (in some sense) reducible to extension (*Principles* i, sect. 53); but extension itself is ultimate – not subject to further reduction. So there is nothing in matter itself which might preclude rigorous mathematical treatment. Since matter is completely shorn of any powers of self-determination, the sorts of vitalism which undermined earlier attempts to systematically treat matter in motion are barred from the outset. Activity in general, motion in particular, cannot originate in extension and if superadded from outside can cause no alteration *within* the matter which is moved. Motion is a purely extrinsic denomination – if we may employ Leibniz's terminology from the outset. It is something which happens *to* bodies on the Cartesian account.

Since motion cannot originate in purely passive extension, Descartes must account for it in some other way. He does so by having God put bodies into motion at the time of creation. God then *preserves* motion through his powers of constant recreation.

Physics, therefore, is on this account nothing more than kinematics. Ultimately, all that happens in the realm of matter may be accounted for in terms of geometry and motion. One of the chief and most useful results which can be derived from this approach according to Descartes, is that God's immutability guarantees the conservation of quantity of motion, where quantity of motion is taken as the product of volume and speed (a scalar rather than a vector). In all interactions among material bodies, the quantity of motion is conserved. Created thinking substance cannot augment or diminish the quantity of motion. So the autonomy of matter in motion – its determinateness and independence from *res cogitans* – is apparently guaranteed. (We should note, however, that this autonomy is compromised almost immediately because, although mental substance cannot alter the quantity of motion in the world, it *can* alter its direction.)

3. FIRST GHOST: GOD HIMSELF

When confronted with the problem of how to get matter and activity back together again once he has separated them so radically, we have seen that Descartes fell back immediately on God. Motion is superadded to matter by divine fiat. God is the ghost who moves the world machine. He does so directly by setting matter into motion at the original creation and by sustaining it ever since through his constant re-creative activity. It appears to follow from Descartes's position that although motion is a mode of bodies, the moving force of bodies is in God rather than in the bodies themselves. Descartes did not draw out this conclusion for fear that his readers would think he was making God into a world-soul. Despite the great difficulties it would create for other aspects of his philosophy, he wanted to leave open the possibility that moving force was *somehow* in bodies as well as in God. Descartes's occasionalist followers, however, did not hesitate to explicitly draw out the conclusion that moving force can exist only in God and not in bodies at all. They maintained that when one body strikes a second, appearing to set it in motion, it is actually God who on the occasion of impact causes the second to move. Bodies themselves are completely without causal power.

Leibniz, himself, early in his career, adopted a strategy very similar in some respects to that of Descartes. In a letter to Jacob Thomasius written in April of 1669, he notes that motion is not derivable from the essential characteristics of matter:

> Hence [he says], there is no motion, strictly speaking as a real entity in bodies. I have demonstrated instead, that whatever moves is continuously created and that bodies are something at any instant in assignable motion, but that they are nothing at any time midway between the instants in motion — a view that has never been heard of until now but which is clearly necessary and will silence the atheists. (Loemker 1969, pp. 93-104)

In the same letter he says that he has "found out...about the perpetual creation involved in motion". In a footnote to this passage, Loemker suggests (Loemker 1969, p. 104) that Leibniz's theory of 'continuous creation' here seems merely to mean that God is the source of all motion and hence that it is very similar to the Cartesian position which he later criticized so vigorously.

Daniel Garber has recently suggested (Garber 1982) that rather more is going on here and that Leibniz is rejecting the reality of motion at least partly because of his view that nothing can be a substance which does not have a principle of action within itself. This view will, of course, play a central role in Leibniz's later matter theory. But Garber suggests that in this early period, Leibniz's view of substance, coupled with his account of motion given above, leads to the consequence that body taken by itself cannot be a substance but that the substance of bodies which lack reason is union with the universal mind, or God.

If this account is correct, then we might say that Leibniz had already begun in the 1660's to forge a tighter link between bodies and the source of activity. We no longer have one substance (God) moving a separate substance (*res extensa*). Rather we find that union with the universal mind is what constitutes the very substance of Leibnizian bodies.

In any event, the later Leibniz lumps these several accounts together and says that they all err in making God the soul of the world and hence that they all collapse into Spinozism by "causing created things to disappear into mere modifications of the one divine substance since that which does not act, which lacks active force...can in no way be a substance." (Loemker 1969, pp. 498-508) The Creator, on this view is not really separate from his creation, and the theological consequences are unacceptable.

Furthermore, we depreciate God's power if we think that the impressions he makes on matter are so fleeting that he must repeat them at every instant. It seems much more appropriate to think that his impressions endure within matter and become "*internal laws* from which their actions and passions follow..." (Loemker 1969, pp. 498-508) The picture here is the familiar one of a God who does not have to intervene constantly in the affairs of matter because he ordered them so well at the outset.

4. SECOND GHOST: THE *ANIMA MUNDI*

When Leibniz argues against the theory of a world-soul, as he very often does, his usual assumption is that the world soul is to be construed as God himself. He sometimes acknowledges, however, (Loemker 1969, pp. 554-560) that one might pursue a slightly different route here and say the universal world soul is not God but a created spirit. He attributes this position to the Cabalists, to Henry More, and to "other newer philosophers, particularly...certain chemists." (Loemker 1969, p. 555)

Now this strategy has one clear advantage over making God himself the immediate cause of motion since it preserves to some extent the distinction between God and his creation. Within that creation, however, all the old problems remain. There would appear now to be just one created substance, the whole world, and no way of individuating bodies within it. Bodies could be differentiated only as modes of a single substance since there would be but a single source of activity among creatures. Individual bodies themselves could not be substances, and even human beings would lack separate immortal souls. Such a notion "degrades the human race" and is theologically heterodox. So, this avenue, too, is closed to Leibniz. If God himself cannot be the ghost in the world machine, then it is scant improvement to postulate a single created ghost to drive the whole mechanism.

5. MULTIPLE CORPUSCULAR GHOSTS

In rejecting the view that there was a single vital principle operating the world machine, Leibniz was rejecting one important view of Nature. It is essential, however, to be quite clear about what he was *not* rejecting. The occasion for spelling out his views about the relationship between Nature and the world machine was presented to Leibniz in his controversy with Sturm.

Robert Boyle had in 1682 written an essay called *Free Inquiry Into the Vulgarly Received Notion of Nature* in which he proposed that the term 'nature' should be avoided and 'mechanism' substituted in its place. John Christopher Sturm of Altdorf defended Boyle's position, maintaining that it is pagan and un-Christian to attribute properties to the 'nature of things' and that nature has no energy or force proper to itself. Leibniz and Sturm corresponded on this issue and Leibniz's reply was published as the essay called "On Nature Itself, or on the Inherent Force and Actions of Created Things". (Loemker 1969, pp. 498-508)

In that essay, Leibniz begins by saying that we cannot decide whether to abandon nature in favor of the machine unless we know precisely what is meant by the term 'nature'. If what is intended is some sort of created world soul or hylarchic principle such as that discussed above, then we must banish Nature immediately. But if that banishment were to leave behind only God and a big passive machine, indifferent to all motion, we would be in still worse shape since now God would have to operate directly and constantly on matter. In other words, we would be back at the position which makes God himself the ghost in the machine.

> Then it would follow that no created substance, no identical soul would be permanent, and hence that nothing would be conserved by God, but everything would reduce to certain evanescent and flowing modifications or phantasms, so to speak, of the one permanent divine substance. And what reduces to the same thing, God would be the nature and substance of all things – a doctrine of the most evil repute. (Loemker 1969, p. 502)

It is necessary, therefore, that nature be preserved – but not as a *single* force governing all matter. Instead agency must be built into matter from the ground up; it must be an intrinsic part of every little bit of matter – not something added from outside. It was in order to accomplish this, of course, that Leibniz developed his ontology of forces.

There are in Leibniz' mature ontology four distinct kinds of force: primitive active, primitive passive, derivative active, and derivative passive. Generally the relations among these forces are cashed out in terms of the monadology. The primitive forces are said to operate at the substantial or monadic level. This is the locus of real activity in the world. But physics deals with matter and motion, which are merely phenomenal, and thus the only forces which concern it are *derived* forces.

Insofar as there is a standard view in the literature regarding the relationship in Leibnizian dynamism between matter and motion, on the one hand, and the ultimate constitutive forces of nature on the other hand, this is it.

There is much merit in the monadic account, and it will be discussed in the next section. But recently Daniel Garber has suggested that it is a big mistake to read Leibniz's later theory of monads back into the writings of the middle period. He maintains that during the 1680's and 1690's when Leibniz was actively engaged in the development of a science of dynamics, he had a different picture in mind.

While the later theory may be properly characterized as a form of pan-psychism, Garber calls the middle theory 'panorganicism'. What he wants to maintain is that the world of Leibniz's middle period is "not a world of souls but a world of organisms, corporeal substances understood on an Aristotelian model as unities of form and matter."(Garber 1983)

So matter on this construal is *not* simply a phenomenal manifestation of the primitive active and passive forces which go into the making up of real substances. Rather, individual corporeal substances *are* "animate machines, living beings, organisms of a rudimentary sort." They are *real*, not mere expressions of an underlying psychic reality.

These physical animate machines, these organisms, are the ultimate constituents of the world. There are organisms within organisms within organisms *ad infinitum*. Inanimate bodies are not themselves organisms but are *composed* of organisms. They are aggregates rather than substances, but they are governed by precisely the same laws and forces that govern the fundamental substances; so they can be treated as true substances for the purposes of physics. The notions of force and law pertain, therefore, not to the later monads but to organisms, the animate machines and their aggregates.

Garber admits that there are a number of difficulties with his view – not the least of which is that on his construal Leibniz's dynamics becomes a sort of mad biology. But it is for this very reason especially useful for our taxonomy of ghosts. It fills in a sort of missing link.

Recall that when we left Leibniz in his controversy with Sturm he had rejected God as a ghost in the world machine and had also rejected the idea of a created *anima mundi* in favor of a sort of pluralism of agency. Each individual substance, in other words should be informed by its own separate principles of activity. What better and more direct way to do this then by making matter itself organic, uniting active and passive principles within it as body and soul are related in human beings?

If Leibniz did adopt this course (that is, if Garber's thesis is correct), then his substantial forms provide instances of the corpuscular ghosts which lent their name to this section of the paper. They drive the world machine pluralistically, each by driving its own little machine. There is on this account, no single *anima mundi*. The world machine is not itself an organism, but its ultimate constituents are.

One difficulty with the pan-organicist view is, of course, that it appears through its hylozoism to threaten the notion of a mathematizable determinate nature and so to fly in the face of the basic Leibnizian project. It puts agency back into matter at too high a price. Physics does indeed become biology and the door is re-opened to all that was bad about substantial forms and organicism understood in the *old* way. The gains of mechanism are lost; the basic mechanical project initiated by Descartes is abandoned.

Leibniz, of course, wanted to *preserve* mechanism with all its determinateness and mathematical rigor; but he also wanted to preserve agency in matter. He wanted both teleology and efficient causation. He finally resolved the tension by saying that the material world is fully governed by the laws of mechanics but that these mechanical laws themselves have a more sublime origin than mere geometry. Their origin lies in a kingdom of ends, a world of teleologically determined monads. Because the two kingdoms do not occupy the same level of being, their laws never come into conflict. Matter is now the phenomenal expression of noumenal ghosts. We should perhaps review here this standard account of Leibniz's fourfold division of forces in terms of the distinction between monadic and phenomenal levels of being.

6. MONADIC GHOSTS

Leibniz's researches into Zeno's problems with the labyrinth of the continuum convinced him that the ultimate constituents of matter cannot be mathematical points because these, though they are exact and have no parts, cannot be real. Nor can "physical points" be the ultimate constituents because these, though they are real enough, have parts; they are not exact. What is required is a point which is both exact and real – a "formal point". Leibniz had in his younger days turned to physical atoms as an alternative to Cartesianism, but soon saw the problems with this approach. He found his answer elsewhere:

> I was compelled to have recourse to a formal atom, since a material being cannot be both material and perfectly indivisible or endowed with a true unity. It was necessary, hence, to recall, and so to speak, rehabilitate the *substantial forms* so decried today, but in a way which would separate the use we should make of them from the abuse that has been made of them. I thence found that their nature consists in force, and that from that there ensues something analogous to feeling and appetite; and that accordingly they must be conceived in imitation of the idea we have of Souls...Aristotle called them *first Entelechies*. I call them, perhaps more intelligibly, *primitive Forces* which do not contain only the *act* or the complement of possibility, but further an *original* activity. (Loemker 1969, pp. 453-459)

Bodies can, therefore, for the sake of analysis (though never in reality) be divided into two parts: an active principle and a passive principle. The passive principle includes notions of impenetrability and natural inertia. These are *real* properties whose repetition gives rise to the *appearance* of extension. This passive principle is sometimes called "primary matter". Primary matter is entirely inert, but it never actually exists because the concept of primary matter is incomplete. What completes the concept is the active principle – force. Primary matter endowed with force is called "secondary matter". This is what really exists in the world.

The forces which we see operative at the level of dynamics are supposed to be *derived from* the original monadic activities, i.e., from the substantial forms of the monads. (Gerhardt 1875, III, p. 457) What we observe is the resultant force of all the monadic activity in the world. The original force is the law governing the exfoliation of each individual monad. The derived force is determined by all the activity of all monads.

Force provides the principle of individuation in the world and guarantees a plurality of substances. The activity of each monad is unique and peculiar to it, though the derivative activity is law-bound:

> Bodies, strictly speaking, are not pushed by others when there is an impact, but by their own motion, or by their elasticity, which again is a motion of their parts. Every corporeal mass, great or small, has already in it all the force which it can ever acquire, but the meeting with other bodies only gives its determination, or rather this determination only happens at the time of the meeting. (Gerhardt 1875, II, p. 116)

Because all the force in the world was implanted in the monads at the moment of creation, and because interaction among bodies neither increases nor decreases the amount of force, but only determines its manifestation, therefore the conservation of force is guaranteed. Leibniz, of course, believed himself to have proven experimentally that the true measure of force is mv^2. Therefore, this is the quantity which is conserved at the phenomenal level. Its conservation is guaranteed by the conservation of force at the monadic level.

Furthermore, forces provide the only thread which can lead out of Zeno's labyrinth of the continuum and render motion explainable, according to Leibniz. Motion is purely phenomenal and entirely relative; but if it is to be subject to the laws of nature, then it must be a *phenomenon bene fundatum*. This is to say that motion must be grounded in something that is real. Leibniz argues, therefore, that there is more to motion than change of place. Change of place cannot be real because space is only relative, not absolute. The additional element, which the Cartesians fail to consider, is motor force.

Motor force is, according to Leibniz, *real*; its phenomenal manifestation is motion. So neither matter nor motion is real according to Leibniz; but both are grounded in the reality of monads. We might say, in other words, that matter and motion have for Leibniz the same status that color and taste have for Newton and Descartes. They are derived characteristics – secondary qualities rather than primary. They are entirely dependent upon there being a conglomeration of ultimate units. A single Leibnizian monad is not material any more than a single Newtonian atom is colored. Just as it would be impossible, however, to take away atoms and still have color or taste, so also would it be impossible to take away monads and still have matter or motion.

Leibniz believed that this theory of matter circumvented the problems which afflicted competitor accounts, especially those of Descartes and Newton. But, of course, this theory faces grave problems of its own, particularly when we try to integrate the foregoing dynamical account of the monad with the psychologistic account developed by Leibniz in his later period. In psychologistic terms, primitive active force should be identified with clarity of monadic perception while primitive passive force should be identified with confusedness of perception. But there seems to be no plausible reason for making these identifications and no coherent way of integrating the earlier and later pictures.

So even if we do not follow Garber in his extreme Aristotelian interpretation of the middle period, he is surely right to stress the dangers we face if we attempt a too-facile identification of the earlier metaphysical points with the later psychologistic monads. I should like particularly to stress how difficult it is to find a proper place in the traditional (amalgamated) framework for derived forces. They are perhaps "less real" than the original monadic activity, but they are also "more real" than the purely mechanical properties of motion and extension. Leibniz stresses again and again that even within the framework of physics itself, there is something absolute about force although motion is purely relative. If we construe the forces of dynamics as mere phenomenal representations of monadic perception and appetition, we run the danger of losing the rather robust sense of reality associated with the dynamic properties of matter during Leibniz's middle period. It is as if we need to create a sort of halfway-house for them somewhere between the psychic activity of monads and the merely mechanical *phenomena bene fundata*. This awkwardness can be explained plausibly by a failure to distinguish between the different metaphysical frameworks of Leibniz's middle and later periods.

The metaphysics of the middle period was created while Leibniz was actively engaged in the development of a new science of dynamics, and it reflects that preoccupation. After 1700 when the psychologistic account of monads came to dominate his thinking, he was no longer actively engaged in dynamical research, and it should not be surprising that his altered metaphysical outlook should provide a less comfortable home for dynamics.

7. *VIS VIVA*: A CLOSER LOOK AT DERIVATIVE ACTIVE FORCE

I should like to retreat here from the big picture to look more closely at the role of derived forces in Leibniz's dynamics. This section will treat *vis viva* (which is derived active force, measured by mv^2), while the next section will examine derivative passive force.

Recall that the biggest event in Leibniz's life as a student of dynamics was his "demonstration" that Descartes had been wrong to identify quantity of motion (mv) with force and that in fact the true measure of force is mv^2. The gist of this demonstration is repeated in several places.[2]

The argument relies on two assumptions acceptable to the Cartesians: (1) that a body falling from a certain altitude acquires the same force which is necessary to lift it back to its original altitude, and (2) that the same force is necessary to raise a body A of one pound to the height of four yards as is necessary to raise the body B of four pounds to the height of one yard.

> There is thus a big difference between motive force and quantity of motion, and the one cannot be calculated by the other as we undertook to show. It seems from this that *force* is rather to be estimated from the quantity of the effect which it can produce; for example, from the height to which it can elevate a heavy body of a given magnitude and kind but not from the velocity which it can impress upon the body. (Loemker 1969, p. 297)

If we follow this line of thought, then force must be estimated by mv^2, the quantity that measures *vis viva*. Thus it is mv^2 that remains constant where the force to raise the bodies is the same.

The quantity mv^2 comes from Huygens who had already found that the sum of the products of mv^2 remains constant for the case of colliding bodies. In addition, Leibniz' energy relations appear to be implied by Huygens' work on the compound pendulum. But it must be stressed that for Huygens the quantity mv^2 had no metaphysical significance whatsoever; it was just a number. Leibniz, on the other hand, invested it with overwhelming ontological significance by making it the measure of something which is substantial, fundamental, and always conserved.

Leibniz explained Descartes's error by saying that consideration of simple machines had led him to equate quantity of motion with force, since "velocity and mass compensate for each other in the five common machines". (Loemker 1969, p. 297) But this works only in statics: if two bodies are in equilibrium on a balance, their possible vertical displacements are proportional to their weights. So the two proposed measures of force happen to co-incide for "dead forces".

"Dead force" is the name given by Leibniz to the infinitely small motion which takes place when a heavy body tries to commence movement but has not yet conceived any impetuosity or living force; he sometimes calls it "solicitation". Its measure is mdv/dt. So it is, in fact, the quantitative equivalent of Newtonian force, though it is conceived in a radically different way.

The living force in a body, on the other hand, is $m\int_0^t v\, dt = m\int_0^t \frac{ds}{dt} dt = ms$ or mv^2. Though Leibniz sometimes said that dead force was to living force as a point to a line, he realized in his more careful moments that the proper relationship is that of a point to a plane. A single integration of dead force over time yields mv (impetus), and a second integration over time is required to yield living force (mv^2): "Hence according to the analogy of geometry or of our analysis, solicitations are as dx, velocities as x and forces as xx or $\int x\, dx$."

Now this probably raises some questions in the mind of the reader. It is sometimes said by commentators that Leibniz's dynamics did not reach its full potential because (1) Leibniz was unable to free himself of the idea of force as something a body possesses by virtue of its motion, an idea based on the impact mode ; and because (2) he did not develop concepts of work and potential energy which would have allowed him to generalize his dynamics of momentum and kinetic energy.

Yet we have seen that the Galilean kinematics of vertical descent provided an alternative to the impact model of force and that it played a key role in his establishment of mv^2 as the true measure of force. And we have seen that Leibniz had in his dead force precisely the mathematical quantity which Newton would exploit so successfully in developing a new concept of force. So why could Leibniz not see that integration of dead force over space in the case of free fall would give him the increment of *vis viva*, exactly what he needed to expand his dynamics in a fruitful way? Why, in other words, did Leibniz not develop the work-energy equation since he appears to have had all the ingredients necessary to do so?

Westfall suggests that Leibniz's failure to do so is based in part on his fidelity to certain metaphysical presuppositions. This, I think, is a plausible suggestion for the following reasons:

> (1) Leibniz's neglect of dead force except as it relates to *vis viva* is understandable when we recall the centrality of activity in his metaphysics of substance. Dead force is not yet active. So it has no real ontological significance except through its relation to motion and *vis viva*. Outside those relationships it is nothing but a number.
> (2) There is no transeunt causality. The monads are windowless; and even at the level of dynamics, bodies do not really interact. Mechanically we may treat two bodies which rebound from one another as if they caused a change of state in one another. But, really, each body rebounds as a result of its own elastic force. So there is no causation across space; bodies dynamically conceived are just as windowless as are monads.
> Newtonian force conceived of as an *external cause* of change of state is, therefore, an impossibility for Leibniz. There are *no* external causes in dynamics. So Newton's 'force' and Leibniz's 'dead force' are two very different things, despite their mathematical equivalence.
> (3) Because causal sequences can unfold only *within* a single monad and (by extension) within a single body, forces can act only through time – not space. A prior monadic state may ground a subsequent state of the same monad and a prior dynamical state may cause a subsequent state of the same body. In fact, each state of a monad recapitulates its past and is pregnant with its future states. Similarly for bodies dynamically conceived.

> It makes sense, therefore, to integrate across time. The integral is the law of the series: the infinite summation of all past, present, and future states. It corresponds exactly to something metaphysically real and so plays a prominent role in Leibniz's dynamics.
>
> Not so the spatial integral. Since there is no transeunt causation, since the law of the series unfolds in time rather than in space, taking an integral of dead force across space is absolutely meaningless. The spatial integral would correspond to nothing real. But dynamical properties must be substantially grounded in order to be permissible in Leibnizian dynamics. Hence, no integrations of dead force across space can be allowed.

It should be stressed here that Leibniz was unequivocally a realist about forces in the sense that "realism" is opposed to "instrumentalism" (though we may debate the extent to which he was a realist in the sense which opposes "realism" to "phenomenalism".) Mv, mv^2, and dv/dt were not for him mere mathematical formulae to be manipulated in any old way that saved the phenomena. Instead, mathematical formulae were intended to describe veridically a non-mathematical reality. (It is true, of course, that Leibniz had modelled his account of that underlying monadic reality on the infinitesimal calculus in the first place. But that is another story.)

8. THE STRANGEST GHOST OF ALL: NATURAL INERTIA

In his youth, Leibniz had accepted the Cartesian doctrine that the essence of matter is extension and had attempted to draw out *a priori* some of the conclusions of that doctrine. (Loemker 1969, pp. 139-142) He had reasoned that mere extension must be indifferent to both motion and rest and that, therefore, when a body in motion, however small, collides with a body at rest, however large, both bodies must move in the direction of the original motion at the original speed. This is a somewhat shocking result, but Leibniz maintained that there is nothing in the purely geometrical conception of matter to provide for a force of resistance or reaction to every action.

This early result was later to be a jumping-off point for Leibniz's attacks on the geometricist conception of extension as the essence of matter. It manifestly violated the metaphysical principle demanding equality of cause and effect, thus making necessary a new account of the essence of body. So Leibniz declared that in all substances active force is coupled with passive force. Recall the outlines of his solution: Primitive active force is the law of the individual monad. Derivative active force manifests itself as dead force, impetus, and living force. Primitive passive force is *materia prima*. Spatial replication of primitive passive force yields the appearance of extension; and impenetrability, too, is a derivative manifestation of passive force. But most importantly, derivative passive force is natural inertia of matter, its resistance to change of state. Derivative passive force is mass.

Just as caution was required earlier not to conflate the Leibnizian with the more familiar Newtonian concept of force, we must take care now not to think of Leibnizian natural inertia as identical to Newtonian inertia. "Inertia" does not refer to the uniform motion of a body in Leibnizian dynamism. It is active rather than passive force that causes a body to continue moving in a rectilinear path. Natural inertia does, however, guarantee the equality of action and reaction: "[E]verything that acts suffers some reaction and consequently a body at rest cannot be set in motion by another without changing something of the speed of the agent." (Leibniz 1691, I, p. 449) The interplay (at the level of phenomena) between active and passive forces underwrites the conservation of force by guaranteeing the equality of cause and effect: If the effect were greater than the cause, we would have mechanical perpetual motion (which is an impossibility). And if the effect were less than the cause, we would not have perpetual physical motion (that is, the world machine would run down). (Loemker 1969, p. 279) Active force guarantees that the quantity of force will not diminish, and passive force guarantees that it will not increase. The presence of both forces in *each* phenomenal body guarantees for Leibniz the determinateness, the mathematizability, of phenomenal interactions. It allows the force ontology to ground a thoroughly mechanistic physics.

But it must be noted just what a dark conception is passive force. Even the name itself seems self-contradictory in much the same way as Newton's "force of inertia".

In my subtitle I refer to it as the "strangest ghost of all", but it is not even clear that it can properly be called a ghost, if we mean by that a source of agency in the world. For natural inertia can initiate no motion whatsoever; rather than being a source of activity, it is a source of *passivity*. But that passivity is itself opposed to the conception of matter as inert. Natural inertia, in other words, is a manifestation of matter's "ertness".

Leibniz recognized very clearly the inadequacies of any concept of matter which made it utterly inactive; and to meet these conceptual difficulties head-on he invented the force ontology. But one of the most important aspects of that ontology had to be a *passive* force to account for the fact that matter is not indifferent to motion. The usual ontological categories seem to break down here. Resistance to motion is in a sense "more active" than is indifference to motion. So passivity, thus understood, is an important aspect of the activity of bodies. The anomalies hidden in the concept of mass are a chief source of the tensions that placed mechanism under strain in the seventeenth and eighteenth centuries.

9. APPENDIX: KANT'S ATTEMPT TO EXORCISE THE GHOST OF NATURAL INERTIA

I should like to review very briefly the structure and intent of Kant's *Metaphysical Foundations of Natural Science* (*MFNS*) (Kant 1970) with a special eye to natural inertia or mass. I shall suggest that although Kant, like Leibniz, was attempting to ground mechanics in a force ontology, he actually failed to ground one of the most fundamental aspects of matter: its mass. There is nothing in Kant's exposition of the original forces of matter which accounts *dynamically* for Newton's third law or for the proportionality between force and acceleration.

I have argued elsewhere (Okruhlik 1983) that throughout his long academic career Kant maintained that the mathematical mechanical approach fails to do justice to the actual concrete existence of physical bodies and that, therefore, it must be supplemented with a metaphysical/dynamical analysis. In this respect, Kant's program was essentially Leibnizian rather than Newtonian. It led him to reject the basic ontology of Newtonian mechanics and to replace it with something quite different – a force ontology or what Kant himself called a "physical monadology". This basic explanatory scheme, this project of grounding mechanics in dynamics survived even the Copernican Revolution; and the critical philosophy was used in the *MFNS* to legitimate Kant's preference for dynamics.

Kant's early attempts to ground mechanics in dynamics had fallen afoul of hylozoism when he posited intensions and other indeterminate force-like entities in matter. In the *MFNS*, however, Kant thought he had pulled off this trick by making forces constitutive of matter while locating the source of mechanical laws (those governing bodies in motion relative to one another) in the systematizing activity of the human mind. The mechanics (ch. 3 of *MFNS*) was still grounded in the dynamics (ch. 2), however, because the bodies treated by the laws of mechanics were precisely those whose concrete natures were described by the dynamics of central forces.

What I want to suggest, however, is that Kant's matter has no inertial mass. Kant did not just repudiate the *phrase* "inertial force" or the idea that matter is innately sluggish or that matter has any power to change its own state. He seems to have gone all the way back to making matter indifferent to motion. He has neglected to give it mass. The original attractive and repulsive powers of matter may give matter its impenetrability, its gravitational attraction, its elasticity and so on. They do not give it any inertial mass. (And one cannot simply pull mass out of a hat as Kant attempts to do in Explication 2 of the Mechanics by saying that matter acts in mass when all its parts move in the same direction and at the same time exercise their moving force externally.)

In an earlier paper (Okruhlik 1983) I pointed out some of the obvious (but often unnoticed) differences between Kant's three laws of mechanics and Newton's three laws of motion. "Inertia" for Kant signifies only the lifelessness of matter – its inability to change its own state. So it is an absolute property, not a property which bodies can possess in various degrees. Kant, therefore, has something like Newton's first law since in the absence of any force, all bodies however large or small, change velocity at the same rate – that is, not at all. Kant does not, however, have anything like Newton's second law since there is nothing in his dynamics which can ground the proportionality between force and acceleration.

Kant's third law, though verbally almost identical to Newton's, in no way draws upon the resistance of matter to impressed force in order to ground the equality of action and reaction. Rather Kant's law says simply that because of the relativity of spatial frameworks, we are always free in a collision to assign velocities in inverse proportion to the masses of the colliding objects, thereby guaranteeing the numerical equality of the resulting momenta though their directions are opposite.

I think these differences can all be explained by the fact that Kant tried to have a dynamics without mass, without the *resistance* of matter. That he should want to exorcise the ghost of Leibnizian passive force is understandable since it seems so fraught with difficulty. In doing so, however, he has also driven out the very massiveness of matter and has robbed the dynamical tradition of one of its original great successes – the invention of a ghost to guarantee the proportionality of cause and effect in the world machine.[4]

University of Western Ontario

NOTES

[1] See, for example, Ernan McMullin (1978) and R. S. Westfall (1971).

[2] "A Brief Demonstration of a Notable Error of Descartes and Others Concerning a Natural Law" (1686), Loemker, pp. 296-302. See also *Discourse* (section 17), the commentary on Descartes (section 36) and the 'Specimen Dynamicum'.

[3] Letter of June 1691, *Ouvres de Leibniz*, A. Jacques (ed.), Paris 1842) I, 449.

[4] I am very grateful to James Robert Brown, Robert Butts, and Howard Duncan for helpful discussions of many of the topics discussed in this essay.

REFERENCES

Garber, D. (1982), "Motion and Metaphysics in the Young Leibniz", ed. M. Hooker, *Leibniz: Critical and Interpretive Essays*. ed. M. Hooker University of Minnesota Press. 160-184.

_________ (1983), "Leibniz and the Foundations of Physics: The Middle Years", Forthcoming in *Leibniz: the Philosophy and Foundations of Science*. eds. J. Brown and K. Okruhlik.

Gerhardt, C. (1875), *Die Philosophischen Schriften von Gottfried Wilhelm Leibniz*. Berlin.

Kant, I. (1970), *Metaphysical Foundations of Natural Science*. trans. J. Ellington. Indianapolis: BobbsMerrill.

Leibniz, G. (1691), *Oeuvres de Leibniz*, ed. A. Jacques. Paris, (1842).

Loemker, L. (1969), *Gottfried Wilhelm Leibniz: Philosophical Papers and Letters*, 2 ed. Dordrecht: Reidel.

NOTE: The page numbers cited within the text will refer the reader to parts of various sections. The section titles are listed below:

"On Nature Itself or on the Inherent Force and Actions of Created Things" (1698).

"Reflections on the Doctrine of a Single Universal Spirit" (1702).

"Studies in Physics and the Nature of Body" (1671).

"On the Elements of Natural Science".

"A New System of the Nature and Communication of Substances, as well as the Union Between the Soul and the Body" (1695).

"A Brief Demonstration of a Notable Error of Descartes and Others Concerning a Natural Law" (1686).

McMullin, (1978), *Newton on Matter and Activity*. South Bend: University of Notre Dame Press.

Okruhlik, K. (1983), Kant on the Foundations of Science", *Nature Mathematized*, ed. W. Shea. Dordrecht: Reidel.

Westfall, R. (1971), *Force in Newton's Physics*. New York: American Elsevier.

R. W. HOME

THE NOTION OF EXPERIMENTAL PHYSICS IN EARLY EIGHTEENTH-CENTURY FRANCE*

One of the most characteristic features of the science of physics, as we know it today, is its highly mathematical form. Physical theories are almost always presented in the guise of extended pieces of mathematical analysis linked to experience through a number of mathematically expressed quantitative functional relationships between physical variables. During the past century and a half, this manner of proceeding has proved immensely successful, to the point where one can now all too easily assume that any worthwhile physics is necessarily quantitative and highly mathematical in its approach. It is therefore important to emphasize that physics has not always been done this way, and that our modern mathematical form of the subject is a relatively recent development whose beginnings are, except in a few special cases, scarcely to be traced back beyond the early years of last century. To be sure, some of the issues that arise whenever one attempts to bring mathematics to bear on the study of natural events had been considered and resolved somewhat earlier in such fields as astronomy, mechanics and geometrical optics. Yet even the very possibility of extending the lessons that had been learned in those cases to the generality of physical phenomena was by no means established, let alone the manner in which, in practice, this might be done.

When this point is recognized, there at once emerges a major task for the historian, namely, to trace the origins of the new mathematical version of physics, as it emerged in the early years of the nineteenth century, from its historical roots in the preceding age. This, however, would be a very large task indeed. In the present paper, my aim is the more modest one of delineating some seldom remarked features of the eighteenth-century background out of which nineteenth-century mathematical physics evolved. It is generally agreed that the new mathematical approach was a largely French invention, the achievement of a remarkable group of investigators based on the Paris Academy of Sci-

J. C. Pitt (ed.), Change and Progress in Modern Science, 107–131
© 1985 by D. Reidel Publishing Company.

ences (or the First Class of the Institut de France, as it was called during the Consulate and Empire) and the infant Ecole Polytechnique. Here, I shall be concerned with the situation in France in the period preceding this development, and more particularly with an influential tradition of experimental physics that emerged there during the first half of the eighteenth century and that came eventually to be associated particularly with the name of the Abbé Jean-Antoine Nollet (1700-1770). I shall show, among other things, that this was a tradition in which not only did mathematical and quantitative considerations play no part, they were actually ruled out, being held to be inappropriate to the subject-matter in question.

The historical importance of this tradition has generally been overlooked in the literature of our subject. Its practitioners do not have major scientific discoveries to their credit, and as a result, few of their names are recognized today. Furthermore, even when they are cited, it is not always sympathetically. Nollet in particular has had a bad press among historians of science, especially those of the English-speaking world. When he is remembered today – if he is remembered at all – it is chiefly as a remarkably successful vulgariser of science. Nollet, we recall, was the man who carried electrical demonstrating to the very pinnacle of French society, the court of Louis XV himself, and amused the King by causing a long line of monks to leap as one into the air when a battery of Leyden jars was discharged through them. His actual scientific work, and especially the theory of electricity he announced in 1745 and defended in a number of subsequent publications, have tended to be dismissed rather scornfully as unworthy of serious attention.

Both Nollet himself, however, and the intellectual tradition he represents, deserve better than this. In the context of his time, Nollet made a major contribution to the science of electricity, and his ideas dominated French discussions of the subject for a generation and more at a time when it was expanding very rapidly indeed.[1] More importantly, his significance for our story is by no means limited to his work in electricity. For a period of thirty years in the middle of the eighteenth century, Nollet was France's leading experimental physicist. His work inspired an entire generation of French investigators, and also led to an important reorientation of physics teaching in universities and colleges throughout the country. Moreover, his influence extended well beyond the confines of France, being felt particularly strongly in Italy, Spain and parts of Germany. Throughout much of Europe his name was for many years almost synonymous, indeed, with the words "physique expérimentale".[2]

There has long been a tendency – especially, once again, among English-speaking authors – to regard French experimental physics in the first part of the eighteenth century as something of an intellectual backwater. The general French caution about embracing the new approach of Newton and his followers has tended to be regarded, with hindsight, as mere blind conservatism prompted by an undue fondness for the *a priori* doctrines of Descartes. That the situation was in fact more complicated than this has recently begun to be recognized.[3] Yet the basic picture of a doctrinaire Cartesianism being slowly overwhelmed by the power of the Newtonian method has as yet scarcely been disturbed. This general outline was first developed by eighteenth-century writers such as Voltaire in discussing the fortunes during the first part of the century of Newtonian celestial mechanics. There are dangers in extending it uncritically to other areas of eighteenth-century science: in fact, our habit of viewing eighteenth-century science as a whole entirely from the stand-point of a debate between Newtonians and Cartesians is often unhelpful, and can sometimes be positively misleading.[4] Nowhere is this more apparent than in the field of general physics, because early eighteenth-century French physics was by no means uncritically Cartesian in outlook.

Nollet – to return to him as our example – was very much an eclectic in his approach. "I do not present myself here under the auspices of any philosopher", he proclaimed in the preface to his immensely popular *Leçons de physique expérimentale*, "it is neither the physics of Descartes, nor that of Newton, nor that of Leibniz, that I have resolved to follow particularly". Though he rendered honor, he said, to great men of all nations,

> I admit nothing on their word, unless it bears the stamp of experiment or is demonstrated according to the rules [i.e. of logic]: in matters of Physics, one ought not to be a slave to authority, and still less to one's own prejudices; one ought to recognize the truth wherever it shows itself, and not affect to be a

> Newtonian in Paris and a Cartesian in London. (Nollet 1743-1764, I, pp. xviii-xix.)

True to his promise, in the body of his book Nollet drew in turn from each of the major intellectual traditions of the day. In so far as he rejected forces acting at a distance as a basis for his physics – he even went so far as to re-word Newton's third law of motion to read "Reaction is equal to compression", so as to avoid any suggestion that there might be situations other than the direct contact of two bodies to which it might apply (Nollet 1743-1764, I, p. 261) - Nollet would have to be classed as anti-Newtonian in outlook. He was firmly persuaded that gravity was brought about by a circulating vortex of aetherial matter, and he was generally content with the standard Cartesian practice of invoking "invisible fluids" as mechanical agents to account for all those physical phenomena where no more obvious mechanical cause could be discerned. On the other hand, he recognized that as yet no satisfactory detailed explanation of gravity had been provided, and so far as aetherial mechanisms more generally were concerned, he conceded that some authors had allowed their imaginations too free a rein by far. (Nollet 1743-1764, II, pp. 79-80, 455) He clearly had no sympathy whatsoever for Descartes' efforts to construct an *a priori* foundation for physics. When it came to the science of motion, it was to Newton that he turned. He took over in its entirety, for example, Newton's concept of inertia, and cited as he did so the Englishman's experimental demonstration, using pendulums, of the proportionality of inertia and weight. (Nollet 1743-1764, I, pp. 179-180) He even accepted Newton's conception of a universal attraction, together with his general analysis of the planetary motions. Nollet insisted, however, that this attraction should be regarded merely as a fact, and not "a principle whose immediate cause is none other than the will of the Creator". On this point, he claimed the support of Newton himself, in opposition to various latter-day authors. (Nollet 1743-1764, II, pp. 411-419; VI, pp. 148-158) Finally, while Nollet simply took over large sections of the *Opticks*, including the whole of Newton's discussions of colored rays, he at the same time firmly rejected the corpuscular theory of light that Newton advocated, in favor of a modified version of Descartes' transmission theory. (Nollet 1743-1764, V, passim)

It is in fact impossible to class Nollet and his fellow French experimental physicists meaningfully as either Cartesian or Newtonian, or to fit them at all comfortably into the story traditionally told of a life-and-death struggle between Newtonian and Cartesian doctrines in eighteenth-century France. Generally in sympathy with the mechanistic spirit of Descartes physics, they were at the same time well aware of the magnitude of Newton's achievement, and drew freely upon the Englishman's work wherever they felt it to be well-founded. In other words, in matters intellectual, they generally adopted a bracingly eclectic approach. Furthermore, in their doctrinaire experimentalism their attitude towards their subject was shared by many eighteenth-century physicists outside France as well, by whom it continued to be widely accepted into the early decades of the following century even after it had been supplanted in France itself. Their style of thinking provides the historical backdrop against which the new mathematical physics constructed by men such as Coulomb, Laplace, Fourier and Fresnel needs to be set. In this paper, then, I shall try to bring out some of the more important features of their approach and of the way in which it evolved.

* * * * * * * * * *

As traditionally studied in universities throughout Europe, physics was not an experimental science. Nor was its subject matter the same as it is today. In both style and content the pattern was set instead by the Aristotelian treatises *Physica, De caelo, De generatione et corruptione, Meteorologica* and *De anima*, the study of which constituted a major part of the traditional undergraduate philosophy course. Consistent with the original meaning of the Greek word φυσις, from which the subject took its name, "physics" dealt with "nature" in general. Traditionally, its primary concern was with very broad questions such as the nature of body, or the conditions determining natural change in general, rather than with particular natural effects.

In France, this conception of the subject still predominated in the early years of the eighteenth century, notwithstanding the well known changes that had taken place in intellectual outlook during the preceding period of the "scientific revolution". In particular, physics continued to be seen, even by the "moderns", as an explanatory science, not a merely descriptive one.[5] Indeed, physics was commonly distinguished from natural history on precisely this basis; that is, natural history was seen as resting content with descriptions of phenomena while physics ascended to their causes. Furthermore, "physics" continued to embrace not only the topics it includes today but also questions that we now assign to the chemical, biological and human sciences and, to some extent, to astronomy. Only very gradually during the course of the succeeding century did the scope of the discipline gradually become confined within the boundaries it enjoys today.

An important change of emphasis can, however, be detected at this period, with ever-increasing amounts of attention being paid to questions of special physics (*physique particulière*) – the branch of the science concerned with "the examination and discussion of the particular bodies or corporeal individuals this universe contains" – at the expense of the more traditional general physics (*physique générale*) that dealt with "the nature, essence and properties appropriate to all bodies".[6] The change is apparent, for example, in Rohault's famous *Traité de physique* (1671), where Cartesian physics is expounded without the metaphysical underpinnings upon which Descartes himself had placed so much emphasis, and is presented instead as an experimentally oriented science chiefly distinguished by its mode of explanation in terms of corpuscular mechanisms in a universally disseminated aether.[7] Likewise, as the Cartesian philosophy at last began to penetrate the curriculum of the University of Paris in the late 1690s, a particular complaint levelled against the reformers by their more traditionally-minded colleagues was that they devoted undue attention to *physique particulière* at the expense of metaphysics and moral philosophy. (Freudenthal 1980, pp. 267-295) A comparison between late seventeenth-century and early eighteenth-century university texts – between, for example, Jean Du Hamel's *Philosophia universalis* (5 vols., 1705) and Pierre Le Monnier's *Cursus philosophicus ad scholarum usum accommodatus* (6 vols., 1750) – tells the same story, with a massive increase in attention to questions of *physique particulière* being evident in the later works.

The idea of experimental investigation was simply inappropriate to old-style Aristotelian "general" physics. With the increasing attention being paid to "special" physics, however, empirical considerations began to play a more and more important role. To some extent this was so even among Descartes' immediate followers, for the master, despite his overwhelmingly *a priori* outlook, had nevertheless allowed experiment a definite place in his science. Rohault, in particular, who was unquestionably the leading Cartesian of his generation, placed great emphasis on experimental inquiry.

Throughout the second half of the 17th century, however, Descartes' supporters remained outside the main institutions of French intellectual life. As is well known, Cartesians were formally excluded, along with Jesuits,' from the early Académie Royale des Sciences, where the experimental approach found its most secure institutional base. One of the Academy's two meetings each week was regularly devoted to experimental research in physics, while the other was given over to mathematical pursuits. To be sure, here, too, "physics" was understood more broadly than it is today, with most of the experimental work done being concerned with questions we would today regard as anatomical, chemical or botanical rather than physical. But men such as Mariotte and Perrault also took up for experimental investigation questions of an unarguably physical kind. At the same time, they adopted a highly sceptical attitude towards scientific theory far removed from the claims of both peripatetic and Cartesian natural philosophy. For both Perrault and Mariotte, and with them the entire *physicien* group in the early Academy, physics was a pragmatic science no longer concerned with either certainty or the ultimate causes of things.

One can, in fact, distinguish at this period three separate groups in France that were concerned in one way or another with "physics", as this subject was traditionally defined. In effect, these groups constituted, for a few years, three separate research traditions, each with its own values, methodological outlook and ontology and each forming, as well, a relatively distinct social group. First, there were the University professors, who took as their task the propagation of traditional learning, not the discovery of new knowledge. Their method was verbal and expository; their theory, Aristotelian; their objective, knowledge that was certain; their primary concern, the elucidation of causes. Secondly, there was the *physicien* group in the Académie Royale des Sciences, committed above all to the discovery of new knowledge by means of experiment, generally inclined to mechanistic theories to explain their observations but fundamentally sceptical about all theory and less and less concerned with the discovery of causes. Finally, there were the Cartesians, whose objective was, as with the professors, certainty and a knowledge of causes, but whose method was experimental and whose theory was programmatically mechanistic. The separation between these groups began to break down in the 1690s. In effect, we are concerned in this paper with the re-grouping of forces that subsequently occurred.

The Académie des Sciences underwent a major restructuring in 1699, one consequence of which was that Cartesians were admitted to the membership for the first time. In due course this had the effect of somewhat diluting the scepticism of the early years. Yet even after its reform, the Academy remained an exclusive and, intellectually speaking, a very inward-looking body. As a result, even though its members did some important work, except for Malebranche they had surprisingly little impact on intellectual patterns in French society more generally. In particular, their empirical outlook had little influence outside the Academy's walls, so that empiricism had to fight its battles anew at the more popular level of the university curriculum and the public lecture, with little assistance from the King's academicians. From this process it emerged in a rather different form that involved, in particular, a more naive view of the relationship between experiment and theory in natural science than had been espoused by, for example, Mariotte. Eventually, with the appointment of Nollet to the Academy, this view, developed outside the Academy's walls, glorying now in the name of *physique expérimentale* but limited in its application to a narrower range of phenomena than included heretofore in the domain of "physics", became the new orthodoxy even within the ranks of that institution.

The change began, however, in the universities where, as I have mentioned, Cartesianism began to make significant inroads into the curriculum in the late 1680s.[8] Though the overall structure of the courses was in fact unaffected by this development, so that general philosophical principles continued to be presented first and then applied in turn to the various traditional domains of special physics – the heavens, and then inanimate and animate nature – increasing attention did begin to be paid to these latter sections of the course. A factor that contributed greatly to the success of the new approach at this period was a very popular course of demonstration experiments, similar in some respects to Rohault's earlier courses for the general public, but now introduced into the University of Paris for the first time. The lecturer was Pierre Polinière, and he offered his course of lectures at the invitation of one of the chief instigators of the Cartesian invasion of the university, Guillaume Dagoumer.

The importance of the demonstration experiment for the history of eighteenth-century physics has recently been stressed anew by Heilbron (1979). In regard to the history of French experimental physics in particular, Pierre Brunet argued many years ago that the example of the Dutch professors 'sGravesande, Boerhaave and Musschenbroek was crucial. (Brunet 1926) Yet in my opinion Brunet has seriously under-rated the significance of Polinière's work, which has yet to be studied with the attention it deserves. Though the Dutch undoubtedly exerted a major influence on French physics, especially on Nollet, I believe that Polinière represents a perhaps equally important native source for a tradition of experimental philosophy.

Polinière commenced his studies at the University of Caen before moving to Paris, where he applied himself particularly to the study of mathematics under the guidance of the celebrated Pierre Varignon. Ultimately this bore fruit in his first publication, a bulky textbook of elementary mathematics that appeared in 1704. (Polinière 1704) Polinière's real love, however, was always physics, not mathematics. He taught himself from books how to repeat the leading experiments of the day, to such good effect that, as has been noted already, he was commissioned by Dagoumer in the mid-1690s to give a course of experimental demonstrations for students at the Collège d'Harcourt. This proved so successful that he was subsequently engaged to repeat the course each year in several of the other colleges as well. He published his lectures in 1709 under the title *Expériences de physique*, and the work proved popular enough to go through two further editions during his lifetime, in 1718 and 1728. A posthumous fourth edition based on Polinière's own manuscript amendments appeared in 1734, a few months after his death, and a fifth, almost identical with the fourth, in 1741. Polinière was of a retiring disposition, and spent most of his life on his estate in Normandy, moving to Paris for only a few months each year in order to give his lectures. Despite this, the lectures achieved a great success in Parisian society as well as in the University, to the extent that the young King Louis XV himself attended the course in 1722.[9]

A consistent emphasis on the dependence of physical theory upon experiment was one of the hallmarks of the experimental physics of Nollet and others later in the century. Something of an evolution towards the views eventually espoused by Nollet and many of his contemporaries can already be discerned, however, in successive editions of Polinière's book.

When Polinière first began to give his course of lectures, it was clearly intended merely to supplement the systematic course of lectures on natural philosophy being taught by his sponsor, Dagoumer. This subsidiary status is reflected in the first published version, in which experiment is accorded a distinctly inferior role in physics to theory. The professors of Paris, we are told, have taken advantage of their students' enthusiasm for Polinière's experiments "to teach them in an agreeable manner what is most curious and useful in Physics, and have applied [the experiments] very ingeniously to the principles they have established in their writings, in order to confirm these and make their truth better understood". (Polinière 1709, preface) By the time the second edition of his book was published nine years later, Polinière had grown more confident, and no longer restricted experiment to the role of merely confirming principles already established in the theoretical writings of the learned doctors. Experiment, he now indicated, could of itself bring students to a knowledge of true physical principles. The professors, he wrote, "advancing step by step have used these experiments to lead their scholars by the hand, as it were, into the truth of the principles they have established by their writings". "These experiments", he went on,

> can even be considered as the principles and fundamentals of an exact Physics (*une Physique exacte*). I bring in the proximate causes of these effects when I believe I understand them, without troubling myself too much about seeking the remote and general causes on which these proximate causes depend. Because these remote causes often not being well known to us, we are in danger of propounding fables. (Polinière 1718, preface)

It has been pointed out elsewhere that in the 1718 edition of his book, Poliniere substantially altered his discussion of light and colours to include an account of Newton's work. (Corson 1968, pp. 402-413) Though he presented Newton's ideas only hesitantly – "I do not pretend that these novelties are incontestable and without difficulty", he said, "I hope only that they will contribute to further deepen research to which the objects we see every day invite us" (Polinière 1718, p. 465) – his readiness to present them as fully as he did at a time when they remained little known in France reveals his determination to keep his presentation as up to date as possible. So, too, does his prompt inclusion in the 1734 edition of an account of the important magnetic researches communicated to the Paris Academy of Sciences shortly before by Du Fay (Polinière 1734, I, pp. 364-366)[10] But what most appealed to him in both cases, it is clear, was the way in which important theoretical advances here so obviously depended upon prior experimental investigation. By the third (1728) edition of his book, Polinière's confidence in himself and in the experimental method he had embraced was complete. Physics, he now maintained, was concerned with "a knowledge of the effects our senses reveal to us, and the study of their causes". (Polinière 1728, p. [viii]) Experiment when properly ordered could lead directly to the fundamental principles of the science, as he believed his own work showed. The proper ordering, however, was important:

> There is here no accumulation of useless, confused and amusing observations of different kinds, without design or consequences. Here are materials investigated with care, prepared and arranged methodically, and which can be considered as the fundamentals of an exact Physics.

Unlike conclusions reached on a merely verbal basis, he said, "the truth I announce will be recognized in all times and places, will always be novel despite its age, and will always gain the admiration of those who begin to have a knowledge of it". (Polinière 1728, pp. [i]-[ii]) Experiment, he maintained, was "the shortest and surest way to know the essence and properties of bodies, and even the economy of the Universe". It is by this means, in short, that we come to know the "immutable laws ... that [God] has imposed on nature".

To insist upon the priority of experiment in physical inquiry was, of course, by the eighteenth century scarcely original. Bacon, whom Polinière cited approvingly from the outset (Polinière 1709, p. 251), had been dead for almost a hundred years, while the Royal Society of London and the Paris Academy of Sciences had long since been successfully established. Nevertheless, I must insist that, in its own time and circumstances, Polinière's stance was significant, as is the evolution in his thinking that is revealed in successive editions of his treatise. Poliniere's lectures constituted the first major infiltration of an experimental outlook into the French educational system. As he himself tells us, once his book was published, the example of the Parisian masters in utilizing his work was imitated in a number of other provincial and Paris colleges. (Polinière 1728, pp. [iii]-[iv]) Initially presented as merely supplementary to the traditional and largely theoretical physics of the schools, experiment became in the third and subsequent editions of Polinière's book the *only* dependable way of arriving at a true physics. In other words, by 1728 Polinière was propounding a complete and thorough-going empiricism, independent of both Aristotle and Descartes; and though many of the explanations he offered for particular effects were modelled after those provided earlier by Descartes, his commitment to them was free of the *esprit de système* that had marked even Rohault's work of a generation earlier.[11]

Of particular interest in the later editions of Poliniere's book is his claim that the principles of physics are directly discoverable by experiment. Such a view was implicit in Bacon's inductivism, but has now been generally discredited. In the eighteenth century, however, it was maintained by many other authors after Polinière. In particular, it was embraced enthusiastically by Nollet (who during his student days in Paris in the early 1720s may well have attended Polinière's lectures), and, following Nollet, by numbers of lesser writers. It became, in fact, for almost a century the leading feature of the experimentalist tradition in physics.

This brings me, then, to Nollet himself. Nollet's commitment to the experimental approach doubtless derived, in part at least, from his personal circumstances of life. Having been born of peasant stock, he had no private income, and so was faced with the need to make a career for himself. Endowed with both a keen brain and remarkable manual dexterity, he found in the physics lecture-demonstration a way of making a living that fitted perfectly with his own talents and interests. He announced his first course of lectures in 1734, so soon after Polinière's death that the timing seems more than coincidental. In these, and later in the published *Leçons de physique expérimentale* that grew out of them, he sought systematically to develop general physical principles upon experimentally demonstrable foundations. Not only was this an excellent pedagogical technique – there is nothing quite so convincing as evidence laid before one's very eyes – it also used Nollet's personal skills to the utmost in the design of spectacular yet meaningful demonstrations.

Already by the time he initiated his course, Nollet had acquired powerful patrons in the Academy of Sciences in the persons of Du Fay and Réaumur, with both of whom he had worked as a resesarch assistant. His lectures, which were aimed at upper-class Parisian society rather than at university undergraduates, were even more successful than Polinière's. Together with the published *Leçons*, they gained enormous public prestige for *physique expérimentale* which culminated in 1753 in the King's creating and naming Nollet as first incumbent of a new chair of experimental physics in the University's Collège de Navarre. This was the first such chair in any French university, and it set the pattern for others established in the course of the next two or three decades.

Nollet, however, went beyond mere pedagogy in his reliance on experiment. Like Polinière, and in a manner similar to Bacon, he maintained that the general principles of any worthwhile physical theory *had* to be straightforward generalizations from experience. "The aim of experimental physics", he announced, "is to know the phenomena of nature, and to show the causes of these *by proofs of fact*." (my emphasis) (Nollet 1753, p. 9) He roundly condemned "all those hypotheses which a too bold imagination takes pleasure in fabricating and multiplying on its own authority to form a body of doctrine, before knowing how they square with the facts that Observation and Experiment can furnish." (Nollet 1753, p. 34) On the other hand, like Bacon, he allowed the *controlled* use of hypotheses a place in his science: "I dare to say", he wrote, "that one can and should allow some, if one is content to conceive of possibilities, in order to submit them to experiment, and learn in that way what reality they can have." (Nollet 1753, p.

33) Yet he insisted that the hypothetical status of such propositions should never be forgotten, and stated plainly that "observation and experiment are the surest means, I would say practically the only ones, that a scholar who seeks to extend the progress of physics can employ." (Nollet 1753, p. 24).

Unlike Polinière, Nollet became a member – indeed, in time, a leading member – of the Paris Academy of Sciences. He was a practising experimental physicist in his own right, and not merely an experimentally oriented physics teacher.[12] His research publications exhibit the same Baconian philosophy, however, as his pedagogical works. When he first presented his new theory of electricity at a meeting of the Academy in April 1745, for example, he acknowledged that what he was proposing was a "system", but insisted that his, unlike certain others, was built upon experimentally established truths: "in forming it", he said, "the imagination has worked only on what has been furnished by experiment, and I make bold to say that it would be a mistake to take it for an assemblage of mere possibilities, or speculations devoid of proof." (Nollet 1745, Mém. p. 108) A few months later, in reporting to a friend that Bose of Wittenberg had advanced ideas very similar to his own, he remarked: "I am not surprised – I flatter myself that the same ideas will come to anyone with a little experience in physics who considers all the electrical phenomena at leisure."[13]

A closer look at the arguments Nollet offered in support of his electrical theory brings out his position even more clearly. Already convinced of the inadequacy of resorting in this instance to unexplained forces acting at a distance, Nollet described a series of experiments which, he said, made manifest the existence of some matter in motion in the region around an electrified body. These included the effect known as the "electric wind" – the cobweb-like sensation experienced when an electrified rod is held near the skin – and the crackling sound, characteristic smell, and luminous discharges associated with a well rubbed tube. "What but a substance", he asked, "is tangible, makes itself heard, has a smell, and is visible? Do not all these features incontestably indicate a matter?" (Nollet 1745, Mém. p. 112) As to the nature of this matter, Nollet's starting point was Herman Boerhaave's conclusion, also reached on what purported to be wholly empirical grounds, that the world was filled with an imponderable elastic all-pervading subtle fiery medium. Nollet had long been of the belief that it was this substance that gave rise, under appropriate circumstances, to the various phenomena of electricity. In the absence of positive proof, however, he had refrained from making his views known. The recent German experiments in which electrical discharges were used to ignite alcohol and other readily inflammable materials gave him the "proof" he had been waiting for, and prompted the presentation to the Academy of his initial paper on the subject.[14]

The naivety of Nollet's empiricism is perhaps most evident of all, however, in his discussion of the means by which the supposed all-pervading Fire brought about the various electrical effects. Fundamental to his entire argument here was the assumption that the various luminous discharges associated with electrified bodies consisted simply of streaming fiery matter rendered visible as a result of being concentrated into a more confined space than usual. The various motions of this matter could thus, Nollet assumed, be ascertained by direct inspection. The explanations of other effects such as the attraction and subsequent repulsion of nearby light objects were to be sought in terms of the motions thus discovered. As is well known, he concluded that any electrified body was surrounded by a double stream of subtle fiery matter, an "effluent" stream flowing out from the body and an "affluent" one flowing inward from the bodies surrounding it.

Nollet's attitude here had a number of interesting consequences. To begin with, the chief desideratum for a researcher who adopted his views became the ever more detailed study of common and well known effects, and not the "torturing" of nature to obtain data about new and unusual situations. One or two seemingly contrary results obtained in special situations could not, according to Nollet, overthrow an understanding gained from multiple observations of more common states of affairs. It was precisely on this basis that he rejected certain difficulties raised by Jallabert against his theory: "it is not by one or two proofs that electricity must be judged", he wrote, "but by the same experiments repeated a thousand times."[15]

It also becomes easy to understand Nollet's bafflement when confronted soon afterwards by the writings of Benjamin Franklin and his supporters. So far as Nollet was concerned, his opponents, in attacking his theory, were denying not a set of abstract propositions but experimentally proven facts. For example, misunderstanding Franklin's doctrine of "plus" and "minus" as an attempt to replace his own theory of effluences and affluences by a theory based on unidirectional streams of electrical fire, Nollet cited again his observation of luminous brushes directed from neighboring objects towards a rubbed glass globe, and could not comprehend how, in the light of this, Franklin's supporters could deny the existence of a second, inward-moving stream. Likewise, to combat Franklin's notion that glass was totally impermeable to electrical fire, he contrived some striking experiments in which luminous discharges appeared inside an evacuated glass chamber when this was electrified on the outside. To his astonishment, however, his opponents remained unconvinced. So far as Nollet was concerned, to continue to assert the impenetrability of glass in these circumstances was tantamount to denying the plain testimony of one's own senses.[16]

To modern eyes – and this brings me back to my starting point – one of the most striking features of Nollet's writings on physics is the complete absence from them of quantitative considerations and mathematical analysis. Nor is this a feature peculiar to Nollet's work. Polinière's treatise is equally non-mathematical, even though, as we have seen, he was a competent mathematician, at least at an elementary level. Economic and social factors clearly played a role here. Both Polinière and Nollet designed their courses of lectures for lay audiences, and would not have been able to assume even the most rudimentary prior knowledge of mathematics. In this respect, the French texts were very similar to those published in Holland and England at about the same time. Even in such a mathematical branch of the science as mechanics, there was everywhere a tendency to offer experimental demonstrations rather than mathematical proofs.

The avoidance of mathematical forms of argument extended, however, even to the research publications of Nollet and many of his fellow eighteenth-century experimental physicists. Some paid lip service, it is true, to the ultimate goal of a truly mathematical physics. But even then they seldom meant by this phrase what we mean by it today. By "mathematical" they usually meant merely "rigorous": they had no conception of a physics in which quantitative functional relationships were manipulated by means of formal mathematical arguments. Indeed, for those who shared Nollet's views of the proper relationship between experiment and theory, such a physics was strictly impossible, since direct observation of nature could never yield precise mathematical relationships between physical variables.

Nollet's published remarks about mathematics were in fact somewhat equivocal. In his 1753 inaugural lecture, he remarked that a knowledge of languages could be useful for a budding physicist. This included, he said, a knowledge of mathematics:

> A language which it is indispensible to learn is that of algebra and geometry; these two sciences have been successfully introduced into physics; everywhere where they can be applied, they bring the exactitude and precision that is peculiar to them; they spread illumination in the mind, they make it reason aright; with their help, it proceeds more rapidly, more surely, and can go further; it is essential if one is to follow authors who travel by the light of these torches. (Nollet 1753, pp. 17-18)

Later in the same work he further emphasized the usefulness, even the necessity, of mathematics for the physicist:

> As it is to be hoped that Beginners who seek to instruct themselves by reading Works on Physics understand the Geometrical and Algebraic expressions that are very commonly used in these today, we should also regard it as necessary for those who want to extend the progress of Physics to possess enough of these two Sciences to assist them in their researches, and to evaluate their discoveries: doubtless there will be many occasions when one is reduced to regretting being unable to make use of them; but even then the geometrical spirit will prevent one from departing from the truth by following false trails, and will make one see the *approximately's* more correctly. Everywhere, moreover, combinations, measurement, calculation will indicate in advance what can be expected from an investigation, will open new views, and will prevent false appearances being taken for reality. (Nollet 1753, pp. 42-43)

Yet this kind of usefulness for mathematical reasoning remains a far cry from the much more crucial role it has since come to play in physics. So far as Nollet was concerned, the principal function of mathematics in physics seems to have been to prevent one from straying into false paths of reasoning. Furthermore, Nollet proceeded immediately to a fierce denunciation of those who had sought to make physics a mere plaything for mathematicians:

> Having recommended in good faith the application of Geometry to Physics, having likewise recognized that the study of Nature only commenced with that happy union to make real progress, would I dare say that it is dangerous for a Physicist to develop too great a taste for Geometry? Living examples will not fail to be opposed to me, who will prove beyond possibility of reply that one can be at the same time an excellent Geometer and a very skillful Physicist; but are these good models always imitated? For a few wise Geniuses, who have not by excelling in an exact Science lost the taste for a study where one almost never finds either precision or complete certainty, and who only have recourse to calculation and Geometrical expressions when the importance of the questions and the nature and necessity of the proofs demands it: how many do we not see who can no longer descend from the high speculations to which they have risen, who disdain everything which is beneath! How many others who could without difficulty reach the common level, amuse themselves by rendering in Algebraic characters truths which would lose none of their value if they were expressed in a manner intelligible to everyone! Such writings, much appreciated, show clearly enough that the small amount of Physics in them has served as pretext for making a parade of a different Science. (Nollet 1753, pp. 43-44)

There is bitterness evident in these remarks, but there is also hidden in them a revealing hint concerning Nollet's attitude towards physics itself. Physics, Nollet says, is a science in which "one almost never finds either precision or certainty." To us, coming after Hume, the lack of complete certainty will come as no surprise; but the claim that physics is generally an *imprecise* science is not something we are used to. Nevertheless, it is entirely consistent with Nollet's naively empiricist approach to his subject. Strictly speaking, even in mechanics and optics, fields where both mathematical and physical styles of reasoning were seen to have roles to play, these were conceived by Nollet and his colleagues as distinct and separate. Thus for many thinkers at this period, mechanics done mathematically was a branch of mathematics; done with nuts and bolts, it was a branch of physics. And though many would have agreed with Nollet that a good mechanic needed a sufficient knowledge of both mathematics and physics, (Nollet 1743-1764, I I, pp. 2-3) they would also have shared his belief, as set out in another letter to his friend Jallabert, that the exactitude of mathematics was, at bottom, simply inappropriate to physics. Commenting on the news that Daniel Bernoulli was trying to unite experiment to mathematical theory, Nollet stressed the need for "Geometrical exactitude not to be disdainful of lowering itself to the *à peu près* inseparable from experimental physics", adding that "I can well imagine that that would come hard to someone accustomed to rigorous demonstrations."[17]

What we have here, I believe, is an eighteenth-century reflection of a very traditional view. "The minute accuracy of mathematics is not to be demanded in all cases", Aristotle had written in the *Metaphysics*, "but only in the case of things which have no matter. Hence its method is not that of natural science; for presumably the whole of nature has matter."[18] In the seventeenth century, van Helmont wrote that "the Rules of the Mathematicks, or Learning by demonstration, do ill square to Nature. For Man doth not measure Nature; but she him."[19] Even Descartes had protested in a letter to Mersenne that "to require of me Geometrical demonstrations in a matter which depends on Physics is to demand that I achieve impossible things."[20] According to Joseph Priestley writing later in the eighteenth century, one of the great achievements of the scientific revolution had been to separate natural philosophy (i.e. in this context, physics) from the mere ratiocination that characterized mathematical reasoning:

> Before this period [following Descartes], we hardly meet with any production among the learned, but such as might have been executed

> without stirring from their closets, except in what respected mechanicks; so that the proper distinction between the natural philosopher and the mathematician was but little attended to. (Priestley 1772, I, p. 132).

As I indicated earlier, however, I believe that so far as Nollet was concerned, the experimental method he adopted was itself sufficient to rule out a quantified mathematical physics. With his over-riding emphasis on direct observation, he could never have reconciled himself to the notion of "experimental error" that is so crucial to the determination of quantitative experimental generalisations. For Nollet, and with him the whole school of eighteenth-century experimental physicists, deviations from mathematical exactness in the data could never be discarded as mere chance departures from a ruling norm. Rather, they were bound up with the irreducibly inexact nature of the subject itself. They were, in fact, of the very essence of a truly experimental physics.

Hence it was that when, in the first years of the following century, mathematics began for the first time to be applied successfully to a wide range of physical questions, those who remained faithful to the old-style *physique expérimentale* quickly found themselves outflanked. Though their predecessors had played a central role in reconstructing the subject on an experimental footing, the too-direct empiricism that I have described, that left no room for a theory expressed in mathematical terms, now in its turn also became outmoded. But that is another story.

University of Melbourne

NOTES

* Research for this paper has been supported by a grant from the Australian Research Grants Scheme.

[1] Cf. Home (1974, II, pp. 269-272); Home (1981, passim); Heilbron (1979, chaps. 11,15).
[2] Torlais (1954). For an account of Nollet's influence even as far away as distant Bohemia, see Smolka (1968, III B, pp. 131-135).
[3] Marsak (1959, pp. 51-60); Hall (1975, pp. 233-250); Guerlac (1977, pp. 479-490); Guerlac (1981, esp. chap. 5). For the more traditional view of French resistance to Newton, Cf. Brunet (1931).
[4] Home (1979a, pp. 235-249); Home (1977, pp. 252-266).
[5] Cf. Rohault (1671, I, p. 1): "This Word, *Physicks*, strictly speaking, and according to the Etymology of it, signifies no more than *Natural*; but we here use it to signify the Knowledge of natural Things, that is, that Knowledge which leads us to the Reasons and Causes of every Effect which Nature produces."
[6] Anonymous (1756, p. 582).
[7] On Rohault, see Schuster (1975, pp. 506-508); also Mouy (1934, pp. 108-138); and McClaughlin (1977, pp. 225-240).
[8] Brockliss (1981a, pp. 33-69); Brockliss (1981b, pp. 131-168); Freudenthal (1980).
[9] These biographical details are drawn from the anonymous "Abregé de la vie de M. Polinière" inserted in the posthumous editions of his *Expériences de physique* (1734, 1741), and from Polinière's own preface to his book.
[10] Cf. Du Fay (1728, Mém. pp. 355-369); Du Fay (1730, Mem. pp. 142-157); Du Fay (1731, Mém pp. 417-432).
[11] Many of Polinière's detailed discussions have been summarized in Hanna (1972, pp. 13-39).
[12] Polinière's only piece of significant research was his study very early in his lecturing career of the phenomenon now known as electroluminescence. On this, see Corson (1968).
[13] Nollet to E. F. Du Tour, 10 December 1745 (Dibner Library, Smithsonian Institution): "je nen suis pas surpris – je me flatte, que les memes idées viendront a tout homme un peu initié en physique, qui aura vû a loisir tous les phenomenes Electriques".

[14] In a letter to the Genevan professor Jean Jallabert dated 19 July 1745, Nollet wrote as follows concerning the new German experiments: "Mais ce qui ma le plus occupé, c'est que jay aperçu dans ces nouveaux phenomenes des preuves assez fortes pour établir un systeme d'explication que je meditois depuis longtemps, mais que je n'osois hazarder sur des raisons foibles ou equivoques". (MS collection, Société d'Histoire et d'Archéologie de Geneve; quoted in Home (1979b, pp. 171-176).

[15] Nollet to Jallabert, 28 July 1747, MS collection, Société d'Histoire et d'Archéologie de Geneve: "Ce nest ny par une ny par deux epreuves quil faut juger de lelectricité, mais par les memes experiences repetées mille fois". Nollet's attitude is well described in Yamazaki (1976, pp. 37-64).

[16] For Nollet's response to Franklin's ideas, see Home (1981, chap. 7).

[17] Nollet to Jallabert, 10 December 1750, MS collection, Société d'Histoire et d'Archéologie de Genève: "Entre les mains dun tel homme cela doit bien aller, pourvû que l'exactitude Geometrique ne dedaigne pas s'abaisser jusquaux *a peu prés* inseparables, de la phys[ique] experimentale: Car je conçois bien que cela doit couter quelque chose a quelqu'un qui est accoutume aux demonstrations rigoureuses".

[18] Aristotle (330 BC, Bk. II, chap 2, 995a 15-18).

[19] Debus (1968, p. 24).

[20] Descartes to Mersenne, 27 May 1638; Descartes (1897-1913, II, pp. 141-142). The translation is from Sabra (1967, p. 23).

REFERENCES

Anonymous (1756), *Memoires de Trevoux 57*.

Aristotle (330BC), *Metaphysics*.

Brockliss, L. (1981a), "Aristotle, Descartes and the New Science: Natural Philosophy at the University of Paris, 1600-1740", *Annals of Science 38*.

(1981b), "Philosophy Teaching in France, 1600-1740", *History of Universities 1*.

Brunet, P. (1926), *Les physiciens hollandais et la méthode expérimentale en France au XVIIIe siecle*. Paris.

_________ (1931), *L'introduction des théories de Newton en France au XVIIIe siècle avant 1738* Paris.

Corson, D. (1968), "Pierre Polinière, Francis Hauksbee and Electroluminescence: A Case of Simultaneous Discovery", *Isis* 59.

Debus, A. (1968), "Mathematics and Nature in the Chemical Texts of the Renaissance", *Ambix* 15.

Descartes, R. (1897-1913), *Oeuvres de Descartes*, eds. C. Adam and P. Tannery. Paris.

DuFay, C. (1728, 1730, 1731), *Histoire de L'Academie Royale des Sciences, avec les memoires*.

Freudenthal, G. (1980), "Littérature et sciences de la nature en France au debut du XVIIIe siècle: Pierre Polinière, l'introduction de l'enseignement de la physique expérimentale à l'Université de Paris et l'*Arret burlesque* de Boileau", *Revue de synthese 101*.

Guerlac, H. (1977), "The Newtonianism of Dortous de Mairan", *Essays and Papers in the History of Modern Science*. Baltimore: Johns Hopkins University Press.

_________ (1981), *Newton on the Continent*. Ithaca/London: Cornell University Press.

Hall, A. (1975), "Newton in France: A New View", *History of Science 13*.

Hanna, B. (1972), "Poliniere and the Teaching of Experimental Physics at Paris, 1700-1730", *Eighteenth Century Studies presented to Arthur M. Wilson*, ed. Peter Gay. N.H.: Hanover.

Heilbron, J. (1979), *Electricity in the 17th and 18th Centuries: A Study of Early Modern Physics*. Berkeley: University of California Press.

Home, R. (1974), "Electricity in France in the Post-Franklin Era", *Actes du XIVe Congrès International d'Histoire des Sciences, Tokyo-Kyoto, 1974* Tokyo.

_________ (1977), "'Newtonianism' and the Theory of the Magnet", *History of Science 15*.

_________ (1979a), "Out of a Newtonian Straitjacket: Alternative Approaches to Eighteenth-Century Physical Science", *Studies in the Eighteenth Century IV*, ed. R. Brissenden and J. Eade. Canberra: Australian National University Press.

_________ (1979b), "Nollet and Boerhaave: A Note on Eighteenth-Century Ideas about Electricity and Fire", *Annals of Science 36*.

_________ (1981), *The Effluvial Theory of Electricity*. New York: Arno Press.

Marsak, L. (1959), "Cartesianism in Fontenelle and French Science, 1686-1752", *Isis 50*.

McClaughlin, T. (1977), "Le concept de science chez Jacques Rohault", *Revue d'histoire des sciences 30*.

Mouy, P. (1934), *Le développement de la physique cartesienne, 1646-1712*. Paris.

Nollet, J. (1743-64), *Leçons de Physique expérimentale* 6 vols., Paris.

_________ (1745), "Conjectures sur les causes de l'électricite des corps", *Histoire de l'Académie*.

_________ (1753), *Oratio habita . . . cum primum Physicae Experimentalis cursum professor a Rege institutus auspicaretur*. Paris.

Polinière, P. (1704), *Elémens des mathématiques*. Paris.

_________ (1709), *Expériences de physique*. Paris.

_________ (1718), *Expériences de physique*. 2nd ed.

__________ (1728), *Expériences de physique*. 3rd ed.

__________ (1734, 1741), *Expériences de physique*. posthumous eds.

Priestley, J. (1772), *History and Present State of Discoveries relating to Vision, Light, and Colours*. London

Rohault (1671), *Rohault's System of Natural Philosophy, illustrated with Dr. Samuel Clarke's Notes*. trans. John Clarke. London, 1723, reprinted New York, 1969.

Sabra, A. (1967), *Theories of Light from Descartes to Newton*. London.

Schuster, J. (1975), Art. "Rohault", *Dictionary of Scientific Biography*. *XI*. New York.

Smolka, J. (1968), "L'abbé Nollet et la physique en Bohême", *Actes du XIIe Congrès International d'Histoire des Sciences, Paris 1968* Paris 1971.

Torlais, J. (1954), *Un physicien au siècle des lumières: l'abbé Nollet, 1700-1770*. Paris: Sipuco.

Yamazaki, E. (1976), "L'abbé Nollet et Benjamin Franklin", *Japanese Studies in the History of Science 15*.

GEREON WOLTERS

SOME PRAGMATIC ASPECTS OF THE METHODOLOGY OF JOHANN HEINRICH LAMBERT

1. In this paper I would like to outline the answers to two questions: (1) What does Lambert mean by 'philosophy'? and (2) What does he mean by 'science'? Both of these very general questions give me the opportunity to present Lambert's positions in sufficient breadth as well as to work out some pragmatic elements in his thought. The former appears to me to be especially desirable because Lambert *as a philosopher* is virtually unknown in English speaking countries. Prior to L. W. Beck's work (Beck 1978, pp. 101-110 and Beck 1969, pp. 402-412) done in the sixties I am aware of only one article written in English on Lambert's philosophy. And that was written in the last century. (Griffing 1893) Much of what has been written on Lambert in German as well as English has concentrated not so much on Lambert's philosophy itself as on the extent of his influence on Kant. Certainly, this is no illegitimate concern. But by this focusing of interest on this single aspect much of Lambert's enormous oeuvre[1] has been neglected. For this reason it generally has not been sufficiently realized that Lambert's philosophical work constitutes only a small part of his total scientific writings, which cover almost all areas of mathematics and the pure and applied sciences. What has most hindered an understanding of Lambert's genuine ideas, however, is the fact that it is also generally unknown that his philosophy was written essentially in the service of science. For this reason I can reduce my initial questions to a single one: What is the relationship between science and philosophy in Lambert's work?

Before I deal with this question or these questions, I would suggest a detour by way of Kant. I take this roundabout way not in order to make the same mistakes others have made, but in order to indulge in this way my passion for biography. I will have to restrain myself so as not to use the available space only telling you anecdotes from Lambert's life. That may indeed be more amusing than

J. C. Pitt (ed.), Change and Progress in Modern Science, 133–170.
© 1985 by D. Reidel Publishing Company.

the paper I have before me, but would undoubtedly conflict with the axiomatic conviction of most that philosophy should only be concerned with something like 'pure thought'. Thus I will carefully restrict the extent of biographical contamination and sketch to only a rough outline of Lambert's life. I am convinced, however, that his life has not had an inconsiderable influence on the *form* of his thought. But first to Lambert and Kant.

2. In contrast to others I consider Lambert's influence on Kant to have been limited. The Kantian System as a whole owes little or nothing to Lambert. Lambert's influence restricts itself on the whole to a single but central concept in the *Critique of Pure Reason*: the problem of the relationship between pure ideas (categories) and objects of experience. Lambert found this problem to be insufficiently handled in section V of Kant's *Inaugural Dissertation* from 1770.[2] Lambert's critique however is a purely internal one and in time Kant himself could have recognized this problem. What is so surprising in this case is the enormous esteem in which Kant held Lambert for so long.[3] It reached its high point in a fragment, in which Kant intends to dedicate the *Critique of Pure Reason* to Lambert. (Kant 1770, *Adak.-Ausq*. XVIII, p. 64). What Lambert fans[4] prefer to overlook, however, is that later Kant's admiration for Lambert cooled off considerably. In the handbook on Kant's lectures on logic edited with Kant's consent by his pupil Jasche in 1800, we find stated succinctly: "One believed that Lambert's *Organon* would represent an important contribution to logic. But it contains nothing more than only subtle divisions, which, like all correct subtleties, sharpen the understanding but are of no essential use."[5] I have the impression that even Kant's praise of Lambert in his letter to John Bernoulli III from November 16, 1781 is thoroughly ambiguous. Bernoulli wanted to publish the correspondence between Kant and Lambert and had written to Kant about it. Lambert had already died in 1777 at the age of 49. In his reply to Bernoulli Kant lamented Lambert's premature death, because he had expected a great deal from his collaboration and criticism.[6] This is good as far as it goes. One reads with astonishment, however, *why* Kant "saw all my expectations for such important assistance suddenly vanish with the unexpected death of this extraordinary genius." (*Akad-Auqs*. X, p. 278). Kant does not complain that Lambert's death deprived him of *professional* assis-

tence. No, Lambert is for Kant just the person "whose bright and inventive spirit through the inexperience in metaphysical speculation makes him all the more free from prejudice and consequently all the more skillful in surveying and appreciating the propositions included afterwards in my 'Critique of Pure Reason'." (*Akad.-Augs.* X, p. 278) That Kant considered Lambert a metaphysical amateure is all the more surprising as Lambert himself had published a two volume opus of almost 1000 pages, which he viewed "as a thouroughly new study of the metaphysical theories", the "Architektonik".[7] At the same time Kant was not completely incorrect with his comment on the "inexperience in metaphysical speculation". I would like to mention two reasons for this opinion. The first reason gives me an excuse for making a few remarks on Lambert's remarkable biography.[8] Lambert was a self-taught person. He never attended university and so had no taint of the metaphysical schools, which dominated and held together German philosophy at that time. He was born in Mühlhausen in Alsace in 1728. At that period Mühlhausen was dependant upon the Swiss city of Basel. After six years of elementary school he worked for his father, whose tailor shop made a very modest income. His attractive handwriting appears to have destined him to higher things. He became a bookkeeper at an iron works at 16 and a clerk with a lawyer and newspaper publisher in Basel at 17. At 20 he had taught himself enough to become private tutor to the influential count von Salis in Chur in the Grisons, in whose service he remained for the next 10 years (1748-1758). In 1759 he became a member of the newly founded Academy of Science in Munchen. In 1761 he was again in Switzerland, where he worked among other things as a land surveyor and hydraulic engineer. In 1763 he began travelling with the Academy at St. Petersburg as his goal. At the beginning of January, 1764 he was in Berlin. His promoters, especially the aesthetician Sulzer, needed almost a year in order to make up for the catastrophic impression Lambert made upon Frederick the Great during a sort of introductory interview for the Berlin Academy.[9] And only the impending appointment in St. Petersburg brought about his acceptance at the Berlin Academy, to which he belonged until his death 1777. Four whole publications from Lambert's enormous oeuvre (see footnote 1) are in a narrow sense specifically philosophical (excluding his writings on logic). Of these four publications only two very short writings were first published in our century.[10] The first of these four works was written in December 1761, the last was completed in October 1764.[11] As Lambert's notes further show, he was occupied exclusively with philosophy only during a few months of these four 'philosophical' years. Up to 1761 he had already made a name for himself as mathematician, astronomer and physicist.[12] After 1765 his interest in philosophy seems to have slackened off.[13] One can draw the conclusion from all of this that Lambert was a 'working scientist' with a broad range of interests, who was interested in philosophy as an academic subject only for a relatively short period of time. To that extent Kant's remark about Lambert's inexperience as a metaphysician is certainly not wrong. It is also not wrong because Lambert pursues neither in the *Organon* nor in the *Architektonik* anything like the prevailing scholastic philosophy nor does he present a philosophical system. Lambert is rather a scientist who reflects philosophically to an uncommon extent on the method of science, also in many non-philosophical works. And with a greater degree of competence than his contemporaries. Before I touch upon Lambert's concept of philosophy and metaphysics, however, I would like to deal with his understanding of science.

I have argued here, that Lambert's influence on Kant was rather restricted. My argument, however, is limited to Lambert's influence insofar it has been admitted by Kant. There remains the fact that there are similarities in other fields, e.g. the "intuitionist" approach to the foundations of mathematics.

3. Lambert is a typical representative of the modern experimental and mathematical sciences and their method. Only a very faint echo can be traced back to the battle Galileo and others had to fight a hundred years before in order to establish the new method. What was accomplished a hundred years before is for Lambert an historical fact and represents a naive epochal consciousness.[14] Galileo still believed that it was necessary to provide his 'programm', namely the "reduction of scientific experience to experience that can be expressed in mathematical terms"[15], with a theoretical, i.e. an ontological justification. The universe *is*, according to Galileo, just as it reveals itself in the laws and experiments of mathematical physics. Sometimes contrary to appearance, it consists of mathematical number and geometric form. This is completely different in Lambert's case.

The question of a justification of mathematical physics and its methods just does not exist for him anymore. I know of no passage in his work where such a question is mentioned, either directly or indirectly. As I will later demonstrate, this is certainly not because Lambert and his contemporaries had convinced themselves that the universe actually is as observations and measurements within the context of scientific law make it appear to be. Mathematical natural science is, as far as Lambert is concerned, (still?) much too weak in those areas he calls 'philosophical'. It (still?) cannot tell us what the universe about us is in truth like. I believe, however, that even if the new method had helped him to develop an elegant ontology, he would probably not have sought therby its cognitive justification. The method of experimental and mathematical physics appears to have been justified for Lambert without question by its practical success. By 'practical success' I mean indeed *practical* success, that is, success not with respect to the contribution a method makes to the enrichment of our 'cognitive assets', but in terms of the effect that experimental mathematical physics has on how we cope with the world about us.[16] As I have already mentioned, I can cite no text reference to support this view of Lambert's as he did not discuss it explicitly. Nonetheless I am convinced that there is sufficient indirect evidence for assuming that Lambert actually held this view. (a) In a draft of the guide-lines for the publication of transactions of the Bavarian Academy of Science (1761) it is stated succinctly "It should depend upon two main principles: 1. The extension of the utilitarian sciences. 2. Their application and propagation throughout the land." (Westenrieder 1784). Most of Lambert's contemporaries naturally also supported this demand of the enlightenment for the practical application and propagation of science. One could say that Lambert advocated an instrumentalist concept of science. The goal of science is accordingly not or not primarily the description of the structure of the world about us but rather the theoretical and practical orientation in it. (b) If Lambert did not stand alone in propagating utilitarian science, one can in any event also say that very few preachers of his day practiced what they preached as faithfully as he. In so far as the use and propagation of science exists in its technical and economic utilization, the decisive borderline between theory and practice can be said to lie in *measurement*. A quick glance at the *Bibliographia Lambertiana* shows that the vast majority of Lambert's writings deals directly with measurement, be it in the more theoretical writings on land-surveying, the measurement of time, weighing, photometry, hygronomy, the measurement of air pressure, of wind velocity and of temperature together with the studies on the theory of error;[17] or in recording series of observations, especially meteorological;[18] or in the design and construction of instruments of measurement.[19] If one also takes into account Lambert's experience as land-surveyor, hydraulic engineer and Prussian senior building commissioner, then his scientific work can be understood for the most part as the application of scientific theory to societal practice. If one further assumes that what Lambert *did* as a scientist depends in his opinion upon how science *should* be conducted, then his practical experience appears to me to be a stronger argument for his instrumentalist conception of science than any treatise on the uses of science 'for life' could possibly be. Lambert's instrumentalism is naturally an instrumentalism within the limits of mathematical natural science. I would like to cite still another reference for practical success as a criterion for the justification of mathematical natural science. This proof was very important for Lambert, who, it should be remembered, was one of the foremost pietists in the already very pious German enlightenment. To us it may rather appear bizarre. Not only that, Lambert, like other enlightenment philosophers promised moral improvement through the scientific elimination of error.[20] Far more then that, in an anonymously published sermon he sees in a dream the mystical union of divine revelation and scientific measurement. On Golgotha the personified "sanctified sciences"[21] appear for worship. Here is just a sample: "Anemometry discovered on Golgotha a whole new and heavenly atmosphere of unusual powers [...] The face of optics was flooded with rays of joy [...] Here is God's brightest mirror, called catoptrics. - Here one doesn't need magnifying glasses, said Dioptry. And perspective drew the divine view toward Zion." (Lambert, 1961a, p.8) This should suffice to show how Lambert justifies the method of mathematical natural science by its success in coping effectively with life, both natural and supernatural.

4. Thus Lambert accepts without further question the experimental and empirical method of modern science because it is successful in a practical sense. What interests him as a philosopher is not its theoretical justification

but the question of what it is that makes this method so successful. Knowledge of the logical structure of this method would be of advantage in two respects: (1) One could apply it still more effectively in the less successful areas of science and (2) it would represent the chance – necessary changes having been made – of transfering the visibly successful scientific methodology to the visibly unsuccessful metaphysics. Lambert appears to have been especially interested in the latter. Such a transfer, of course, cannot function *directly* because the objects of scientific and metaphysical investigation are different. Lambert was thus faced with a two-fold task: (1) The analysis of the structure of the mathematical method, i.e. the development of a methodology for the natural sciences and (2) proof of the applicability of this methodology to metaphysics. A metaphysics, it should be mentioned, requiring a new definition of its object.

In order to characterize Lambert's methodology more closely, a number of comments on his concept of science are first necessary. Two aspects of science as a complex of propositions i.e. science in theoretical perspective, are important for Lambert: a more *'semantic'* aspect referring to what scientific propositions are talking about and a more 'syntactic' and *'systems-theoretic'* aspect concerned with the connection between scientific propositions. I would like to begin my analysis of the *semantic* aspect (I am using the term 'semantic' for want of a better) with Lambert's classification of the various types of knowledge.[22] This classification is traditional and owes much to *Christian Wolff*. Together with Wolff, Lambert distinguishes three hierarchically ordered types of knowledge: 1. 'historical' knowledge, 2. 'philosophical' knowledge and 3. 'mathematical' knowledge. 'Philosophical' and 'mathematical' knowledge, taken together, constitute the ideal instance of science in a sense that I will explain later. Lambert's use of 'historical' and 'philosophical' differs from our present usage.[23] 'Historical', similar to Locke, is the merely passive empirical knowledge of isolated facts.[24] 'Philosophical' knowledge on the other hand is – in the sense of Aristotelian natural philosophy – knowledge of the essence (Wesen) and the causes of phenomena, while 'mathematical' knowledge consists of its quantitative expression.

> The doctrine of gravity may serve as an example. The philosopher scarcely knows what he should look for with respect to its theory.

> The mathematician on the other hand compares the space transversed in falling with the elapsed time and the velocity and finds the relationships between these three dimensions to be constant and independant in free space on the mass and the weight of the object and in this way he demonstrates for the philosopher more exactly how his theory of the cause of gravity should be like, because he gives him propositions which must be deducable from this theory if it is to be correct. Kepler's laws of planetary motion are based on just this sort of touchstone of theories that had been developed previously in order to explain the corresponding mechanism.[25]

The (theoretical) pre-eminence of mathematical knowledge so rests on the fact that 'philosophical' theory must not contradict the quantitative, '*law-like*' expression of phenomenal experience[26] and that the dimensions used in the equations of mathematical physics determine the conceptual framework of the respective 'philosophical' theory. A natural philosophy that is not based on quantifiable and consequently measurable concepts is as far as Lambert is concerned not scientific. "One will not be able to pronounce philosophical knowledge a completely scientific knowledge if it is not at the same time thoroughly mathematical [...]. Then the simple ideas that the philosopher must look for [...], are just those which the mathematician can and does use as dimensions as soon as the philosopher has found them."[27]

The paradigm of a successful method is, according to Lambert, Euclidian geometry. "Geometry has always given the philosopher the chance to enrich and impróve his theory of whatever can be called method." (Lambert 1771, p. 683) What pleases him so much in Euclid is the fact that he finds solved what I have elsewhere (Wolters 1980) termed a methodological 'basis problem'. For Lambert science is ideally *axiomatic* as we shall see when we take a closer look at the syntactic and systems-theoretic aspect of science. Axiomatic science raises two methodological questions for Lambert as it already did for Aristotle: 1. Securing the deductive connection of scientific propositions (the problem of deduction) and 2. securing the beginnings of deductions (Deduktionsanfänge) i.e. determining the meaning of those predicates found in axioms (the basis problem). Wolff in Lambert's eyes had already accomplished a great deal and "had achieved at least half of the method".[28] Wolff failed with respect to the second question, however, because he believed it possible to construct an axiomatic system on the basis of nominal definitions. (Lambert 1771, p. 685) Lambert admits being plagued at the beginning by "despondency" because of this problem: "The questions where should one begin and where should one stop defining? How is it possible to avoid the great and terrifying vicious circle [...] Does the analysis of concepts have an end or has it no end? And if characteristics (Merkmale) of characteristics develop ad infinitum, where will we have to remain standing and are we standing securely or only hypothetically?"[29] But then he stumbles, more through chance, on the solution

> I had first read Euclid long after I had read Wolff, and consequently not so much with the intention of acquainting myself with his theorems than in order to judge the esteem Wolff himself expresses for him in his writings on logic because of the strict order and connection of propositions. I knew already pretty well what scholastic and mathematical method were and in spite of that Euclid's first proposition astonished me. I just thought he would begin with the first theorems of the comparison of angles.[30] But he takes angle, side and figure all at once and instead of beginning with a theorem, he begins with a problem. What, I thought, mustn't the theory come first before one moves on to the practice? But Euclid had apparently thought still further [...]. This problem, which I viewed as a guide to practical geometry, served to place the whole theory beyond all doubt [...]. Should not this means also be applicable for proofs of ideas which can be taken for fundamental ideas?:[31]

Let me summarize Lambert's text. The methodological solution of the problem of deduction as the logical connection of *propositions* had already been worked out by Wolff. The deductive method must fail (Aristotle had seen this as well) with the proof of the basis of deduction. On the whole it is not the proof of *propositions* that is really important for the 'basis of deduction', but only the justification of meaning and use of *concepts*. Then concepts are the semantic element on which all knowledge is based. This is precisely what Euclid succeeded in doing through a 'practical' justification of basic concepts, just as Lambert finds it with Euclid.

Euclid, as Lambert sees it, begins with *definitions*. Lambert is not interested in those definitions which we can find listed under the name οζσι at the beginning of the first book of Euclid's *Elements*: "That Euclid begins with and amasses definitions, this is really only a nomenclature."[32] Such a listing of definitions is 'hypothetical". And it remains hypothetical as long as the "possibility" of corresponding concepts has not been proven.[33] The concept of possibility, which Lambert brings now into play here, is extremely important. Lambert also terms it "real" or "positive categorical" possibility as well as, and this is particularly interesting, doability or feasability (Thulichkeit).[34] This doability is expressed in *postulates* that, as Lambert interprets Euclid, state an unconditioned and unrestricted possibility of performing certain actions. – The positive posibility of concepts *somewhere* in a conceptual system is in principle relatively easy to prove. Namely through reduction to the positive possibility of the characteristics of one such concept. The analytical method used here, however, is of no further help once the analysis has been terminated; for then a direct proof, of positive possibility is required.[35] *What* such a direct proof looks like and *what* it proves or is supposed to prove, this Lambert finds in Euclid. Simple logical consistency ('Gedenkbarkeit' i.e. thinkability) is as incapable of characterizing 'positive possibility' as simply infering from facts to possibility. (Lambert 1771, p. 19f) When Euclid, with just this inference in mind, establishes his above mentioned 'nomenclature', then he is "doing nothing more than what in his opinion a clockmaker or some other artist does, when he begins to make his apprentice acquainted with the names of his tools. For Euclid it is enough if one grants that such figures could exist, even if there were only one."[36] Such ostensive definitions can by no means guarantee what Lambert calls the "unconditioned possibility" or "doability" (Thulichkeit). Geometry, as a theory of invariant size and position, cannot be based on the accidental size or position of a single or of several ostensibly safe examples. An "example (serves) only as a demonstration of one or more possibilities. Experience or examples alone are not capable of showing immediately how far the possibility extends." (Lambert 1771, p. 20) Expressed in modern terms: The universality or the use of universal quantifiers in the propositions of the synthetic Euclidian geometry already requires a methodological justification at its base. This justification exists in the proof of the 'doability' or feasability, of an in principle *unrestricted possibility of action*. Whereas simple ostensibility is sufficient for Euclid's definitional 'nomenclature', "he requires in his postulates the 'unconditioned possibility' of straight lines and circles of any size and in any position as a basis for the whole geometrical theory.

"Et hoc si dederis danda sunt omnia."[37] After Euclid has stated these "universal, unconditioned [...] or *simple* possibilities or doabilities" (Lambert 1771, p. 12) in the form of postulates, his methodological undertaking is still by no means ended. In the course of the rest of Euclid's geometry it becomes evident to Lambert just what a methodological fox Euclid was. In the longer passage quoted above Lambert is astonished that Euclid begins with the problem of constructing an equilateral triangle, i.e. with a special case. And he does this, although he has made universality his particular aim. "But Euclid had thought still further." For if he began with the construction of a general triangle,

> then one could doubt the universality [...]. There would have been nothing easier than to have placed before him three lines, one of which would have been greater than both of the others together. He should just try to make a triangle out of these lines. The matter would have soon miscarried. I could see from this now why he began with the possibility of an equilateral triangle. He only needs one line to do it. The trick thereby lies in the fact, that the possibility of an equilateral triangle shows itself, as it were, to be self-evident. One cannot better refute whoever believes something like this impossible than by showing him how he himself could put it into practice. (Lambert 1915, p. 79)

In his construction Euclid pursues further this method of securing the universality of theory through possibilities of action. If the "doabilities" were still unrestricted with respect to the "simple possibilities" of straight line and circle, this is no longer the case in the further construction of the theory. "Restrictions are met with by the combination of such simple possibilities and these Euclid determines for the most part by means of his ninth and twelfth axioms."[38] Let me summarize: If one regards science as a deductive complex of propositions, then history shows geometry to have had a paradigmatic position. The exemplary success of geometry – exemplary also for 'philosophical' knowledge, rests neither (empiristically) on a distinctive relationship to reality, nor (rationalistically) on the intuition of ideal entities. The paradigmatic character of Euclid's geometry rests for Lambert methodologically on the justification of the *universal* application of their fundamental concepts through recourse to 'unrestricted' human possibilities of action ('postulates'). Only certain restrictions of these possibilities of action formulated in 'axioms' must be considered. This justification of the fundamental concepts of a theory in terms of possibilities of action is oriented on Euclid's actual procedure. I would like to call it its *'operative justification'*.

The operationality of the fundamental concepts in this sense warrants their universality. What has remained open up to now, is the essentially semantic aspect, the relationship to facts, the meaning of the fundamental concepts, or, as Lambert prefers to call it, the "matter of knowledge".[39] I will first touch upon this more closely in the section after next. If we just assume for a moment that the 'material' aspect of the fundamental concepts has been clarified, then operationally justified fundamental concepts are 'true' concepts. Although a correspondence theoretic aspect of the concept of truth doubtless comes into play in this manner, this aspect is, as we shall immediately see, not conclusive for Lambert.

5. It would like to turn now to the more syntactic, systems-theoretic aspect of science as a complex of propositions. With respect to the semantic aspect Lambert, as we have seen, distinguished propositions as scientific by the specification of quantified dimensions and ideally by causal proofs. Knowledge derived from *experience* expresses itself for him principally in singular proposi-

tions.[40] Only complicated logical procedures permit their universal formulation. But even knowledge derived from experience that is formulated in universal propositions, in the way in which Lambert understands it, refers to directly sensually perceivable facts and remains as well within the realm of 'common' or 'historical' knowledge.[41] What distinguishes scientific knowledge before knowledge derived from experience is that scientific propositions as opposed to empirical propositions are not isolated logically.

> Scientific knowledge is based accordingly on the *dependancy* of one knowledge or another and examines how the one could be determined by the other. Hence in this way it is opposed to common knowledge, because the latter views every proposition, every idea as subsisting of itself and without any connection or at most finds them harmonizing only through individual inferences and comparisons. (Lambert 1764, p. 605)

True scientific propositions and concepts taken together form the "system or the realm of truth", which is admittedly not known to its full extent and which should probably be understood more as a regulative idea in Kant's sense.[42] The systematic nature of science determines at the same time its cognitive and practical use: "To make experiences superfluous and consequently to determine in advance what one would still have to experience."[43] Lambert illustrates what he means by this with a series of examples: So geometry makes possible the determination of longitudinal positions, whose actual measurement is *practically* impossible. Or: "Newton (remained) in his room and determined the shape of the earth from a few truths known to him."[44] With this, scientific knowledge transcends to a certain extent (See ftn.41) the realm of sensual knowledge. This transcendence occurs, (1) in the sense just mentioned, through the prediction of empirical facts possibly unknown up to that point; predictions however, which follow logically from true propositions. (2) Through the possibility of stating *hypotheses*: "An hypothesis is an arbitrarily assumed idea of a thing, with which one wants to explain it. One gives the thing a certain structure, attributes, mechanism etc. in order to explain and derive what experience teaches us."[45] "Physical" hypotheses come into play at that point where a well-founded theory is still missing.

Their purpose is the same as that of a well-established theory: Explanation and prediction.[46] On the one hand hypotheses have in Lambert's writings the negative aftertaste which they have with many of his contemporaries in the wake of Newton's "hypotheses non fingo". Science proper should have a non-hypothetical form such as Euclid's geometry. On the other hand, Lambert knows only too well the state of the sciences, in order not to know that for the time being hypotheses are indespensable for scientific progress and the practical application of science. I would like to discuss briefly a few aspects of Lambert's theory of hypotheses:

(1) Hypotheses don't have existential import. "What gives a possibility the so despised name of hypothesis is not that it is a possibility, but the unproven presupposition that this possibility might exist." (Lambert 1770, p. 86) (2) Hypotheses are "conjectures based on insufficient reasons."[47] (3) As conjectures, hypotheses are not true, but only have a certain "degree of probabiity"[48] and can only be included ad interim[49] in a scientific system of propositions. (4) For the maintenance of hypotheses it is necessary that they be a) internally consistent, b) consistent with empirical propositions and c) consistent with the remaining propositions of the scientific system under consideration.[50] (5) An hypothesis "is either confirmed or not".[51] (6) Hypotheses approach the truth according to their degree of efficiency a) in predicting the existence of still unknown phenomena and b) through experimental confirmation.[52] (7) Methodological analogies for the correction of hypotheses offer a) the rule of false position, b) the art of decoding and c) approximation methods for the solution of algebraic equations.[53] (8) A logical procedure for the correction of hypotheses or for their transformation into true propositions is based on the 'analytical' discovery of 'identical propositions'. (Lambert 1764, p. 404ff) (9) From Lambert's coherence theory of truth it does not follow that a hypothesis, 'falsified' in the Popperian sense, is a false proposition.

The aspects 1-6 cited above correspond essentially to what Laudan (See ftn. 53) relates about the renaissance of the hypothetical method in the 18th century. With a view toward Hartley (and also Le Sage) as well as toward the methods of approximation cited here under points (7a) and (7c), Laudan complains: "Unfortunately, Hartley himself did not indicate how we can apply such mathematical methods to the natural sciences [...]. (He) simply took it for granted that one can in a more or less straightforward fashion, import these mathematical techniques into the logic of the natural sciences." (Laudan 1981, p. 232). Lambert is more thorough here. He sees differences between the use of the rule of false position in mathematics and in the natural sciences: 1. The rule of false position, just like the procedures for solving equations and the methods of decoding, have "only to fulfill a few conditions. On the other hand, it is not enough for the physical hypotheses that they satisfy only *some* experiences, they must satisfy *all* of them."[55] For a complete test of an hypothesis, at least a complete classification of the phenomena in question would have to be specified. Only in a few cases, "more mathematical than physical", does he view this as given. He gives as an example the law of optical refraction, where one would only have to examine all degrees of angels of incidence. 2. The rule of false position utilizes indefinite magnitudes in its singular steps of approximation. This is not acceptable for hypotheses, because they cannot be recursively and so to speak necessarily approximated to the truth. For *hypotheses* one needs much more "a mean to correct the erroneous in an assumed hypothesis with the help of that, which possibly follows from this hypothesis contrary to truth."[56] That is, the correction of an hypothesis requires a new idea, whereas the next step in the mathematical method of approximation is determined by the method itself. If under (1) a complete classification of the *phenomena* is required which can be explained through an hypothesis, under (2) it is necessary for the explanation of a phenomenon or a class of phenomena that a complete class of alternative *hypotheses* be tested. Lambert gives as examples among others Kepler's hypotheses of elliptical planetary motion, with which he was able to improve upon the Copernican hypothesis of circular motion. (Lambert 1764, p. 571) Circles and ellipses actually form a complete disjunction of the class of closed curves from conic sections. In short, according to Lambert the application of the rule of false position requires a complete disjunction of classes of hypotheses and phenomena for the methodically complete correction or testing of hypotheses.[57]

I would now like to turn to the analytical treatment of hypotheses cited under (8)[58]. The example given by Lambert for this method is, for want of a better, not overwhelming. The point of departure for the syllogistic analysis in the mood 'barbara' is the empirically (more or less) true proposition: The earth casts a round shadow (EaSh). The questionable hypothesis: The earth is a sphere (EaSp). If we put both of these propositions in 'barbara' in such a way that the empirically true proposition EaSh is the conclusion and the hypothesis EaSp the minor premise, then we obtain:

$$\begin{array}{l} ? \\ \underline{\mathit{EaSp}\quad} \\ \mathrm{EaSh} \end{array}$$

In the next step, the correct major premise must be introduced. In purely formal terms we obtain (*):

$$\begin{array}{l} \mathrm{SpaSh} \\ \underline{\;\mathit{EaSp}\quad} \\ \mathrm{EaSh} \end{array}$$

The major premise is thus: Spheres cast round shadows. The hypothesis, i.e. the minor premise is, of course, not proved by this. Lambert now asks himself what must happen to the major premise so that the hypothesis can appear as the conclusion of a correct inference:

$$\begin{array}{l} ?\ \mathrm{SpaSh} \\ \underline{\mathit{EaSh}\quad\quad} \\ \mathrm{EaSp} \end{array}$$

One sees immediately that the inference is valid when the major premise is conversible (thus SpaSh *and* ShaSp), i.e. when the grammatical subject and predicate have the same extension. In terms of predicate logic: $\bigwedge x\, Sp(x) \leftrightarrow Sh(x)$. In Lambert's words: The major premise in (*) must be an "identical proposition". In the example: The spheres and only these cast shadows. This 'analytical' method for building and proving "identical propositions" is viewed by Lambert as the most important method for the substitution of hypotheses through true propositions. Logically, this involves proof of the equality of extension of classes. The more the proof of this equality advances, the more probable the hypothesis becomes.[59]

Lambert's theory of truth corresponds to the systematic nature of science. It represents a combination of an 'intuitionistic' and a coherence theoretic concept of truth.[60]

The 'intuitionistic' concept of truth is applicable only to a few concepts, namely the simple ones, and those propositions formed from these concepts by using synthetic methods both strict and without gaps. According to Lambert most scientific proposition cannot be infered deductively and without gaps from 'intuitionistically' true propositions. In any event, one looks in vain in Lambert's work for a definition of truth. To adopt a distinction made by N. Rescher: Lambert holds no "definitional", but rather a "criterial theory of truth".[61] Whereas the intuitionistic concept of truth, with its relation to the facts represents a "guaranteeing criterion" for the truth of proposition, the coherence theoretic concept of truth supplies an "authorizing criterion", that means a rational warrant for accepting a proposition. Lambert's authorizing criterion:

> What evidently is contrary to the, as far as I know, most correct concepts and propositions [according to the guaranteeing criterion], that I can withhold my approval until the contradiction has been removed or the connection been proven. And on the other hand, what is in harmony with my most correct propositions, that I would deny my approval without any reason, until the opposite had been proven to me. I don't believe that we hold geometric truths for irrefutable in any other way. They are connected correctly and their opposite can always be reduced to a contradiction that makes it inconsistent. However, both of these basic rules, to which our approval directs itself, still does not reach up to truth proper.[62]

Now the authorizing criterion in the case of geometric propositions is practically a guaranteeing criterion as well because of the strictly deductive connection of geometric propositions with the fundamental ideas proven as true according to the guaranteeing criterion. This however is not true generally, because the system of scientific propositions is full of "gaps" and "leaps". These also make hypotheses necessary. What is decisive is, that the acceptance of most scientific propositions is not determined by its correspondence to facts, but by its coherence relation to other propositions. This has an interesting consequence with respect to the rejection of propositions, especially the falsification of hypotheses. Lambert objects strongly to simply eliminating hypotheses that have not stood the experimental test. "For it seldom or never happens that an hypothesis, particularly when it is part of a larger context, should be completely wrong. One would therefore proceed inconsistently, if one wanted to reject along with this hypothesis itself also the propositions interwoven into it."[63] Thus, with respect to falsifyability Lambert supports a view, which in its overall structure corresponds to those which Duhem, Neurath and Quine later held, though with in part different and highly elaborated arguments. Lambert stands here in opposition to the correspondence theoretician and 'Popperian', Wolff.[64]

6. Let me turn now to the last question. What does Lambert, inexperienced as he is in this area, really mean by 'metaphysics'? As we have already seen, Lambert considers the question of the nature of physical objects and the causal connections between them as 'metaphysical'. Of course, not every question concerning cause and effect is metaphysical. For example, the proposition 'This body moves because I pushed it', is not a metaphysical, but an empirical proposition. This is the case for the simple reason that cause and effect here are observable events. Metaphysical research into the nature of things and their not manifest causes relates in the Wolffian tradition to that part of the universe that is not directly observable, that is to something like theoretical entities. In this tradition metaphysical propositions are propositions with a high degree of universality. By A. G. Baumgarten, a Wolffian Lambert often refers to critically, it is stated that general metaphysics is the "science of the more general predicates of the thing"[65] which means the word 'thing' as grammatical subject. Correspondingly, special metaphysics as 'cosmologia generalis' investigates the nature of bodies. General as well as special metaphysics thus use very general concepts. *This* universality is determined by abstracting according to genus and species and ends with the highest order of genus, 'ens' or 'thing'. If abstracting in this manner could be conducted in a methodically correct way, then the solution would be quite elegant. "The universality that should reign here (i.e. in metaphysics) leads, as it were, to omniscience."[66] But that is not the way it is, as everyone knows. "The process of abstraction necessarily presupposes special knowledge and, so viewed, resembles the

ontology of a juggler's box, in which is previously placed whatever one wants to pull out of it."[67] In other words, as long as the sciences, on which an abstracting metaphysics is based, are not adequately founded, this metaphysics (i.e. metaphysics in the Wolffian tradition) will remain nothing more than a cheap trick. Yet Lambert does not completely reject metaphysics, although in his view it had still accomplished nothing up to that point. His conception of metaphysics rests on a new understanding of simplicity. The traditional Wolffian metaphysics equated the degree of universality with the degree of simplicity. Accordingly, 'thing' is the most universal and at the same time the simplest concept. But Lambert does not consider this to be true. Only the method of conceptual analysis makes it appear simple. Lambert, on the other hand, uses an 'anatomic' or 'Euclidian' concept of simplicity as a point of departure. He attributes its discovery to Locke.[68] Not 'analysis', according to genus and species, but the (iterated) *division* of concepts into their (intentional) components ('Merkmale') leads finally to simple concepts. Seen in this way, simple concepts are the material basis of all knowledge. In this *epistemological* question Lambert follows Locke up to the point where he feels he would only have to "copy down for the most part" what Locke had written. (Lambert 1764, p. 29) There is, however, an essential difference. Locke, as Lambert sees him, didn't take the trouble to construct a system of science on the basis of his simple ideas. But this is precisely what is so important to Lambert, because *epistemology should not just be a theory of knowledge* per se, but a theory of its most outstanding part, scientific knowledge.

> Locke, namely, is content to base his whole work on empirical propositions and proceeds in this way thoroughly a posteriori because he simply takes things as they are [...]. Now this is not our intention here. We have seen in the *Dianoiologie* why scientific knowledge, as far as it is able to proceed *a priori*, is superior to common and merely historical knowledge [...]. It is not enough to have selected simple ideas [as Locke did]. We must also see where we can get universal possibilities in order to combine these simple ideas (Lambert 1764, p. 29)

As we can see, even when he is establishing a metaphysics, Lambert refers to the Euclidian paradigm, whose application (see section 4) had already secured for the exact sciences both a factual basis as well as universality. This is admittedly not an analytic, abstractive universality, but rather an operative one founded on possibilities of action. In the present context I can only sketch Lambert's procedure in general terms.[69] Lambert begins with an arbitrarily chosen concept gained from experience and continues to divide it iteratively into its component parts until he arrives at components which are no longer divisible. These components are the simple ideas. Lambert does not say much about this procedure. In any event the result is more or less the same simple ideas that Locke came up with. Once one has simple ideas, they immediately lose their empirical character that had accompanied their discovery.[70] Simple ideas, *per definition*, cannot have component parts at all and therefore, *a fortiori*, no contradicting ones. It is for this reason that they are 'possible' and can be conceived of at any time without repeatedly recurring to the sensual experience in which they originated. This is the reason why they are *a priori* for Lambert. At the same time they are also 'true'. This aspect of the Lambertian concept of truth begins with 'intuition', but is not dependant on it. I would like to call it an 'intuitionistic' concept of truth. In this way he believed he had a basis for synthetically producing scientific knowledge, which has something to do with facts, but at the same time is independant of the actual state of affairs. Lambert now attempts once more to solve the problem of the universality of simple ideas, in terms of a universality that is both operative and founded on possibilities of action, just as in his eyes Euclid had solved it with respect to geometry. Once again postulates and axioms establish 'universal possibilities' (Lambert 1761a, p. 76ff) for fundamental concepts, i.e. their universal meaning and use.

At the same time the postulates and axioms specify the possibilities of combining simple fundamental concepts in order to form complex concepts and thus propositions. As far as Lambert is concerned, experience does not come into play here as well. The result is that *a priori* sciences come into existence. "So we have for example three sciences, which are in the strictest sense a priori; namely Geometry, Chronometry and Phoronomy." (Lambert 1764, p. 658) Other disciplines such as dynamics or the theory of matter are in some way dependant on the existence of our real universe and are *a priori* only to a lesser degree. What remains of metaphysics, you might ask? What about God and immortality? The answer is simple. That was already the Lambertian metaphysics. Its function was to provide a framework of categories for the exact sciences. A framework of categories, which secures and explains both the relationship of scientific knowledge to experience and, as far as it is possible, their universality and *a priori* nature. Thus metaphysics is essentially philosophy of science. And what about the 'nature' of bodies? Here Lambert has a harsh reply at hand for the traditional metaphysicians.

> A philosopher asks immediately: What is a body? In doing this he immediately makes a nominal definition, which is in itself superfluous, because everyone understands the word, and which specifies little or practically nothing of what a mathematician can measure with respect to a body. The philosophers should trouble himself more with discovering everything that can be measured regarding bodies in every respect. And then one would know in detail and in all respects and with respect to each part, what a body is.[71]

Traditional metaphysics is thus "superfluous". Long live science! Long live Methodology!

But what about the question of God? In those philosophical writings Lambert published himself, God appears as *obiter dictum* on *circa* 7 out of 2000 pages. And it looks as if Lambert has harnessed even God in the advancement of science. He distinguishes between two languages, the "physical language" of the sciences and the "teleological language", which concerns the laws of nature as an expression of "divine perfection". "In this connection we only note that both languages name the same thing and consequently should differ at the same time only in the naming." Whereas the physico-theological translation from the physical into the teleological language – very common during the enlightenment, I might add – "properly speaking only serves to edify", the reverse translation leads to hypotheses and thus to the discovery of natural laws. Lambert's edification herewith is "disproportionately larger and more extensive".[72]

Let me summarize. In a manner that is extremely unusual for his time, Lambert's epistemology is completely oriented towards the foundation of scientific knowledge. It does this, on the one hand, by providing science with material categories in an 'operatively' established and universal form, and on the other, by reflecting methodologically on scientific method with Euclid's axiomatic and synthetic procedure in the *Elements* as a paradigm. Measured according to *this* paradigm of scientific knowledge, metaphysics has not accomplished and in a certain sense can not accomplish anything. Thus it finds itself in the following dilemma. Its function would be to tell the sciences something about the nature of its respective domains. Metaphysics would have to be able to specify and explain the dimensions used in equations of mathematical physics. Such an abstract exploration, however, would have to presuppose the existence of the respective sciences and the dimensions used in their equations as a basis of abstraction. In Lambert's view, this means that metaphysics is only possible when it has already become superfluous. Nevertheless, Lambert does not arrive at an explicit verdict on metaphysics, because he lacks an incisive logical and epistemological instrument such as e.g. the Kantian concept of *a priori* synthetic judgements. So he decides pragmatically that metaphysics is not wholly impossible, but rather "superfluous", because it is of no use. Under the aspect of use, Lambert also sees science as a whole. Purpose and goal of metaphysics is not an ontologically realistic *knowledge* of the universe, i.e. *theory*, but the provision of instruments for coping with individual and social life. This general precedence of practice over theory already has a remarkable correspondence in the foundation of theory itself. Euclid's geometry shows in exemplary fashion how certainty and success of scientific knowledge do not rest, as it were, '*realistically*' on a distinctive relationship to reality. It is rather human action and the careful analysis of possiblities of action which form the foundation of scientific theory. Even the truth of propositions is not the result of a comparison between language and reality. It exists rather in its logical harmony with the propositions most justified. This 'liberal' concept of truth, which crosses over smoothly into the concept of probability, leaves systematic room for hypotheses. Hypotheses are parts of a comprehensive theoretical network of relations and are thus not isolatedly falsifiable. As a result Lambert can be placed in the vicinity of the 'conventionalist' Duhem-Quine thesis.

If I were to label Lambert's philosophy, I would choose the much used and long-suffering word *'pragmatic'*. In this way I want to characterize the syndrome that the overall conception of philosophy is not ontological, but instrumentalistic; that the foundation of scientific theory is itself not theoretic but has to do with acting, i.e. is 'poietic' in the Aristotelian sense of the word; that scientific truth rests not upon the comparison of language and reality, but on the comparison of propositions; that scientific progress results not only from the expansion of absolutely certain knowledge, but also from the continuous correction of conjectures; in short, that human *knowledge* is above all human *action*.

That is not to assert, that Lambert has presented a thoroughly pragmatist conception in the contemporary sense. My claim is and can only be more modest. I have dealt with the fact that at certain, and usually important places in Lambert's philosophical thought pragmatic ideas play an important, if not decisive role. Certainly Lambert is not yet able to connect these ideas to an overall pragmatic network that excludes completely ontological realism. So there are only important pragmatic *elements* or *aspects* in Lambert's philosophy.

In my view philosophy, German philosophy in particular, would have done well by adopting these Lambertian elements and intentions immediately. Unfortunately, for more than half a century it tried unsuccessfully to become "omniscient".

Universitat Konstanz

NOTES

[1] Cf. Steck (1970), *Bibliographia Lambertiana*. In his career as scientific writer which lasted only 22 years (1755-1777) Lambert published more than 30 monographs and more than a hundred papers in learned journals. Besides that there is an enormous correspondence and hundreds of reviews.

[2] The "Inaugural Dissertation" in: I. Kant, *Gesammelte Schriften* [refered to in short as: *Akad.-Ausg.*] II, pp. 385-419. Lambert refers to that problem in his letter to Kant from October 13th, 1770 (cf. Kant, *Akad.-Ausg.* X, pp. 103-111, 108). Cf. Beck (1978). Lambert's critique of the concept of time as expounded in the "Inaugural Dissertation" did not impress Kant very much. He rejects it (1781) in the first *Critique* (A 36-40) without naming Lambert explicitly, who had died four years before (1777).

[3] Decisive for Kant's estimation may well have been the fact that Lambert was the only one of his German contemporaries who combines to a high degree both scientific and philosophical competence. A biographical circumstance may also be important in this context. In 1765 the successful academy member, Lambert, took up a correspondence with the considerably older unknown "Privatdozent" at Königsberg and suggested a scientific collaboration. Furthermore, he informed Kant that he would endeavor to secure him a position at the Academy in Berlin (Kant, *Akad.-Ausg.* X, p. 54). Nothing came of this project, and in 1766 Kant took up his position as sub-librarian in Königsberg.

[4] W. S. Peters, (1968). Peters does not seem to know the assessment that I am going to quote. He seems to take as Kant's last word on Lambert, what Kant writes in a letter to John Bernoulli III. This also applies essentially to L. Braun who on the other hand opposes vigorously the habit of referring to Lambert only "de précurseur, d'épigone, de phénomène de transition" etc. (Braun (1977)).

[5] *Immanuel Kants Logik*. Ein Handbuch zu Vorlesungen, *Akad.-Ausg.* IX, pp. 1-150, 21. *Organon* means: *Neues Organon oder Gedanken über die Erforschung und Bezeichnung des Wahren und dessen Unterscheidung von Irrthum und Schein*, I-II, Leipzig 1764 (Reprint in: Johann Heinrich Lambert, *Philosophische Schriften* [referred to for brevity's sake as: P.S.], III, Hildesheim 1965. Volumes I, III, VI, and IX contain good introductions by the editor, H. W. Arndt). The title *Neues Organon* has to been understood as an homage to the highly esteemed Francis Bacon.

[6] In his first letter to Lambert on New Years Eve 1765, Kant confesses that he holds Lambert "for the foremost genius in Germany, who is capable of making an important and lasting improvement in the same sort of investigations, with which I am at the moment preoccupied" (*Adak.-Ausg. X*, p. 54). Lambert, stated more precisely his offer of scientific collaboration in a letter written on October 13th,

1770. He has in mind the founding of a "private society" "where everything would be avoided that can all too easily ruin public academic societies. The so e members would be a small number of selected philosophers, who would also have to be proficient in physics and mathematics [...]. Members of this society would talk abcut their writings or at least give an adequate idea of them, in order to gain assistence if need be. If, however, each member were to retain his own opinion, then each of them could have his opinion printed, but with the necessary modesty and with the understanding that everyone could be mistaken." (*Akad.-Ausg*.X, p. 104). Much is ant cipatory here of the spirit of the later Vienna Circle.

7 *Anlage zur Architectonic, oder Theorie des Einfachen und des Ersten in der philosophischen und mathematischen Erkenntniß*, I-II, Riga 1771 (Reprinted as: P.S., III-IV, Hildesheim 1965). The passage quoted is in the "Vorrede" p.III. Kant was, by the way, invo ved in providing a publisher for the *Architectonic* (cf. *Akad.-Ausg*. X, p. 52ff.).

8 There is no comprehensive Lambert biography. Useful is Humm (1972). Humm also touches briefly upon Lambert's remaining stations in life. Previous b ographers especially take pleasure in Lambert's exceedingly odd personality, warranted by many anecdotes. Of particular interest is the Lambert biography presumably written by J. C. Lichtenberg (1778), which has been reprinted in Steck (1943).

9 Fredrick to d'Alembert (Jan. 1765): "One almost forced me, as it were, to take the most horrible creature in the universe and appoint him to our Academy. He calls himself Lambert. And although I can testify that he lacks sound common sense, people claim he is one of Europe's greatest geometricians. But because this man does not know the language of mortals and only speaks in equations and algebraically, I do not intend so soon to have the pleasure of conversing with him. I am – as compensation – very satisfied with monsieur Toussaint, whom I procured. His science is more human than that of the latter. Toussaint is an inhabitant of Athens, Lambert however a Carribean, or a savage from the Kaffir's coast. In spite of this the whole Academy, even Euler, is on its knees before him. And this animal, completely covered with the filth of the most uncouth pedantery, accepts this homage like Caligula that of the people of Rome, from whom he wanted to be accepted as God. I beg you not to let these little anecdotes from

our Academy out of your hands." Translated from Harnack (1900, 1.1 p. 366). But Frederick changed that somewhat bizarre assessment in the course of time: After several years Lambert was appointed senior royal building commissioner ("Königlicher Oberbaurat") and his salary was doubled. And shortly after Lambert's death we find in a letter to the very same d'Alembert the following: "You talk about a question to be presented to the academy. Alas! We have lost recently the poor Lambert, one of our better subjects. I do not know who could be able to treat the question if it is permitted [to the Sovereign] to deceive people. I believe Béguelin will be the only one capable of treating that question philosophically (Translated from *Oeuvres posthumes de Fréderic II, Roi de Prusse*, Vol. XI, Berlin 1788, p. 266).

[10] With the exception of the already mentioned *Organon* and *Architectonic* there are only two additional, posthumously published works devoted exclusively to philosophy: *Abhandlung vom Criterium veritatis*, ed. K. Bopp, Berlin 1915 (Kant-Studien, Erg.-Hefte 36) and: *Über die Methode die Metaphysik, Theologie und Moral richtiger zu beweisen*, ed. K. Bopp, Berlin 1918 (Kant-Studien, Erg.-Hefte 42). The latter is a fragment drafted for a prize competition of the Berlin Academy. Kant answered the question sent out by the Academy in his (1764). Moses Mendelssohn won the prize. Unfortunately, as far as I know, no one has as yet undertaken the attractive task of comparing the various answers.

[11] The dates are taken from a scientific diary which Lambert kept until his death: *Johann Heinrich Lamberts Monatsbuch*, ed. K. Bopp, München 1915 (Abhandlungen der Königlich Bayerischen Akademie der Wissenschaften, Mathematisch-Physikalische Klasse, XXVII.6).

[12] Up to that time Lambert had discovered in mathematics the well known series named after him (Observationes variae in mathesin puram [1758], in Lambert (1770, pp. 16-51).. He had also published important contributions to the theory of perspective, Lambert (1759). His reputation as an astronomer had been established by his (1761c). In this book he had put forward the astronomical theorem named after him which states that in a parabolic orbit the time needed to move along an arc of that orbit is dependant only from the corresponding chord and the sum of the radius vectors of the extremes of the arc. One should also mention his popular (1761b) where he, among other things,

developed a theory of the various milky way systems which is in accordance with modern theories in several essential points. (English translation: *Cosmological letters on the Arrangement of the world edifice*, ed. S. L. Jaki (Edinburgh 1976). As a physicist he was one of the founders of photometry, (1761d); partially translated into German: *Photometrie*, I-III, ed. E. Anding, Leipzing 1892 (Ostwald's Klassiker, Vol. 31-33)).

[13] Cf. the remark in the preface of the *Architectonic* (P.S. III, XIII) "that only in the first two years after having finished the manuscript did I often think about inserting new ideas occuring to me occasionally".

[14] His terminus a quo for his scientific epoch is the "renewal of the sciences" (cf. (1764) Dianoiologie 573, P.S. I, p. 365)

[15] Cf. Butts (1978).

[16] This Lambertian conception of the primacy of practice in justifying cognitive methods corresponds to newer systematic considerations on the same topic. Cf. J. Mittelstrass (1972) and N. Rescher, *Methodological Pragmatism (1977)*.

[17] Cf. O. B. Sheynin, (1970/71, pp. 244-45) and L. Tilling (Thesis), (unfortunately I have not yet seen this study). According to Sheynin "Lambert schould be regarded as the main predecessor of Gauss" (Sheynin, 1970/71, p. 244).

[18] Lambert promised himself much from its uses for farming. He himself kept a meteorological diary, whenever he lived at one place for a longer period of time. One can judge how seriously he took the matter by the fact that he, although severely ill, still read his meteorological instruments on the day of his death and made entries in his notebook. Cf. Steck (1970, p. 95).

[19] The whole of volume III of Lambert's is devoted to Lambert's correspondance with the ingenious Augsburg designer and builder of scientific instruments, Georg Brander. During his time at the Munich Academy Lambert lived in Brander's house.

[20] E.g. in a letter to his (calvinist) pastor in Mühlhausen (cf. Lambert to pastor Rißler Nov./Dec. 1750, in: Lambert, 1782, pp. 7-11,8) *Deutscher gelehrter Briefwechsel* II. Lambert's religious practice provoked altogether lots of mockery in Berlin which at that time was heavily influenced by French religious libertinism.

[21] Cf. Lambert (1761a, p. 9).

[22] *Architectonic* (455, 681ff, P.S. III, 77f., 302ff). The superiority of mathematical knowledge over phlosophical knowledge can be taken as a further confirmation for Lambert's instrumentalism expounded in section II.
[23] For the historical development of these terms, cf. the excellent sketch in: F. Kambartel (1968, Ch. 2.)
[24] Historical knowledge "is simply reporting what actually is in nature, and what happens with things nature puts before us". Lambert (1764), Dianoiologie 600, P.S. I, p. 387).
[25] *Architectonic* (455, P.S. IV, p. 77f). This text may be understood as Lambert's version of the "hypothesis non fingo".
[26] For Lambert's conception of hypothesis, see below. As a precaution I would like to remark that neither here nor elsewhere does Lambert use the "sloppy mode of reasoning" that L. Laudan (1981, p. 228) has found in the works of several Cartesians and Newtonians of the 17th and 18th centuries. For Lambert successful prediction is never "prima facie evidence that the theory was a proven truth" (Laudan, 1981, p. 228), but only a necessary condition for its truth. Cf. particularly Lambert (1764), Alethiologie 175, P.S. I, p. 544; Phanomenologie 166, P.S. II, p. 333f.
[27] *Architectonic* (683. P.S. IV, p. 304). The subordinate clause, "as soon as the former [the philosopher] found it", is probably negligence on the part of Lambert. It stands in clear contradiction to the text quoted above (455) as well as to the fact that philosophy has accomplished nothing in the area to which Lambert has relegated it. The success of mathematical natural sciences is thus not dependant on the respective 'philosophical' theories. Compare also the invariance of the *measurement* of light with respect to the various *theories* of light in Lambert (1761d, p. 3).
[28] Lambert to G.J. Holland on April 27th, 1767, in Lambert (1788, P.S.IX, p. 1894). Wolff treated virtually everything axiomatically, including for example the construction of Latrines. Cf. Wolff (1750, p. 480f.). In addition it is interesting to note that Lambert placed so much emphasis on Wolff's merits. It was certainly not unknown to Lambert that Wolff had not invented formal logic. What is decisive is that he applied it, i.e. that he practiced it consistently.
[29] Lambert to Holland (April 21st, 1765), (P.S. IX, 21). Lambert's correspondence with Georg Jonathan Holland provides the best short version of Lambert's methodology.

Holland was first extraordinary professor at Tübingen and at the same time tutor to the princes at the Württembergian court in Stuttgart. Later he became a member of the St. Petersburg Academy.

30 As Wolff had done after having defined straight line and circle (cf. C. Wolff, *Elementa matheseos universae* I, Halle/Magdeburg ²1742, p. 135ff., (ch.2, coroll. 2ff.)).

31 *Criterium veritatis* 79, loc.cit. 45f. – I have corrected a wrong reading of the published text from the manuscript (University Library Basel).

32 Lambert to Holland (April 21st, 1765), P.S. IX, p. 29.

33 Lambert to Holland, op. cit. p.30: "In the proofs Euclid uses the expression *per definitionem* in no way different to the expression *per hypothesin*. For as long as the possibility of the concept has not been proven, the definition is still only an hypothesis. If it is evident in itself or only through a single example that at least some of these figures do exist which the definition indicates, then the definition may be stated first, as mere nomenclature. The conditions of its possibility, however, must follow as from principles and *postulates* [...] This is in my opinion how Euclid deals with definitions and concepts. It should also be possible in metaphysics." See in this context Lambert's letter to Kant (P.S. IX, p. 337f). This text is not included in the Academy Edition of Kant's letters. I can only touch here upon those variants of Lambert's concept of possibility which are important for the present context.

34 Cf. *Architectonic* (243, P.S. III, p. 215 (Cf. *ibid.* 19, p. 16); *ibid.* 67, p. 53; *ibid.* 70, p. 55). See with respect to the passage quoted here: Lambert (1895, p. 7). "The term *per definitionem* means nothing different for him (Euclid) than the term *per hypothesin*. A closer look reveals, for example, that he does not take the categorical in his theorems from the *definitions*, but, understood properly, he takes it particularly from the *postulates*. It holds for these in a preeminent way, when Cicero says: *Si dederis danda sunt omnia* (*De finibus bonorum et malorum*, lib. V. p. 83)." It is by the way only a small step from Lambert's *Theorie der Parallellinien* to non-Euclidian geometry.

35 In his (1761) Lambert had still regarded the Leibniz-Wolffian style of analysing concepts as useful in a methodological context. But in *Organon* and *Architectonic* it became clear to him that it was of no great use (cf. Wolters, op.cit. p. 58f.).

[36] Lambert to Holland (April 21st, 1765), (P.S. IX, p. 29).

[37] Lambert to Holland, (loc. cit., cf.ftn. 33). The latin quotation: Once one has admitted [the postulates and axioms], one is also obliged to admit everything else [i.e. the whole of geometry].

[38] Lambert, (1771). The "ninth and twelfth axioms" in old editions of the "Elements": The whole is larger than the part, and: Two straight lines do not encompass an area. In our context the shortcomings of Lambert's methodological analysis of Euclid's geometry are of course of no particular interest. As seen through Pasch or Hilbert the basic concepts of geometry would be 'point', 'straight line', 'plane', 'between', and the like.

[39] Cf. Lambert (1771, preface, P.S. III, XXII, XXVII); Lambert (1764, Alethiologie 1, P.S.I, p. 453); Lambert to Kant (Febr. 3rd,1766), *Akad.-Ausg.* X, p. 64f.

[40] "In nature and consequently in experiences, observations, and experiments, everything is individual and determined in time and space" (Lambert, 1764, Dianoiologie 588, P.S. I, p. 376).

[41] "And this goes so far that it becomes a rooted prejudice in people who have nothing more than common, or, at the most, something of the slightly better historical knowledge: they believe one would not be able to think beyond what one knows through the senses; and what one could not experience immediately, that would be beyond the limits of human knowledge, and would be impossible for us to know." (Lambert, 1764, Dianoiologie, 601, P.S. I, p. 387f.)

[42] Lambert (1764, Alethiologie 160, P.S. I, p. 538. As the "Alethiologie" further shows a lot of work is needed to establish something systematically coherent from "fragments", "cahos" (as Lambert constantly misspells), "gaps" and "leaps". A detailled analysis of these terms is not possible here.

[43] Lambert, (1764, Dianoiologie 678, P.S. I, p. 434; cf. 633, P.S. I, p. 411; 604, P.S. I, p. 389).

[44] Lambert (1764, 606f., P.S. I, p. 390f).

[45] Lambert (1764, 567, p. 359f). This paragraph is followed by a short analysis of the concept of hypothesis. More on hypotheses in: "Von der analytischen Methode und den Voraussetzungen", and in: "Von den Vermuthungen"; both of these texts in Lambert (1782) (repr. as P.S. VI, Hildesheim 1967); additions to the above texts in: *Logische und philosophische Abhandlungen* II, ed. J. Bernoulli, Ber-

lin/Leipzig 1787, pp. 32-39 (repr. as P.S. VII, Hildesheim 1969); Lambert (1761, 71ff., loc.cit. p. 41ff.). Apart from the concept of hypothesis expounded here Lambert distinguishes two other ones which are completely unproblematical: (1) Conventional signs and divisions, e.g. arithmetic figures or the division of the circular arc into 360 degrees, (2) The acceptance of approximate values in mathematics, e.g. $\Pi = \frac{22}{7}$. Cf. Lambert (1764, Dianoiologie 152, P.S. I, p. 98f).

46 A well founded theory is in Lambert's terminology an 'a priori theory'. More on this in the next section. "We have, however, [...] already remarked that we will make use of this mean [i.e. the formation of hypotheses] in the natural science only because of the lack of more a priori knowledge." Lambert (1764, Alethiologie 244, P.S. I, p. 578). See (P.S. VII, p. 32f) with respect to explanation and prediction.

62 "Because nothing more is required for conjectures than insufficient reasons, and because, moreover, most of our knowledge rests only on such reasons, thus not only conjectures have an extremely broad range, but also very many and very general kinds of inferences can be applied." (P.S. VI, p. 338)

47 (P.S.VI, p. 339): "So conjectures become the more certain the higher their degree of probability. Therefore the theory of conjectures is very closely akin to the logic of the probable."

48 *Ibid.*

49 Lambert (1764, Phanomenologie, p. 52, P.S.II, p. 247). The same formulation, by the way, in: Chauvin, (1713, p. 297), repr. Dusseldorf 1970, 297: "Hypothesis [...] loco veri quaesiti ad interim supponitur".

50 Cf. (P.S. VII, p. 33). For 'consistency' in (c) see the following remarks on Lambert's theory of truth.

51 *Ibid.*

52 Lambert (1761d, 4, op. cit.4): "Then under the principal and most certain criteria for an hypothesis approaching truth, one has to include the case, when one can predict the incidence of new phenomena from a system of scientific propositions and when one can infer propositions from it, with which the experiments agree that were conducted for this purpose."

53 Cf. Lambert (1771, p. 293ff., P.S. III, p. 283ff). Due to restrictions of space and time I cannot here touch upon the distinction (loc.cit.) between 'logical' and 'metaphysical' truth and several other related distinctions.

[54] Lambert (1764, Dianoiologie p. 568, P.S. I, p. 360f). Laudan (1981) has demonstrated convincingly how, apart from the better known epistemological trends, the hypothetical method, which had fallen into disfavor from about 1720 on, came back into favor among a few scientists (among others, Lambert) around 1750. A turning point, by the way, which the professional epistemologists recognized only with some delay. Wolff, by the way, never adopted this verdict on hypotheses. (Cf. Wolff (1740 p. 126ff; further: p. 606ff., p. 1162f.)

[55] Lambert (1764, Dianoiologie p. 569, P.S. I, p. 364). E.g. In order to determine the zero of a monoton function $f(x)$ two points $(a_K, f(a_K))$, $(b_K, f(b_K))$ are chosen on the graph of f, so that $f(a_K) < 0$ and $f(b_K) > 0$. The chord connecting the two points intersescts the x-axis in a_{K+1}. The tangent to the curve in $(b_K, f(b_K))$ intersects the x-axis at b_{K+1}. Again $f(a_{K+1}) < 0$ and $f(b_{K+1}) > 0$. The values of a_{K+1} and b_{K+1} can be expressed recursively. The continuation of this procedure (nest of intervals) delivers the zero of the function. Indeed, 'only few', precisely stated 'conditions' must be fulfilled in this approximation procedure.

[56] Lambert (1764, Dianoiologie p. 571, P.S. I, p. 363). In Lambert (1787, P.S. VI, VII) as well as in the other works of Lambert's there are frequently historical remarks about how occured "lucky inspirations" (glückliche Einfälle).

[57] Oddly enough Prevost uses the very same example as evidence for his thesis that the mathematical rule of false position has *nothing* to do with correcting hypotheses in the natural sciences. Cf. Laudan (1981, p. 233).

[58] As to the logical structure of this procedure cf. Neemann (1977).

[78] Lambert (1764, Dianoiologie p. 415, P.S. I, p. 271). Lambert himself is quite aware of the fact that this procedure has not yet shown significant results. But he expects great things from it for the future. (Cf. loc.cit. pp. 407, 266f.).

[59] Cf. (P.S. VI, p. 290f.).

[60] Cf. D Rescher (1973). From Rescher's book (p. 4) I adopt also the distinction between 'guaranteeing criterion' and 'authorizing criterion'.

[61] According to Rescher (1973, p. 1), a *definitional* theory of truth attempts "to provide a definition of the conception 'is true' as a characteristic of propositions", whereas a *criterial* theory consists in "the attempt to specify the

test-conditions for determining whether or not there is warrant for applying the characterization 'is true' to given propositions". For Lambert's adoption of the 'criterial route' cf. Lambert (1761), (1764, Alethiologie, P.S. I, p. 453ff.); (1771 pp. 289-306, P.S. III, pp. 280-298).

[62] Lambert (1761, 15, p.16). Cf. (1764, Alethiologie 179, P.S. I, p. 546).

[63] Lambert (1764, Dianoiologie p. 567, P.S. I, p. 360). What Lambert says on pages 380 and 247 does not contradict the passage just quoted, because it relates to the deductively complete system of geometry.

[64] Cf. Wolff (1740, pp. 609, 449): "Si hypothesi philosophicae vel unum, quod observatur, phaenomenon repugnat, falsa as est".

[65] Cf. Baumgarten (1749, para. 4): "scientia praedicatorum entis generaliorum."

[66] Lambert to Kant (February 3rd, 1766, *Akad.-Ausg.* X, p. 62).

[67] Lambert in a review of the anonymously published book *System der Wesen*, enthaltend die metaphysischen Prinzipien der Natur (in: Allgemeine Deutsche Bibliothek XI.1, Berlin 1770, p. 273). As to the universality of metaphysics with regard to the sciences cf. Lambert (1764, Dianoiologie p. 50, P.S. I, p. 33).

[88] Lambert to Holland (April 21th, 1765), (P.S. IX, p. 33). I cannot discuss here the reasons which brought Lambert to this assertion.

[68] Cf. (P.S. IX, pp. 34, 348) (to Kant); Lambert (1764, Alethiologie p. 29, P.S. I, p. 472); Lambert (1771, p. 7, P.S. III, p. 4ff.).

[69] More detailed in: Wolters (1980, p. 67ff.).

[70] "Because the possibility of a fundamental idea arises at the same time with the perception, it becomes completely independant of experience, such that, even if we have to thank experience for it, this only gives us, as it were, the occasion of its consciousness. Once we are, however, conscious of it, then it is no longer necessary to look for a reason for assuming its possibility from experience, because the possibility is already there wth the mere perception. Accordingly, the possibility becomes independent of experience. And this is a requisite of a priori knowledge in the strictest sense." Lambert (1764, Dianoiologie p. 656, P.S. I, p. 421f.). Cf. Kant (1781, B1) "Although all our knowledge begins *with* experience, it does not follow that it all arises *of* experience". Cf. further *Organon* (Alethiologie

p. 21, P.S. I, p. 465f.). – Lambert's theory of fundamental ideas has surprising similarities with the theory of William Whewell; for example the idea that fundamental ideas constitute necessarily true propositions and so become the foundation of science. Further, that necessary truths originate from sensual perception and so are "matters of experience prior to apprehending them as necessary" (Cf. Butts' 1968) – I do not know, however, if Whewell had read Lambert's philosophical writings.

[71] Lambert to Holland (August 19, 1765), (P.S. IX, p. 84). The verdict in Lambert's published writings does not come out that clear. But they show through their example what is stated here explicitly.

[72] Lambert (1764, Phanomenologie p. 231f., P.S. II, p. 391ff). The connection between teleology and the theory of hypothesis has been noticed with regard to Lambert's *Cosmologische Briefe* by R. Wahsner, "Die philosophische Position Johann Heinrich Lamberts", in: Sternzeiten (1977, pp. 27-41, 31).

REFERENCES

Baumgarten, A. (1749), *Metaphysica*. Germany: Halle; repr. New York: Hildesheim, 1963.

Beck, L. (1978), "Lambert and Hume in Kant's Development From 1769 to 1772", *Essays on Kant and Hume*. New Haven/London. (First published in German in *Kant-Studien*, 1969.)

Beck, L. (1969), *Early German Philosophy*: *Kant and His Predecessors*. Cambridge, Mass.

Braun, L. (1977), "Lambert et Kant", *Colloque international et interdisciplinaire Jean-Henri Lambert*, Mulhouse, 26-30 Septembre 1977. Paris 1979.

Butts, R. (1978), "Some Tactics in Galileo's Propaganda for the Mathematization of Scientific Experience", *New Perspectives on Galileo*. eds. R. Butts and J. Pitt.

Papers deriving from and relating to a workshop on Galileo held at Virginia Polytechnic Institute and State University. Boston/Dordrecht.

Butts, R. (1968), "Introduction," *William Whewell's Theory of Scientific Method*. Pittsburgh.

Chauvin, S. (1713), *Lexicon philosophicum* repr. Dusseldorf 1970.

Frederick II (The Great) (1788), *Oeuvres posthumes de Frederic II, Roi de Prusse*. Berlin

_________ (1900), *Geschichte der Königlich Preu ischen Akademie der Wissenschaften zu Berlin*. trans. A. Harnack. repr. New York: Hildesheim, 1970.

Griffing, H. (1893), "J. H Lambert. A Study in the development of Critical Philosophy", *Philosophical Review*. vol. 2. pp. 54-62.

Harnack, A. (1900), *Geschichte der Königlich Pren ischen Akademie der Wissenschaften Zu Berlin*. Berlin. repr. New York: Hildesheim, 1970.

Humm, F. (1972), *J. H. Lambert in Chur. 1748-1763*.

Kambartel, F. (1968), *ErFahrung und Struktur*. Frankfurt.

Kant, I. (1781), *Kritik der reinen Vernunft*.

Kant, I. (1770), *Gesammelte Schriften* [*referred to in short as*: *Akad.-Ausq*.]

Kant, I. (1764), "Untersuchung über die Deutlichkeit der Grndsatze der naturlichen Tehologie und Moral", *Akad.-Ausq*. x, 273-301.

Lambert, J. (1918), *Über die Methode die Metaphysik Theologie und Moral richtiger zu beweisen*. ed. K. Bopp. Berlin.

_________ (1915), *Johann Heinrich Lambert's Monatsbuch*. ed. K. Bopp. München.

_________ (1895), *Theorie der Parallellinien* in: *Die Theorie der Parallellinien von Euklid bis auf Gauss*. Leipzig.

_________ (1787), *Logische und philosophische Abhandlungen II*. ed. J. Bernoulli. Berlin/Leipzig.

_________ (1783), *Deutscher gelehrter Briefwechsel*. ed. J. Bernoulli. Berlin.

_________ (1782), *Logische und philosophische Abhandlungen I*. ed. J. Bernoulli. Berlin.

_________ (1771), *Anlage zur Architectonic oder Theorie des Einfachen und des Ersten in der philosophischen und mathematischen Erkenntni*. Riga.

_________ (1770), *Beytrage zum Gebrauche der Mathematik und deren Anwendung* in: *Opera Mathematica*. ed. A. Speiser. Zurich, 1946.

_________ (1764), *Neues Organon oder Gedanken über die Erforschung und Bezeichnung des Wahren und dessen Unterscheidung von Irrthum und Schein*.

_________ (1761d), *Photometria sive de mensura et gradibus liminis, colorum et unbrae*. Augsburg.

_________ (1761c), *Insigniores orbitae cometarum proprietates*. Augsburg.

_________ (1761b), *Cosmologische Briefe uber die Einrichtung*. Augsburg.

_________ (1761a), *Die Wissenschaften. Ein Traum*.

_________ (1761), *Abhandlung von Criterium veritatis*. ed. K. Bopp. Berlin.

_________ (1759), *La perspective of Franchie de l'embaras du plan geometral*. Zurich.

Laudan, (1981), *Science and Hypothesis*. Boston/London: Dordrecht.

Mittelstrass, J. (1972), *Das praktische Fundament der Wissenschaft und die Aufgabe der Philosophie*. Konstanz.

Neemann, U. (1977), "Die Bedeutung der identischen Satze in der Wissenschaftstheorie Johann Heinrich Lamberts", *Colloque international et interdisciplinaire Jean-Henri Lambat*, Mulhouse. 26-30 Septembre 1977. Paris 1979, pp. 181-186.

Peters, W. (1968), "I. Kants Verhaltnis zu J. H. Lambert", *Kant-Studien 59. pp. 448-453.*

Rescher, N. (1973), *The Coherence Theory of Truth*. Oxford.

__________ (1977), *Methodological Pragmatism*, Oxford.

Sheynin, O. (1970/71) "J. H. Lambert's Work on Probability", *Archive of the History of Exact Sciences 7.*

Steck, M. (1943) (Editor) Johann Heinrich Lambert. Schriften zur Perspecktive. Berlin.

Steck, M. (1970), *Bibliographia Lampertiana*. Hildesheim.

Tilling, L. (Thesis), *The Interpretation of Observational Errors in the 18th and Early 19th Centuries*. Ph.D. Thesis, University of London.

Wahsner, R. (1977), "Die philosophische Position Johann Heinrich Lamberts", *Sternzeiten* East Berlin.

Westenrieder, L. (1784), *Geschichte der baierischen Akademie der Wissenschaften I*. München, pp. 482-531.

Wolff, C. (1740), *Philosophia rationalis sive logica*. Frankfurt/Leipzig.

__________ (1742), *Elementa mathesos universae I*. Halle/Magdeburg.

__________ (1750), *Anfangsgrunde aller Mathematischen Wissenschaften*. Frankfurt.

Wolters, G. (1980), *Basis und Deduktion*. Studien zur Entstehung und Bedeutung der Theorie der axiomatischen Methode bei J. H. Lambert (1728-1777). Berlin/New York.

DANIEL M. HAUSMAN

CLASSICAL WAGE THEORY AND THE CAUSAL COMPLICATIONS OF EXPLAINING DISTRIBUTION*

1. INTRODUCTION

For at least half a century classical political economists were largely in agreement on wage theory. The demand for labor was determined by the so-called wage fund or wages fund, while Malthusian population theory accounted for the supply of labor. Classical wage theory attempted to explain wages in a given period, to identify those factors which would influence the trend of wages over time and to account for the eventual level of subsistence wages in the approaching "stationary state" in which economic growth would cease.

Modern economists have repudiated virtually all of classical wage theory. They no longer regard the supply of labor as determined by the responsiveness of the laboring population to sub- or super-subsistence wages. Nor do they accept the existence of a wage fund. Instead they subscribe for the most part to a marginal productivity theory of demand for labor and to more complicated theories of the supply of labor. Although I do not want to defend classical wage theory, I think that there are some interesting methodological lessons to be learned from it and that the marginal productivity theory which succeeded it, considered as a theory of average wages and of the distribution of income, is no improvement. As we shall see, there are formidable difficulties in the path of any (causal) explanation of the distribution of income. We shall also see that, in constructing and assessing explanations in economics, it is important to pay attention to the causal structure of such explanations.

In section 2, I shall introduce classical wage theory, while in section 3 I shall talk in general terms about supply and demand explanations. In section 4, I shall show how supply and demand explanations of wages run into difficulties which, as we shall see in section 5, classical wage theory, although unacceptable for other reasons, escapes.

J. C. Pitt (ed.), Change and Progress in Modern Science, 171–197.
© 1985 by D. Reidel Publishing Company.

Finally in Section 6 I shall have something to say about John Stuart Mill's famous recantation of classical wage theory in 1869. What is most interesting about classical wage theory is its causal structure – the example it provides of an alternative to modern equilibrium analysis.

2. CLASSICAL WAGE THEORY

Classical wage theory was a theory of "average wages" and of the distribution of income between workers and non-workers, not a theory of wage differentials. The classical political economists assumed that they could talk about a single input into production called "labor", and I shall follow them in this. Nominal wages are the money earnings of laborers. Real wages are the goods that workers and their families consume.

According to classical wage theory, in each production period the number of laborers (which was equated with the supply of labor) is fixed, and so are the stocks of wage goods, which are largely foodstuffs harvested annually (but see section 6 and footnote 7). Wages are just a quotient. As Nassau Senior put it in 1830, "... the rate of wages depends on the extent of the fund for the maintenance of labourers, compared with the number of labourers to be maintained." (1830, pp. ii-iv). Half a century later, Henry Fawcett writes, "... wages in the aggregate depend upon a ratio between capital and population." (1883, p. 131) Note that Senior and Fawcett are not just insisting on the arithmetical truth that average wages equal aggregate wages divided by the number of laborers. This arithmetical truth might enable one to *calculate* average wages, but it certainly would not enable one to *explain* wages. Although the stock of wage goods available at the beginning of the production period (which is what the classical economists called "the wages fund") comes to equal the aggregate wages paid in the period, the wage fund is also one of the factors which *determines* aggregate wages.

Aggregate real wages clearly cannot be larger than the wages fund, but they also cannot (in "equilibrium") be smaller. To see why aggregate wages cannot be less than the wages fund, let us follow the classical economists and simplify by supposing that capitalists advance to workers stocks of wage goods and that employers of servants and unproductive laborers also pay them in kind. Since there are profits to be made by putting laborers to work or pleasures to be gained from hiring servants, excess stocks of workers' consumption goods will give rise to a competition among employers and an increase in wages (Mill 1871, Bk. II, ch. XII, sec. 1, p. 362; Longe 1866, p. 21). But until the composition of output and thus the stocks of worker's consumption goods can be changed, aggregate wages are fixed. Individual wages are the quotient of the wages fund (demand) divided by the number of workers (supply).

In the longer run, neither the number of workers nor the wages fund is fixed. Both are growing. Wages will increase or decrease, depending on whether the supply of labor or the wages fund increases more rapidly. Classical wage theory should, I believe, be regarded as fundamentally a dynamic theory of temporary equilibrium, which is devoted to identifying those factors which influence the changes in wages over time. The wages fund increases when investment increases and when inventions increase the productivity of labor, particularly in the wage-good sectors. An increased demand for servants would have an ambiguous effect on the wages fund. If it went along with a decrease in savings and investment, the output of wage goods and thus the wages fund would decrease. But a demand for servants, instead of luxury goods, would increase the demand for worker's consumption goods and would lead to an increase in the wage fund. In any event the classical economists were in fact in the habit of first mentioning direct demand for labor services and then (as in the quotation above from Fawcett) ignoring it. I shall follow them and simplify by assuming that all workers are employed by capitalists. With some possible, but unlikely exceptions, the wage fund increases with investment and improvements in productivity.

Classical political economists were convinced that limited supplies of land must eventually choke off this economic growth (Ricardo 1817, ch. 6, pp. 139-40). And even while such growth could be sustained, economists feared or expected that the labor force (the supply of labor) would at least keep pace with the wages fund. In Mill's words, "We must, therefore, in considering the effects of the progress of industry, admit as a supposition, however greatly we deprecate as a fact, an increase of population as long-continued, as indefinite, and possibly even as rapid, as the increase of production and accumulation." (1871, Bk. III, ch. XXIV, sec. 4, p. 699) Unless the rate of increase of the labor force could be held down, increases in the wages fund would not improve the condition of the average worker.

Classical wage theory thus has apparent immediate practical implications. It says that there are only two ways to increase wages: increase the wages fund (by increasing saving and investment or by technological innovations) or slow the growth of the labor force. To quote Mill once more, "The condition of the [laboring] class can be bettered in no other way than by altering that proportion [between capital and population] to their advantage: and every scheme for their benefit, which does not proceed on this as its foundation, is, for all permanent purposes a delusion." (1871, BK. II, ch. XI, sec. 3, p. 350) The nineteenth century political economists believed that the most practical way to increase the rate of growth of the wages fund was to eliminate as many of the obstacles in the way of untrammeled competition as was possible. Schemes to raise wages by legislation or trade union agitation, which in effect increase these obstacles, must inevitably fail. In Francis Walker's words, "The amount of it [the wage fund] cannot be increased by force of law or of public opinion, or through sympathy and compassion on the part of employers, or as the result of appeals or efforts on the part of the working classes." (1976, p. 138, but see footnote 7 below)

Interferences with the workings of the market which attempted to benefit workers by regulating or supplementing wages were held to be particularly injurious. Not only would the consequent decrease in profits slacken the pace of accumulation, but the higher wages would spur the growth of population. Although thus powerful anti-union ammunition, it is worth noting that classical wage theory was in fact seldom used to attack unions (perhaps because economists doubted whether unions were powerful enough to make a serious attempt to raise average wages).[1]

Instead in Mill's writings and to a great extent in the writings of both his contemporaries and his immediate predecessors, concerns about population growth were predominant. To ease the pressure of population, economists urged "restraint" and emigration. John Stuart Mill also argued that the increase in population might be slowed if women were emancipated, since they would have other occupations and would not consent to bear and to care for so many children (1871, Bk. IV, ch. VII, sec. 4, p. 760). In his more optimistic moods, Mill hoped that a fairly lengthy improvement in the circumstances of laborers (which might be brought about by emigration) might accustom workers to living better and provide them with the education and the prudence needed to limit their numbers and thus to preserve and perhaps even to increase their gains (1871, Bk. II, ch. XIII, sec. 3, 4, pp. 380-82).

With some exaggeration, Francis Walker summarizes the practical implications of classical wage theory as follows:

> If the workman would not give up when told to enlarge his dividend [to increase output and thus the wage fund], he was struck dumb on being informed that his only alternative was to lessen his divisor [the number of workers]. The divisor aforesaid being flesh and blood, with certain attachments to home and life, and with a variety of inconvenient affections, was not to be lessened so easily. If the workman turned from words to blows, and went out "on strike" with a view to better his condition, it was regarded as the act of an irrational animal whose instincts, unfortunately, were not politico-economical. (1876, p. 143)

Among the classical economists, there was little investigation of the details concerning the demand or supply of labor (Taussig 1896, p. 163; Blaug 1978, p. 195). What really mattered about classical wage theory was that aggregate wages would grow as the economy grew, but that permanent improvements in the circumstances of workers depended on a limitation of their numbers. The bogeyman of overpopulation so fascinated or terrified most economists that they paid scant attention to the more immediate factors which might affect wages.

The above, in a nutshell is classical wage theory and its policy implications. By the end of the nineteenth century, this theory was in complete disfavor. Mill had recanted his previous espousal in 1869. Others (for example Cairnes 1874, Part II, ch. 1-3 and even Taussig 1896) continued to defend a wage-fund doctrine, but with so many qualifications that little content remained. By the end of the first decade of the twentieth century, James Bonar could remark that "The wages fund theory is the crowning instance of an untrue abstraction ... and it has probably done more injury to the reputation of economic theory than any other generalization ever received into economics textbooks and then expunged from them." (1911, p. 75; quoted in Breit 1967b, p. 509).

Merely to make explicit the heroic assumptions upon which classical wage theory relies is enough to explain this repudiation. But before saying anything more about classical wage theory, I want to consider in some detail the difficulties which confront any attempt to explain wages and the distribution of income in terms of the supply and demand for labor. After these difficulties are recognized, we will be in a much better position to appreciate the merits and deficiencies of classical wage theory.

3. SUPPLY AND DEMAND EXPLANATIONS OF WAGES

The amount of labor that workers will be willing to supply is, of course, a function of many variables besides the wage; and so is the amount of labor that employers demand. One can assume, however, that these various influences are roughly independent or separable, so that one can treat supply of and demand for labor as functions of a single variable – the wage rate – , although, of course, these functions will shift with changes in the other determinants of demand and supply. According to classical wage theory, the situation can be represented by the following graph (figure 1):

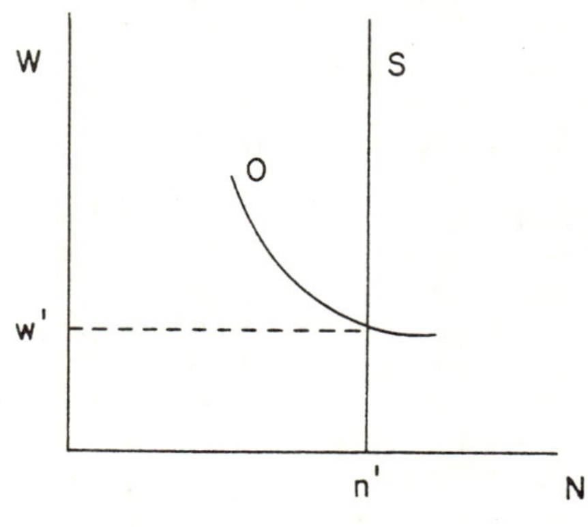

D is a demand curve for labor, a graph of the demand function. It is a downward sloping hyperbola with unitary elasticity, since the total amount of wages paid is a constant equal to the wages fund. S is the fixed ("perfectly inelastic") supply of labor. In general, of course, demand curves need not have unitary elasticity, nor need supply curves be perfectly inelastic, although economists do assert that in ordinary circumstances demand curves will be downward sloping, while supply curves will be upward sloping or horizontal. In the general case one will, of course, have the familiar Marshallian "cross".

It is conventional to measure the quantity demanded or supplied along the horizontal axis, although quantity (which in this case is number of units of labor) is in fact the dependent variable. Wages are measured along the vertical axis. At the intersection of the demand and supply curves the amount of labor sought by employers equals the amount offered by workers. The "equilibrium" amount of employment will be n' and the equilibrium wage w'. In supply and demand theories, of which classical wage theory is a special case, one explains wages and employment, in principle at least, in terms of the supply and demand of labor and in terms of the market processes that adjust wages to equalize the amount of labor supplied and the amount demanded. Such causal explanation requires that the demand and supply functions show how much labor would be demanded and supplied at various prices, that demand and supply be determinable prior (in at least a causal sense) to the determination of the wage, and thus that there is a general, but not necessarily perfect, separation between the forces that affect demand and the forces that affect supply (Friedman 1949, p. 57).

Of course, as everybody knows, in the real world there is often involuntary unemployment (excess supply) as well as occasional excess demand. The supply and demand apparatus does not attempt to explain what wages will actually prevail at any moment. Instead one has a sketch of an explanation of the *equilibrium* level of wages around which actual wages will fluctuate and toward which actual wages will tend. In classical wage theory one only has such equilibrium accounts of wages in a particular period (or of wages in the eventual stationary state). The course of wages over time requires a dynamic theory of the trends in such successive temporary or one-period equilibria.

In modern economics this concern with developments over time has been muted. When contemporary economists consider what will happen with a shift in supply, for example, they do not usually ask the dynamic question: "What will be the chain of consequences?" Instead they are concerned with the (comparative statics) question, "If equilibrium were reached, what would be the new levels of wages and employment?" The fact that supply and demand theories are theories of economic equilibrium and not of actual (disequilibrium) states of the economy is, as we shall see below, of central importance.

As mentioned above, the supply of and the demand for any commodity or service depend on many other things besides the price of that commodity or service. Demand for hair pins, for example, will depend on the price of hair pins, but it will also depend on fashions, on incomes, and on the prices and properties of other commodities like hair sprays or permanents. In writing a simplified function that portrays demand for hair pins as a function of just one variable, the price of hair pins, one is assuming that these other causal factors are not significantly influenced by changes in the particular price.[2] There will always be *some* influence of the price and quantity of hair pins on the other variables that affect demand for hair pins, but it is reasonable to treat the influence of these other variables as separable. These remarks apply, of course, equally to supply functions. The amount of any commodity or service that producers or owners will supply is obviously affected by changes in techniques of production or by changes in the prices of other inputs.

What one holds constant or treats as a separable causal factor affects the nature of the equilibrium explained. Marshall introduced the useful, but potentially confusing distinction between short and long-run equilibrium prices (1920, Bk. V., ch. III, sec. v). The difference is mainly on the supply side. In the short run, the number of hair-pin factories, for example, is fixed. Supply can only be increased by measures like working over-time or double-shifts. Economists thus argue that short-run supply curves will be upward sloping – only at a higher price will more be produced. In the "long run", on the other hand, much less is taken as fixed, even factory capacity can adjust. Long run supply curves should thus slope upwards (if at all) much more gently.

The interesection of a demand curve and short-run supply curve thus indicates what the price and the quantity exchanged will be, holding constant or treating as separable, factors like plant capacity, while long run equilibrium price and quantity are determined holding constant or regarding as independent only the basic givens (like tastes and technology) and factors which are assumed to be largely independent, like prices of unrelated commodities. "Long-run" thus has very little to do with real time. Over any real long-run of time, tastes and technology and other prices will change considerably. Long run equilibrium explanations are not devoted to saying what will *happen* over any long time period, but to saying what prices and quantities would be, assuming full adjustment of the production process (within the limits of the given knowledge of technological possibilities). Carrying Marshall's distinctions with them, contemporary economists have in fact sometimes misunderstood classical wage theory. Equilibria are important in classical wage theory: there are both temporary (one-period) equilibria and the Marshallian long-run equilibrium of the eventual stationary state. But the main concern of the classical economists was with the "long run" in an entirely different sense. They were interested in trends over time, not fundamentally with long-run equilibria.

Demand and supply explanations are thus causal explanations, in which for any price causally prior supply and demand functions give rise to amounts demanded and amounts supplied. Differences in amounts demanded and amounts supplied are translated through the (prior) mechanism of the competitive market to price changes. The equilibrium price is the price that equalizes demand and supply. Modern demand and supply explanations are not concerned with any actual sequence of events over either a short or a long period of time, but with equilibrium prices and quantities. Both demand and supply for any commodity or service, including labor, will depend on a great many other things than just the price of that commodity or service, but these other factors are assumed not to vary significantly with prices in order that one can separate out the causal influence of an individual price on demand and supply from the causal influence of other prices, incomes, tastes, technology, and so forth.

4. THE CAUSAL COMPLICATIONS OF EXPLAINING DISTRIBUTION

There is nothing wrong in principle with asserting that demand and supply of hair pins are, *ceteris paribus*, functions of the price of hair pins – where the other things assumed unchanging include other prices, incomes, tastes and technology. But there is something wrong in principle

– indeed something self-contradictory – about asserting that demand and supply of labor are, *ceteris paribus* functions of the wage rate. It is well known that not only the price level, but also relative prices depend significantly on wages. I am in fact concerned here exclusively with the dependence of relative prices on wages. One cannot in general sensibly consider what demand would be, were the wage larger than it is, prices being what they are, because if the wage were larger, relative prices would not be what they are. Of course, prices may not *change* immediately with changes in wages, but remember that what is being attempted is an explanation of equilibrium wages. The *ceteris paribus* condition built into the supposed labor demand and supply functions can never be satisfied. The causal structure is incoherent: the demand and supply functions depend on wages and employment as well as *vice versa*. Those functions do not and cannot correctly state how much labor would be demanded or supplied at various wage rates.

One might argue, however, that with a slightly different construal of demand and supply functions, these difficulties disappear. Why not take the demand for (supply of) labor function to state how much labor would be demanded (supplied) at a given wage, at the relative prices which correspond to that wage? Instead of holding prices constant, let them adjust.

This suggestion has the drawback that it divorces the demand for (supply of) labor function from the projections of individual employers and laborers concerning how much labor they would seek or offer at different wage rates, since these individual projections do not take price changes into account. But a more serious difficulty remains. Relative prices depend not only on wage rate, but also on the *amount* of employment. Both the size and composition of the demand for goods and the relative cost of production depend on the size of the employed labor force and its total expenditures. The amount of labor which would be demanded or supplied at any given wage, w*, depends, among other things, on what prices are. But the prices depend not only on the wage but on how much labor is actually employed. Demand and supply can only be defined *after* wages and employment are already given and cannot be used to explain wages and employment.

To clarify this point, suppose that the labor market were a particularly simple sort of market somewhat like that depicted in figure 1 with the supply of labor some fixed quantity, n'. In this case there is no problem defining a supply of labor function or drawing a supply curve. Since supply does not depend on wages or prices, we need not worry about the dependence of prices or wages and on total employment. Consider now what happens when one tries to define a demand function in which, unlike the special case of the wage fund doctrine, aggregate wages are not determined by the output of the previous period. One way to define a demand function would be to take employment as a given, n'. For each wage rate one can then calculate what prices would be, given that employment is n'. Given the particular wage and those prices, one can say what the demand for labor would be.

But the resulting function would not correctly state the demand for labor if the supply of labor shifted. Since this putative demand curve is derived on the assumption that employment is n', it will typically have to shift. One thus cannot derive equilibrium wages and employment from a new given supply of labor and this putative demand curve. Such a curve has no role in an explanation of wages or employment. Only one point on the curve – the actual wages and employment before the shift in supply – is of any significance.

But there is one other possiblity. For each given amount of labor supplied, let us suppose that there is one and only one wage rate for which the amount demanded will equal the amount supplied. (There is, incidentally, no reason why this supposition should be true). One can then trace the locus of all of these pairs of equilibrium wages and employment. This graph may slope downward to the right, although it need not. Why not regard this curve, which Don Patinkin calls a "market equilibrium curve" (1956, p. 45), as the demand curve?[3] The intersection of this curve and any given supply does correctly "determine" equilibrium wages and employment.

A market equilibrium curve and a supply curve *may* jointly enable one to *deduce* equilibrium wages and employment, but they do not enable one to *explain* them. Indeed they are often not even sufficient for the deduction of equilibrium wages and employment, since when supply is a variable function of price the "supply" and the "demand" curves will usually be the identical market equilibrium curve. Market equilibrium curves do not show how much labor would be demanded or supplied in a given market at wages other than the equilibrium wage. They show instead what equilibrium employment is at other wages. Market equilibrium curves are given only *after* equilibrium prices and quantities are determined. They are not causally prior to equilibrium prices or quantities, and they thus have no role in an explanation of *how* equilibrium prices and quantities are determined. The mere fact that market equilibrium curves can sometimes be used to *deduce* equilibrium prices or quantities does not establish that they have any explanatory role.

There is in general no way to define supply and demand functions that can be used to explain equilibrium wages and employment. The basic problem is that labor is such a ubiquitous input that just about everything else in the economy depends on how much of it is employed and how much it is paid. There is no way, in Herbert Simon's (1953) terminology, to "causally order" the determination of the values of various economic variables in such a way that labor supplied and demanded can be known while wages and employment remain to be determined (see also my forthcombing b). One cannot isolate some small set of proximate causes of equilibrium wages and employment in the way that one can isolate the main causal factors upon which depend the price of hairpins and the quantity of hairpins exchanged. The only equilibrium account of wages and employment which could be acceptable would be a general equilibrium account in which wages and employment were determined by tastes, technology, the initial endowments of commodities and resources and the basic institutional constraints of the market. But, as I have argued elsewhere (1981), current general equilibrium theories are at best sketches of such explanations or promisory notes for such explanations.

Let me mention two criticisms that the argument above faces. First one might object that although prices are sensitive to both wages and employment, this fact makes supply and demand explanations of wages impossible only if this sensitivity is appreciable. If demand and supply functions for labor which ignored price changes were not too far off, one could still offer supply and demand explanations of wages and employment.

This objection is in principle perfectly reasonable, but the onus is on the objector to show that the sensitivity of prices to wages and employment is sufficiently low that one can continue to define approximate supply and demand curves. I have seen no such demonstrations, and I have elsewhere used some simple models to show how significant the dependence of prices on wages and employment is like to be (forthcoming c). It is also worth pointing out that this defense of supply and demand theories of wages turns out to be a defense of the labor theory of value. If prices are not sensitive to wage changes (and thus to changes in the rate of interest), then they are well approximated by labor values.

Second, one might object that the entire criticism rests upon my taking the notion of 'labor' as a single input too seriously. Once it is recognized that there are many varieties of labor, one recognizes that explaining the various wages these varieties of labor are paid presents no special problems. But the difficulty with this objection is that once one retreats to offering separate explanations for the compensation paid to different kinds of labor – which may be an entirely sensible move to make – one has also abandoned the attempt to offer any simple unified theory of the distribution of income between workers and non-workers (which may also be quite sensible). In so far as one is considering supply and demand theories of wage determination as theories of the distribution of income, one is forced to theorize in terms of a single ubiquitous input called "labor".

5. THE CAUSAL STRUCTURE, VIRTUES AND VICES OF CLASSICAL WAGE THEORY

There is thus a general problem in explaining distribution. To offer a causal explanation, one must take certain features of the economy as causally prior to the distribution of income between capitalists and workers – as already determined while the wage and the amount of employment remain to be determined – where this precedence is at least causal, if not temporal as well. But supply and demand for labor do not in general possess such causal priority. Note that this is not just the familiar point that economic variables are simultaneously determined. The "givens" in a typical general equilibrium model, although not in any sense earlier in *time*, can nevertheless be regarded as *causally* prior to the variables whose values are determined in the general equilibrium solution.

It is interesting to notice, however, that classical wage theory can be regarded as an answer to this critique. If it were otherwise acceptable, it would show that supply and demand for labor are indeed both temporally and causally prior to the determination of wages. To clarify this point, let me present a fragment of a mathematical restatement of classical wage theory that I have developed elsewhere:

> In an extremely simple one-commodity model where capital consists entirely of advances of some consumption commodity, "corn", to workers, one can write the following highly simplified "supply" and "demand" functions for labor (see Hausman forthcoming a, section 1):
>
> $$\text{(S)} \quad N_{t+1} = N_t \left[1 + c_t (w_t - w^*_t) \right]$$
>
> $$\text{(D)} \quad w_{t+1} N_{t+1} = w_t N_t \, (1 + s_t r_t)$$

(S) is a simplified restatement of Malthusian population theory. It says that the current (t + 1) supply of labor depends on the supply of labor in the last period times a growth factor which reflects the difference between actual wages in the last period, w_t, and subsistence wages, w^*_t, and the responsiveness of population to this difference, c_t, which will be some positive number. (D) states that the current wages fund is equal to the wages fund in the last period plus net investment, where net investment equals net output or income (the rate of profit, r_t, times capital, $w_t N_t$), times the capitalists' propensity to save or invest, s_t. Although the classical economists had little to say about the extent to which wage changes would lead to the substitution of other inputs for labor, it is possible to include substitutability (and considerations of marginal productivity) in the model (Hausman forthcoming a, sec. 3; Hahn 1972, p. 82; Hollander 1968; Samuelson 1951, p. 328).

To discuss how wages will change over a long period of time requires that one say something about how c_t, w^*_t, s_t and r_t will change over time and about the extent to which they depend on contemporary or earlier values of one another and on contemporary or earlier wages and employment. But all these questions may be postponed in considering wages in any given period. Aggregate wages and labor supply are entirely determined by features of the previous period. Aggregate wages are simply the wages fund. Average wages are the quotient of this wages fund divided by the given supply of labor.

Classical wage theory thus purports to explain in principle what wage will clear the labor market in the given period. If the labor market does not clear, wages can be higher or lower than the wages fund divided by the number of laborers. So we have an explanation of *one-period equilibrium* wages in terms of supply and demand of labor. No causal difficulties arise, however, because both supply and demand are fixed in the period before. On the view of causation that I have defended elsewhere, causes need not necessarily precede their effects in time; but causation remains, of course, an asymmetrical relation. The temporal precedence of the determinants of the wage rate in classical wage theory neatly and decisively establishes this causal asymmetry.

Although classical wage theory is a theory of the market-clearing wage, there is no implication of stationary or general or long-run equilibrium. Other markets need not clear. Both supply and demand in succeeding periods may (and typically will) differ. Since population and the wages fund are fixed when the period begins, they may be regarded (in combination with the competitive processes of the labor market) as determining wages and employment in the particular period. The causal structure of classical wage theory in its one-period application is thus simple and sound.

This conclusion may appear completely uninteresting, since the assumptions of classical wage theory, thus far left implicit, are so hard to swallow. The goods that workers consume are not harvested only once a year; they flow on to the market in streams of varying steadiness throughout the year. The portion of output that workers may consume is not fixed. Aggregate wages can be immediately increased by lessening the consumption of non-workers. The amount of labor supplied (as opposed to the population of laborers and their dependents) is not independent of wages. In times of business crises, as even classical wage theorists like J. S. Mill pointed out (1871, Bk. II, ch. XI, sec. 2, p. 345), capitalists may hoard rather than invest, and total wages paid may be less than the wages fund. Finally, workers do save and may thus be paid partly in claims to future output.

Although I sympathize with the above objection, I should point out that there are ways in which one can continue to defend a highly qualified (and rather uninteresting) short-run wages-fund theory. Since workers in fact save comparatively little, it is perhaps reasonable to ignore their savings. It may be reasonable to separate theories of business crises from theories of what would happen in their absence and thus to assume full employment and complete distribution of the wages fund (Pigou 1949, p. 174 points out, however, that, contrary to classical wage theory, inventories of consumer goods, after an initial increase, *fall* during recessions). When concerned with the distribution of income, we may be interested in wages per member of the labor force, the population of which one can regard as more or less fixed at any given time. Given the number of workers compared to the number of non-workers and the differences in much of their consumption, the flow of goods that workers consume, while not fixed, cannot vary both rapidly and appreciably. Finally, even though there is a flow of wages, rather than a stock, production is a roundabout and time-consuming process. To increase the flow of wage goods will require a considerable time period. And, in any case, as will be discussed further in Section 6, in its long-run applications, classical wage theory does not rely on the wage fund being rigidly fixed anyway. What remains after all the needed qualifications are made are the less than exciting truths that to increase real wages will affect investment, and to do so appreciably without reducing employment takes time (Taussig 1896, Part I, esp. pp. 82-96).

But the classical political economists were not particularly interested in what might be possible within any given period. Their attention was instead often focused on the long-run – not on any Marshallian long-run equilibrium, but on what was likely to happen over the succeeding fifty or one-hundred years. Book IV of Mill's *Principles of Political Economy* is appropriately titled "Influence of the Progress of Society on Production and Distribution". How would the wages fund, the size of the labor force and the amount of wages change? Apart from theories of the stationary state, the classical economists offered no theory of a long-run equilibrium wage or of a long-run equilibrium amount of employment (see Hahn 1972, pp. 88-89). Wages and employment would change from year to year – there is, in their quite sensible view, no single wage or level of employment to be explained. Their inquiry was directed instead toward the factors responsible for changes (particularly general trends) in wages, profits, rents, employment, output and so forth.[4]

One might also propose a proportional growth model that bears a superficial resemblance to classical wage theory. Dropping all time references from (S) and (D), except those on N, and setting supply and demand equal to one another, one can solve for an equilibrium wage and for an equilibrium rate of employment growth (which is equal to the rate of growth of the wage fund).

But such a theory has little to recommend it. To take (S) and (D) as stating supply and demand for labor, *ceteris paribus,* where the other things taken to be unchanging include subsistence wages, the rate of profits or interest and capitalist's propensity to save, commits even more blatantly exactly the same causal error that oversimplified purported explanations of stationary equilibrium wages in terms of supply and demand commit. Moreover, such a proportional growth model is completely out of step with the aims, interests and convictions of the classical economists. They believed that the rate of profit or interest is not only a decreasing function of the wage rate, but that it must decrease (ignoring technological improvements) with increases in the labor force (which demand more extensive or more intensive agriculture which lowers profits). The propensity of capitalists to save was taken to be a decreasing function of the rate of interest, reaching zero at some small positive interest rate.

At the end of the development lay the stationary state – an economic state in which the rate of profits was so low (because of diminishing returns in agriculture) that there was no net investment and no economic growth. As J. S. Mill especially appreciated, the economy does not, however, march off relentlessly toward some inevitable stationary state whose character we are powerless to influence. The stationary state that society arrives at depends on the subsistence demands of workers. These may be large or small depending on the extent to which economic growth is soaked up by an increasing labor force. Classical wage theory was a schema for identifying the crucial factors influencing the growth of capital and population and the trend of wages. It is a dynamic causal theory of long-period developments, not a theory of long-run equilibrium.

6. MILL'S "RECANTATION" OF CLASSICAL WAGE THEORY

One of the most remarkable events in the history of classical political economy occurred in 1869 when, in a review of a book by William Thornton, Mill restated some of Thornton's criticisms of classical wage theory and retracted his life-long espousal of it (1869, pp. 642-46). Francis Longe also published a confused attack on classical wage theory in 1866, but he seems not to have influenced Thornton or Mill. Mill's recantation is a fascinating episodue for both the historian and the philosopher of science. The fact that it is generally labelled a "recantation" itself raises questions: Does one *recant* one's commitment to a scientific theory?

What makes the episode particularly fascinating is that the doctrine is one that is central to Mill's economics, and the criticisms Thornton offered seem to virtually all commentators weak and confused. To add to the puzzlement one should feel at Mill's retraction is the fact that in the last revision of the *Principles of Political Economy*, which Mill made about a year after his recantation, he made few changes in his exposition and defense of classical wage theory![5]

There are lots of theories about Mill's actions. Some have felt that Mill simply blundered in his retraction, and that his blunder can be explained in part by his desire to help Thornton (whom he liked) or his distress at becoming fully aware of anti-union and inhumane applications of classical wage theory – at becoming aware that, in Francis Longe's words, "...it excludes altogether the influence of liberal principles from that field of social action, where it is for the interest of society that they should be ever most influential." (1866, p. 8) Others have tried to find reasons for Mill's change of heart, although not necessarily very good ones: that he confused monetary and real analysis (Ekelund 1976, esp. p. 67), that he came to have doubts about the existence of a fixed period of production (Schwartz 1972, p. 95), or that he realized dimly that the wages fund doctrine was inconsistent with the assumption of fixed input coefficients (Hollander 1968).

Since a large part of Thornton's critique (and Longe's too) was devoted to questioning supply and demand analysis, one might suspect that confusions or worries about the causal structure of such theories might have a role in this famous episodue. In fact, with great generosity one can extract from Thornton's bewildering and mediocre discussion one interesting and germane point. Thornton may argue that, since actual trading takes place typically at non-equilibrium prices, there is no point to theories of equilibrium (1870, pp. 47n, 66-67n, 81-82).[6] In other words, Thornton seems to criticize classical wage theory not because it is too little like the neoclassical theories which succeeded it, but because it is too much like them. It relies on the notion of an equilibrium, albeit a partial and temporary one. Thornton seems to argue against all such theorizing – there is no way to abstract from the detailed day-to-day influences on wages (1870, esp. pp. 81-82).

Although Mill did not explicitly glean this particular point from Thornton's book (which is, given the confusions in Thornton's exposition, hardly surprising), he seems to some extent to have accepted it with respect to the labor market (*not* with respect to markets in general) (1869, pp. 645-46, 657-58). The main point in Mill's recantation, which is independent of his confusion of real and monetary analysis, seems to me to be the following:

> The less he [the capitalist] expends on the one [his labourers], the more may be expended on the other [himself], and *vice versa*. *The price of labour, instead of being determined by the division of the proceeds between the employers and the labourers, determines it*....
>
> ... and the law of wages, on the side of demand, amounts only to the obvious proposition, that the employers cannot pay away in wages what they have not got. On the side of supply, the law as laid down by economists remains intact. (1869, p. 645, my emphasis)

The portion of the output of consumption goods in period t that constitutes the wages fund of the next period is not fixed in period t and not independent of wages in period t+1. One cannot regard the wages fund as causally prior to wages. They are mutually determining in the one-period labor-market equilibrium, and thus no real explanation of wages in terms of the supply and the demand for labor can be given.

It is important to realize that Mill is not only arguing that wages can be increased in a given period at the expense of capitalists' consumption (the details of which claim Marshall 1920, p. 825 ably criticizes). To argue only that it is not absolutely impossible for unions to raise wages is not to concede much (Gordon 1973, p. 18); it still might be universally inadvisable for them to do so.[7] One can still argue that action to increase wages will slow accumulation, increase population and thus harm the workers. As one of the lesser lights of nineteenth-century political economy put it:

> There is only a certain produce to be divided between capitalist and labourer. If more be given to the labourers than nature awards, a smaller amount will remain for the capitalist; the spirit of accumulation will be checked; less will be devoted to productive purposes; the wage fund will dwindle, and the wages of labourers inevitably fall. For a time, indeed, a natural influence may be dammed back; but only to act, ultimately, with accumulated force. In the long run, God's Laws will overwhelm all human obstructions. (Stirling 1869, pp. 26-27; quoted in Breit 1967b, p. 512)

But Mill is not only conceding that wages can be increased above their "natural" aggregate market-clearing or "equilibrium" level. He is claiming instead that classical economics has no good explanation for wages at all. He is questioning whether there is any single equilibrium wage and whether, if there were, real-world competitive labor markets insure that such a wage is paid. Moreover, even if workers were paid "equilibrium" wages, Mill appears to doubt that there is any causally prior demand for labor which can figure in an explanation of wages. The policy conclusions drawn on the basis of classical wage theory condemning collective action to raise wages are thus completely unjustified. Mill would certainly concede that there is a trade-off between wages and economic growth (although he does not stress the point in his review of Thornton); but there is no reason to believe, given the imperfections of the labor market and the dependence of worker's subsistence demands on their actual wages, that efforts to raise wages are always or even usually harmful to the interests of workers in general.

Homilies to the working class concerning the harmfulness of attempting to increase wages are as unjustified as are homilies concerning the absolute impossibility of doing so. But (and here I am speculating), Mill is not sure what one should replace the wages-fund doctrine with. He did not change his exposition in the *Principles of Political Economy,* I suggest, simply because he did not see how to reformulate his whole theory of distribution. Mill's sympathies for the laboring classes led him to see that some of the abstractions of classical wage theory were unreasonable; but the very same sympathies (along with his unshakeable Malthusianism) made him cling to its general analysis of the long-run course of the economy.

Classical wage theory did not collapse because of the argument I have attributed to Mill, which no one at the time formulated very explicitly or attempted to answer. It collapsed because the abstractions it involves are extreme and perhaps unhelpful, because it failed to say anything about the contribution of additional labor to additional output (although this deficiency could have been remedied), and because economists both lost interest in the dynamic questions that it attempted to answer and became engrossed in the new puzzles thrown up by the emerging marginal productivity analysis.

It is ironic that, if my interpretation is correct, Mill recanted his espousal of classical wage theory on the basis of criticisms that apply much more forcefully to its successor. Classical wage theory did, to be sure, rely on supply and demand to explain a temporary equilibirum in the labor market; and one may reasonably doubt whether one can define causally prior demand and supply functions even within a single period. But twentieth century theory attempts an explanation in terms of supply and demand of wages in a stationary or proportional growth equilibrium. These demand and supply theories are incapable of explaining wages and employment because supply and demand for labor functions cannot be fixed prior to the determination of wages and employment. I am proposing no resurrection of classical wage theory – its difficulties are many and apparent. But its dynamic and straight-forward causal structure is interesting in its own right and as a contrast to twentieth-century equilibrium theory.

University of Maryland

NOTES

*I would like to thank Professor H. S. Gordon for his careful and helpful criticisms.

[1] This claim is argued especially by Blaug (1978, p. 195); Schwartz (1972, p. 71) and Taussig (1896, pp. 104, 194-95) and is borne out by my own reading. The claim is disputed by Breit (1967b, pp. 512-13) and Gordon (1973, pp. 21-23). Applications to trade unionism seem to become more frequent in the second half of the Nineteenth Century.

[2] There is some disagreement about what other causal factors one should hold constant. See particularly Friedman (1953). Both Friedman (1953, pp. 48, 59-63, 90) and Buchanan (1958, esp. p. 159) insist that *ceteris paribus* stipulations must be consistent with general equilibrium analysis. From the Marshallian perspective that seeks "an engine for the discovery of concrete truth" (1885, p. 159; quoted twice in Friedman 1953, pp. 56, 90), it is not clear (to me at least) why "approximate consistency" would not often be enough.

[3] In the course of validly criticizing Patinkin's construction of an aggregate demand curve for money (1958, p. 263), Buchanan mistakenly argues that a market equilibrium curve "is the only meaningful demand curve for money which may be derived for the whole economy" (1958, p. 263). If he is right, then there is in fact *no* meaningful economy-wide demand curve for money. His grounds for regarding the market equilibrium curve as a demand curve is that it enables one to predict what will happen if supply changes. But he overlooks the explanatory role of demand curves, which, as I argue in the text, market equilibrium curves cannot play.

[4] This construal of wages-fund theory was suggested by comments of Joseph Schumpeter's (1954, esp. pp. 666-67, 669). Yet in his long footnote 50 on p. 667, he suggests an equilibrium interpretation that has been explicitly developed by William Breit (1967b, pp. 523-28). I think that economists have generally failed to take sufficient notice of the dynamic nature of classical wage theory. I argue the point in Section 2 of my "A Modern Restatement of Classical Wage Theory" (forthcoming a).

[5] In the "Preface to the Seventh Edition," Mill remarks, "Since the publication of these [earlier editions], there has been some instructive discussion on the theory of Demand and Supply, and on the influence of Strikes and Trades Unions on wages, by which additional light has been thrown on these subjects; but the results, in the author's opinion, are not yet ripe for incorporation in a general treatise on Political Economy." (p. xxxi) The most substantial change I notice in the body of the *Principles* occurs in Book V, ch. X, sec. 5. In discussing unions, Mill had originally written "But if they aimed at obtaining actually higher wages than the rate fixed by demand and supply – the rate which distributes the whole circulating capital of the country among the entire working population – this could only be accomplished by keeping a part of their number permanently out of employment." In the seventh edition, the sentence is replaced by "They would also have a limited power of obtaining, by combination, an increase of general wages at the expense of profits. But the limits of this power are narrow; and were they to attempt to strain it beyond those limits, this could only be accomplished by keeping a part of their number permanently out of employment."

[6] Longe does not attack supply and demand theories in general, but argues that the supply and demand apparatus is misapplied in the case of labor. There is, perhaps some echo of Thornton's argument (1866, pp. 24-27); but Longe's discussion is at least as confusing as Thornton's. He argues that demand and supply are misapplied because in wages fund theory they are not variable functions of wage rates and because "demand" is a fund (of money on Longe's interpretation) (1866, pp. 30-35). Taussig (1896, p. 252) seems to think that Longe has a valid point here. I disagree. My discussion of Mill's recantation might be regarded as an elaboration and a clarification of Schumpeter's claim that Longe's and Thornton's argument boils down to the claim "that there is little point in inserting aggregate wages as a 'proximate cause' that, as such, plays a role of its own". (1954, p. 670)

[7] In fact a number of historians of economics have argued that the classical economists did not really insist on a wage fund which was *fixed* even in a single period (Taussig 1896, pp. 214-15; Schumpeter 1954, p. 666; and Breit 1967b, p. 510). Gordon has disagreed (1973, p. 15n). Mill vacillates. In the original sentence quoted in footnote 5, he is committed to a fixed wage fund (and see also 1871,

Bk. II, ch. xiv, sec. 7, p. 404). Yet in at least one passage (1871, Bk. II, ch. xii, sec. 1, 2, pp. 362-63), Mill seems to regard an increase in aggregate wages within a single period as entirely possible. Thornton may have regarded his own critique as mainly directed against the fixity of the wage fund, since he makes the following large concession to classical wage theory, "If the increase of wages obtained trenched very far upon profits, the growth of capital would infallibly be checked, and more or less of the capital already invested might very probably be withdrawn from business." (1870, p. 301)

REFERENCES

Blaug, M. (1978), *Economic Theory in Retrospect*. 3rd. ed. Cambridge: Cambridge University Press.

Bonar, J. (1911) *Disturbing Elements in the Study and Teaching of Economics*. Baltimore.

Breit, W. (1967a) "Some Neglected Early Critics of the Wages Fund Theory," *Southwestern Social Science Quarterly*, 48:53-60.

__________. (1967b) "The Wages Fund Controversy Revisited," *Canadian Journal of Economics and Political Science*, 33:509-29.

Buchanan, J. (1958) "*Ceteris Paribus*: Some Notes on Methodology," *Southern Economic Journal*, 24: 259-70.

Cairnes, J. (1874) *Some Leading Principles of Political Economy*. rpt. London: Harper and Brothers, 1900.

Ekelund, R. (1976) "A Short-Run Classical Model of Capital and Wages: Mill's Recantation of the Wages Fund," *Oxford Economic Papers*, 28:66-85.

Fawcett, H. (1883) *Manual of Political Economy*. London: Macmillan.

Friedman, M. (1953) "The Marshallian Demand Curve," in *Essays in Positive Economics*. Chicago: University of Chicago Press, pp. 47-99.

Gordon, H. S. (1973) "The Wage-Fund Controversy: The Second Round," *History of Political Economy*, 5: 14-35.

Hahn, F. (1972) *The Share of Wages in the National Income. An Enquiry into the Theory of Distribution*. London: London School of Economics and Political Science.

Hausman, D. (1981) "Are General Equilibrium Theories Explanatory?" in J. Pitt, ed. *Philosophy in Economics*. Dordrecht: Reidel, pp. 17-32.

__________. (forthcoming a) "A Modern Restatement of Classical Wage Theory."

__________. (forthcoming b) "Are There Causal Relations Among Dependent Variables?" *Philosophy of Science*.

__________. (forthcoming c) "Does the Wage Depend on the Marginal Productivity of Labor?"

Hollander, S. (1968) "The Role of Fixed Technical Coefficients in the Evolution of the Wages Fund Controversy," *Oxford Economic Papers*, 20: 320-41.

Longe, F. (1866) *A Refutation of the Wage Fund Theory of Modern Political Economy as Enunciated by Mr. Mill and Mr. Fawcett*. rpt. in J. Hollander, ed. *A Reprint of Economic Tracts*. Baltimore: Johns Hopkins Press, 1903 (original pagination).

Marshall, A. (1920) *Principles of Economics*. London: Macmillan.

Mill, J. S. (1869) "Thornton on Labour and Its Claims," *Fortnightly Review*, ns. V, pp. 505-18, 680-700. rpt. in *Essays on Economics and Society*, vol. V, of *The Collected Works of John Stuart Mill*, ed. J. Robson. Toronto: University of Toronto Press, 1967.

_________. (1871) *Principles of Political Economy with Some of their Applications to Social Philosophy*, 7th ed. rpt. and ed. W. Ashley. London: Longmans, Green, and Co., 1909.

Patinkin, D. (1956) *Money, Interest and Prices*. Evanston, Ill.: Row, Peterson & Co.

Pigou, A. (1949) "Mill and the Wages Fund," *Economic Journal*, 59: 171-80.

Ricardo, D. (1817) *Principles of Political Economy and Taxation*. rpt. and ed. R. Hartwell. Harmondsworth, Middlesex: Penguin, 1971.

Samuelson, P. (1951) "Economic Theory and Wages," in D. M. Wright, ed. *The Impact of the Union*. New York: Kelley and Millman, pp. 312-42.

Schumpeter, J. (1954) *History of Economic Analysis*. Oxford: Oxford University Press.

Schwartz, P. (1972) *The New Political Economy of John Stuart Mill*. Durham, NC: Duke University Press.

Senior, N. (1830) *Three Lectures on the Rate of Wages*. London: John Murray.

Simon, H. (1953) "Causal Ordering and Identifiability," rpt. in *Models of Discovery*. Dordrecht: Reidel, 1977, pp. 53-80.

Stirling, J. (1869) *Trade Unionism with Remarks on the Report of the Commisioners on Trade Unions*. Glasgow.

Taussig, F. (1896) *Wages and Capital: An Examination of the Wages Fund Doctrine*. New York: D. Appleton and Co.

Thornton, W. (1870) *On Labour, Its Wrongful Claims and Rightful Dues, Its Actual Present and Possible Future*. 2nd. ed. London: Macmillan.

Walker, F. (1876) *The Wages Question: A Treatise on Wages and the Wages Class*. New York: Henry Holt.

JACQUES VONÈCHE

GENETIC EPISTEMOLOGY IN THE CONTEXT OF EVOLUTIONARY EPISTEMOLOGY

1. INTRODUCTION

In the end, all genetic epistemologies fall into three main categories: naturalistic, culturalistic and logistic. In this paper I will take one example for each: James for the naturalistic approach, Ernest Cassirer for culturalism and Piaget for logicism. The main feature of logicism would be, in this view, its attempt at blending in the appropriate proportions, the biological and the sociological components of knowledge into a terminal product that would be so abstract as to depend on logics only in the final analysis. Originally, this paper was aimed at comparing Piaget's genetic epistemology to evolutionary epistemologies that somehow set the pace for Piaget: James's, Dewey's and the Chicago School (especially Mead), Cassirer's and above all Baldwin's. I would have liked to add to the above list of names a few German ones such as Wundt, Simmel, Sigwert and British ones such as Hobhouse. These additions would have made the picture more complete and more complex. In fact, too complex a picture for the present, hence, my more limited scope.

2. WILLIAM JAMES

Evolutionary epistemology explains knowledge using a Darwinian model of analysis. This attempt is part of a wider enterprise which views knowledge as the outcome of biopsychological and sociohistorical causes. In the strictest sense, knowledge is treated as any other biological species or physical organ. The essential question remains the same. Can the evolutionary model (random variations and selective retention) account for the range of knowledge processes so far identified as well as for the surprising but necessary fit between knowledge (often limited to scientific theories) and reality?

J. C. Pitt (ed.), Change and Progress in Modern Science, 199–232.
© 1985 by D. Reidel Publishing Company.

In 1881, Thomas Huxley did not hesitate to answer: "The struggle for existence holds as much in the intellectual as in the physical world. A theory is a species of thinking, and its right to exist is coextensive with its power of resisting extinction by its rivals", echoing in the intellectual spheres the massive repressive actions in the sociopolitical realm of his fellows in the ruling class.

In one of the 1880 issues of the *Atlantic Monthly*, James had published an essay entitled "Great Men, Great Thoughts and the Environment" in which he argued in detail for a direct parallel between the Darwinian conception of evolution and what he called "mental growth of the race". This would later be known as ethnogenesis, sociogenesis or phylogenesis of mind. James's fundamental distinctions are evolutionary in the most biological sense, since they are Darwinian in origin.

The first distinction drawn by James separates spontaneous variations from natural and sexual selection. This distinction parallels, more or less, the contemporary complementarity between chance and necessity. Spontaneous variations represented the chancy side of the couple, whereas selection represented its necessity aspect.

It is worth noticing that this conviction that chance was an important factor of evolution, which James shared with Darwin, goes along with the implicit assumption that the space of errors described by Nature was what mathematicians call "amorphous". This assumption was not to be shared by Baldwin, Piaget and Bertalauffy. For these thinkers, the space of errors is structured with the revolting consequence to strict Darwinists that Nature either remembers its mistakes or even learns from them. This difference leads to some basic epistemological choices. Whereas the believers in chance tend to remain strictly biological and somewhat mechanical in their approach to the acquisition of knowledge, the structuralists usually lean upon logic.

The second important distinction made by James opposes production mechanisms to preservation ones. The production side stems out of spontaneous variations. As such it depends on chance. A preservation mechanism is caused by natural and sexual selection.

It is important to observe that the tendency of Nature towards spontaneous variations which are physiologized in Darwin's model are usually ignored by him in favor of environmental selection. In James's epistemology this tendency towards variations is conceptualized somehow as metamorphoses of the same form (Species). These metamorphoses should not be confused with Goethe's idea of metamorphosis, which is structuralistic in nature. This is clearly explained in the work of one of Goethe's contemporary followers on this specific point, Chomsky. James is closer, here, to Lamarck's metaphysical scale of beings. This position is coherent with James's general functionalistic approach to philosophy and his idea of groping in thinking.

Another noteworthy point about this opposition between chance and necessity is that variation and selection are independent processes in the Darwinian evolutionary epistemology. This mutual independence of the production cycle of causation from the conservation cycle reflects an epistemological position on knowledge: it prevents any form of direct knowledge from existing. In other words, it assumes that the world is mysterious for human observers. Remember James's famous hypothetical baby bombarded with stimulations of all sorts that appear, at first, as complete chaos.

In this line, James appears much less impressed than Piaget by novelties. This difference, once again, can be explained by a different stand about Darwinism. Piaget, who was critical of Darwinism and neo-Darwinism (he confused the two all too often) needed new forms and new species all the time. Only such novelties would have guaranteed the creative aspect of the production cycle. To James this was not so important, because, in the end, all variations are variations of the same basic form, as in "The Varieties of Religious Experience".

According to James, the causes of variation "are molecular and invisible; inaccessible, therefore to direct observation of any kind" (James 1880, p. 444). Thus, speculation about this production cycle of causation is unadvisable. One stays with the observable range of variations.

> As a matter of fact, the new conceptions, emotions, and active tendencies which evolve are originally *produced* in the shape of random images, fancies, accidental outbirths of spontaneous variation in functional activity of the excessively unstable human brain, which the outer environment simply confirms or refutes, adopts or rejects, preserves or destroys, –

> *selects*, in short, just as it selects morphological and social variations due to molecular accidents of an analogous sort. (James 1880, p. 456)

Once again here is the idea, dear to James, that the brain is "a seathing caldron of ideas, where everything is fizzling and bobbing about in a state of bewildering activity" (James 1880, p. 456).

The next question is that of the specific mechanism of selection. James leaves this question open for one excellent reason: he is not equipped to answer it and prefers to say nothing than to speculate.

Despite such obvious gaps in his explanation of knowledge, James remained faithful to the Darwinian position. This is especially clear in the way he dealt with the Kantian *a priori* synthetic categories of knowledge. James accepted Kant's idea that without *a priori* categories of understanding experience would be impossible. But he was bothered by the *a priori* aspect of these categories, because *a priori* validity was in contradiction with an evolutionary explanation of the mind. Thus, he imagined a way out of this dilemma by assuming that the Kantian categories might be "pure *idiosyncrasies*, spontaneous variations, fitted by good luck (those of them which have survived) to take cognizance of objects (that is to steer us in our active dealings with them), without being in any intelligible sense immediate derivations from them" (James 1890, II, p. 632). According to this theory, internal processes instead of environment stimuli are the proximate causes of *a priori* mental structure.

His discussion of the construction of scientific theory is exemplary:

> I do not see how any one with a sense for the facts can possibly call our systems immediate results of 'experience' in the ordinary sense. Every scientific conception is in the first instance a 'spontaneous variation' in one's brain. For one that proves useful and applicable there are a thousand that perish through their worthlessness. Their genesis is strictly akin to that of the flashes of poetry and sallies of wit to which the instable brain paths equally give rise. But whereas the poetry and wit

> (like the science of the ancients) are their 'own excuse for being' and have to run the gauntlet of no further test, the 'scientific' conceptions must prove their worth by being 'verified'. This test, however, is the cause of their *preservation*, not that of their production.... Thousands of years ago men started to cast the chaos of nature's sequences and juxtapositions into a form that might seem intelligible. Many were their ideal prototypes of rational order: teleological and aesthetic ties between things, causal and substantial bonds, as well as logical and mathematical relations. The most promising of these ideal systems at first were of course the richer ones, the sentimental ones. The baldest and least promising were the mathematical ones; but the history of the latter's application is a history of steadily advancing successes, whilst that of the sentimentally richer systems is one of relative sterility and failure... when you give things mathematical and mechanical names and call them just so many solids in just such positions, describing just such paths with just such velocities all is changed. Your sagacity finds its reward in the verification of nature of all the deductions which you may next proceed to make... What makes the assumption 'scientific' and not merely poetic, what makes a Helmholtz and his kin *discoverers*, is that the things of Nature turn out to act as if they *were* of the kind assumed. (James 1890, II, pp. 665, 666, 688)

The themes and problems outlined by James have remained central to all subsequent discussions in evolutionary epistemology: the role of chance, the objectivist notion of science as leading to instead of proceeding from epistemological certainty, as well as the craving for more technological explanations. There are no irreducible elements of sense-data as in the empiricist tradition, no subjectively guaranteed certainty as in the Cartesian cogito, no Kantian transcendental deduction of epistemological *a priori*. All that remains is ordinary experience, scientific knowledge and the hypothetico-deductive method.

Piaget was influenced by some aspects of this biological approach to question epistemology. But he rejected some aspects of it: the role of chance, the strict adherence to Darwinism and the notion of species. In the beginning of his career, Piaget opposed Darwinism essentially for two main reasons. First, he had been socialized in a rather Lamarckian form of transformism and a field of study, natural history of molluscs, that was largely insensitive to Darwin's breakthrough. Second, Piaget did not like the moral consequences of Darwinism: struggle for life and elimination of the misfits as prerequisites for evolution. As a generous young man (who had thought at one point of becoming a Protestant minister) Piaget expected evolution to lead to increasing cooperation, support, and compassion in the more advanced forms of life. His paper on "Biology and War" (Gruber & Voneche 1977) was emblematic of this view. Later on, when he had lost interest in morality, Piaget resisted Darwinism because of the excessive role given in it to chance, and because experimentally established variations were all intraspecific instead of interspecific. In sum, for Piaget, Nature did not agree with its Darwinian and neo-Darwinian model. Hence, he could not follow the line of evolutionary epistemology introduced by James in the Genevan circles of Flournoy and Claparède with which he was familiar.

Hence, as one would have expected given Piaget's interest in morality as well as Swiss proximity to Germany, there was a re-direction of Piagetian thinking towards the socio-cultural perspective being developed by Cassirer at the time, a form of epistemology apparently adequate to Piaget's purpose.

3. CASSIRER

Ernst Cassirer's epistemology was essentially a reaction against an empirically oriented picture of Mankind explicit in its faith in the coming of a better world. This view was frequently associated with the implication that there is some necessary connection between a predominance of individuals with I.Q.'s over 150 and an age of great cultural efflorescence.

Cassirer was preoccupied with the problem of mind and its place in nature. He opposed French naturalism and adhered explicitly to a form of idealistic naturalism that he shared with Dewey, Woodbridge, Mead and, to a lesser extent Charles Peirce and William James. He belonged to that group of German philosophers who had been disillusioned by Hegelian absolutism at the same time as they refused Kantian dualism between mind and matter.

Dewey had an early behavioristic analysis of mind as experience. But he moved increasingly towards a position according to which experience was too subjective a word to stand for dominant features of mental phenomena. In the end had replaced experience by culture. Cassirer took over at this point in Dewey's personal evolution and defined culture, in its manifold forms, as "mind's place in nature". Such a view conformed to the ideas developed by George Herbert Mead in *Mind, Self and Society* But Cassirer differed from the Chicago School in that he had arrived at his cultural interpretation of mind through Dilthey, the German philosopher of history. Dilthey also had moved from the Hegelian position of *Geistwissenschaften* to *Kulturwissenschaften*.

In order to understand fully this passage from *Geist* to *Kultur*, the reader should be reminded of the philosophical landscape at the time of Cassirer's writing. In Germany and its Hinterlands, Richard Avenarius's popular empiricocriticism had influenced such diverse people as Ernst Mach, Vladimir Ulianov Lenin, and William James. It led Mach to his phenomenalism and James to his radical empiricism. It was this radical empiricism of James that Cassirer criticized for its extreme subjectivism. At the same time, British philosophy, except for a few neo-Hegelians such as Bradley, had little inclination to identify mind and culture. Even Whitehead was only concerned with the relationship between mind and nature. This view was grounded in James's pure and immediate experience that led to the popular idea of the existence of a gulf between minds.

Dilthey, on the contrary, divided human nature into two components: 1) personal experience or mind in the concepts of subjective idealism and 2) sociology of knowledge or culture as historically conditioning of mind (here, the influence of Max Scheler is evident). Based on this distinction, Cassirer elaborated an analysis of mind according to which personal experience was either a mere natural process or a mere sensation. On the other hand, all *cognitive* aspects of mind (or meaning by opposition to direct experiences mediating meaning) were *social* and, if lasting and communicable, *cultural*.

According to Cassirer, the produce of cognition is *science*. But the notion of science in German philosophy is not the mere factual description plus exact lawful relationships obtaining between natural events. It is a larger concept. *Wissenschaft* covers any systematic ordering of concepts (logic) and it is not divorced from daily experience, since it is conceived only as a more profound and more adequate knowledge of what we already know in our daily encounter with the world. Hence, mythology and poetry are part of this enquiry. Hence, *Kulturwissenschaften* correspond to the English category of humanities analyzed from this logical standpoint. For Cassirer, the logic of humanities amounts to the study of disciplined perception. It is the logic of concrete universals by opposition to abstract universals that form the logic of natural sciences by a process of isolation of generalized aspects. It is a logic of styles by opposition to the study of uniqueness in particular personalities, but it should not be conceived with the study of streams of history. Styles are concepts upon which we build our understanding of particular persons and of historical events. The logic of styles does not enable us to predict personal behavior and historical events. Rather it brings us to a deeper knowledge not of the *how* but of the *what* of human nature. The logic of styles refers to a critical study of such type and style concepts and not to a supposed methodology of historical predictions or of formulas for artistic creation. *Kultur* in this view is not the cultivation and education of the human race (Lessing, Herder, Schiller, Goethe, Wilhelm v. Humboldt, Hegel). Instead, it is the conception of human nature as the potentiality for self-definition, for teaching itself without welding it to theories of historical evolution in order to give it "scientific" rigor, as did Cassirer's German predecessors beginning with Leibniz.

From this description of Cassirer's approach, it is clear that Piaget could not agree with it for several reasons. First, the emphasis Cassirer placed on humanities instead of natural sciences was unacceptable to someone like Piaget who was trained in natural history. Second, Piaget suspected a lack of scientific rigor in Cassirer's approach, since it was not grounding its logic in a biological theory of evolution. Third, Cassirer's approach was centered around knowledge and understanding, whereas Piaget was looking for explanations and know-how. Fourth, in spite of his claims to the contrary, Piaget favored a causal model of explanation, whereas Cassirer proposed a heuristic model of knowledge. A causal model brings one from situation A to situation B by anticipation. A heuristic model short-circuits the causal loop by breaking up the analysis into two complementary components: a situation and its correction as exemplified in bicycle-riding and the law of balance.

4. WHAT IS GENETIC EPISTEMOLOGY?

The phrase "genetic epistemology" was coined, in 1906, by James Mark Baldwin in his famous book: *Thoughts and Things or Genetic Logic* (Baldwin 1906, 1908, 1911). For Baldwin, genetic epistemology "concerns itself with the implications of reality at each of the greater stages of conscious process, from the most primitive to the most derived" (Baldwin 1911, vol. 3 p. 16). Epistemology, for Baldwin, is equivalent to "instrumental logic" or the science of truth by the control of facts by opposition to "axiology" or the science of value or worth covering the whole range of "normative" disciplines. This dichotomy is founded on the fundamental distinction between the two possible controls (according to Baldwin) mediated in the thought content. If the control of fact, then we have epistemology. If the control of the inner sort, then we have axiology, that is to say "the relative selection and utilization of facts through the mediation of the thought system as means" (Baldwin 1908, vol. 2 p. 383).

What is essential here, is what Baldwin called "genetic modes" according to which nature presents "genetic series" as such, movements of progressive change, which are qualitative, irreversible and non-mechanical. These genetic series constantly place in evidence the rise of new and progressive "genetic modes" or sorts of organization which are *each for itself novel, sui generis, and creative*. In passing from one genetic mode to another, nature achieves a real evolution; there is an actual production of novelties" (Baldwin 1911, vol. 3 p. 258). The logic of this general movement is stated in the "axioms" of genetic sciences (Baldwin 1902) and is interpreted for the movement of mental process in the "canons of genetic logic" (Baldwin 1906, vol. 1 ch. 1). This movement is externally guided in accordance with the Darwinian principle of natural selection.

The concept of genetic epistemology was not Baldwin's exclusive property. One might mention, from Germany, W.

Wundt (Wundt 1880), C. Sigwart (Sigwart 1894, 1895) and E. Cassirer (Cassirer 1903, 1920). In the Anglo-Saxon tradition, there were B. Bosanquet (Bosanquet 1888) and L. T. Hobhouse (Hobhouse 1888, 1912).

Thus, the tacit procedural assumption that "genetic epistemology" is, or ought to be, identified with Piaget's specific undertaking constitutes a total misapprehension of the domain. On the contrary, the concept of "genetic epistemology" should be disembarrassed from the unhappy fusion it has undergone, in recent times, with Piaget's specific attempt to orient that discipline along certain lines.

The goal of epistemology is to provide a theory of knowledge in whatever form it occurs. This aim has traditionally been conceived as a matter of logic rather than psychology, of abstract norms rather than concrete facts, as a normative discipline, regulative of inquiry, rather than a positive discipline shaped by the results of empirical investigation. But the demarcation of epistemology from empirical inquiry involves a decision and not a description of "the way things are". This decision is not regulated simply by the character of the concept to be defined, but depends also on the soundness of the interest it is designed to serve, so that the proceeding is partly normative. On the other hand, between the conception and the creation of such a logic falls the shadow of a presupposed psychology, because every epistemology is shaped by underlying conceptions of the mind and conduct of which cognition is a product.

This dual characteristic of thinking as both a process and a product was clearly understood by Baldwin and by most of the early genetic epistemologists. Hence, the recurrence in Baldwin's writings of such reversals of terms as "worths as thoughts" and "thoughts as worths". This passage from facts to norms and vice-versa is indeed central to any epistemology and especially to genetic epistemologists who agree in recognizing that any form of knowledge is triple: 1) a social system; 2) a group of psychological behaviors, and 3) a system *sui generis* of signs and cognitive activities. But they do not all agree on Piaget's solutions to these two main issues: the relationship of facts to norms cleverly eschewed as "normative facts" (faits normatifs) or the consideration of science as the form of knowledge *par excellence*. Cassirer, for instance, would consider Piaget's emphasis upon the scientific forms of knowledge as extremely limitative, and Baldwin as simply preposterous. The "fusion" of norms and facts under the heading of "normative facts" would be considered as muddy by most of the genetic epistemologists mentioned here.

Besides limiting genetic epistemology to the sole attempts of Jean Piaget, there is another confusion generally made by philosophers and scientists alike, that of fusing Piaget's epistemology with his psychology. Numerous examples can be given of people who sincerely think that genetic epistemology is the title Piaget bestowed upon his theory of intellectual development in human ontogeny, although it should have seemed unlikely that a person of Piaget's considerable historical erudition and critical acumen would conflate individual developmental psychology and genetic epistemology.

In fact, Piaget defines genetic epistemology as "the study of the mechanisms of the growth of knowledge" (Piaget 1957, p. 14). In addition to this general and broader form of genetic epistemology, there is a more specific notion of genetic epistemology:

> Under its limited or special form genetic epistemology is the study of successive states of a science as a function of its development. Thus conceived, genetic epistemology could be defined as the positive science, empirical as well as a theoretical, of the becoming of positive sciences *qua* sciences. (Piaget 1957)

Once again, other genetic epistemologists would disagree with the idea that genetic epistemology could be a positive science only. They would include in it a normative aspect. As we have seen, they would not share the same fascination for science.

Piaget's own fascination is easily explainable in terms of the original purpose of genetic epistemologies: to see any form of knowledge as a triple. Science is indeed so social that it has become a social institution, so psychological as to engender specific forms of behavior and such a complex and coherent system of signs and cognitive activities that it requires a long and difficult training to be practiced. The coherent system of signs and cognitive activities evolves with time and has to be explained. Piaget assumes that the explanation of this development of scientific knowledge comes from its psychological and sociological aspects which are thus indissociable from the epistemic aspect. Indeed, Piaget goes so far as to maintain that a

> systematic study of the development of any sector of scientific knowledge will necessarily be led, in attempting to disengage the sociogenetic and psychogenetic roots of this form of knowledge, to push the analysis of its formative mechanisms to the terrain of prescientific and infra-scientific common beliefs in the history of societies ..., in the development of the child and even to the frontiers of physiological processes. (Piaget 1957 p. 13)

The implicit assumption of this statement by Piaget is that knowledge can be attained through the presuppositions, methods, procedures, and resultant findings of empirical sciences such as history of ideas, history of science, sociology of knowledge, developmental psychology, and even psychophysiology. No special place is given, in this context, to epistemology itself as a separate way of knowing or of assessing, at least, the validity of claims to knowledge. Thus, it may be argued that such positive inquiries into the history, sociology and psychology of epistemic claims to knowledge presuppose, incontrovertible resolutions of ontological and epistemological issues. Piaget explicitly stated this several times in his paper "De la psychologie génétique a l'épistemologie génétique" (stupidly translated into English as "Genetic Psychology and Epistemology" (Piaget 1952); see especially the two examples of number concept and conservation). Piaget can be apodictic about these resolutions, because he believes that the experimental method renders findings incontrovertible.

> From the University of Louvain to Soviet laboratories, psychologists agree today upon a number of issues (perception, habit formation, intelligence development, etc....) without making it possible while reading an experimental report, to recognize the philosophy of its author. (Piaget 1947, p. 118)

5. IS GENETIC PSYCHOLOGY RELEVANT TO EPISTEMOLOGY?

Since it is beyond our scope, here, to examine the relations of all genetic psychologies to all conceptions of epistemology, we will limit ourselves to two questions dealing with Piaget's genetic psychology: 1) is this psychology relevant to philosophical epistemology, and 2) is Piaget's psychology of intelligence relevant to Piaget's epistemology?

Piaget's answer to the first question is clear:

> Genetic psychology is a science whose methods are more and more closely related to those of biology. Epistemology, on the contrary, is usually regarded as a philosophical subject, necessarily connected with all the other aspects of philosophy and justifying, accordingly, a metaphysical position. In these circumstances, the link between the two subjects would have to be considered either as illegitimate or, on the contrary, as no less natural than the transition from any scientific study to whatever form of philosophical thought, less by way of inference than by inspiration and involving, moreover, the addition to the latter subject of considerations beyond its scope. (Piaget 1952 p. 51)

This disclaimer takes part in Piaget's general depreciation of philosophy (see: "Insights and Illusions of Philosophy" Piaget 1971a). As an empirical researcher, Piaget distrusts arm-chair psychologists: "The unfortunate thing for psychology is that everybody thinks of himself as a psychologist. As a result, when an epistemologist needs to call on some psychological aspect, he does not refer to psychological research and he does not consult psychologists; he depends on his own reflections." (Piaget 1970a p. 8)

As far as the second question is concerned and to the extent that the reader accepts Piaget's claim of not being himself a philosopher, Piaget has pointed out many times that "specialists in genetic psychology, and especially in child psychology, do not always suspect what diverse and fruitful relationships are possible between their own subject and other more general kinds of research, such as the theory of knowledge or epistemology." (Piaget 1952)

The preliminary condition for their "diverse and fruitful relationships" is indeed the acceptance by philosophers and developmental psychologists of Piaget's own idea of what epistemology really is all about. This answers positively the second of our two questions, but leads to a new question, in turn. What is epistemology for Piaget? As we have already seen, the central task of genetic epistemology is the study of "knowledge, and in particular scientific knowledge, on the basis of its history, its sociogenesis, and especially the psychological origins of the notions and operations upon which it is based" (Piaget 1970a, p. 1). The common denominator of the history, sociogenesis and psychogenesis of knowledge is, for Piaget, the stages of scientific thought which are the answer to what Piaget considers as the fundamental question of a truly experimental epistemology, namely: "By what process does a science pass from one determinate form subsequently held to be inadequate to another determinate form afterwards held to be superior by the common agreement of the experts on this subject?" (Piaget 1952, p. 51).

As one can see, this particular form of epistemology rules out any form of philosophical criteriology external to the scientific inquiry itself and rules in the reflection of scientists themselves upon their own disciplines. It calls for an epistemology internal to the various scientific disciplines and it justifies, in advance of any empirical testing, the fundamental categories of a discipline as defined by the practitioners of the discipline. This position renders the status of scientific breakthroughs rather hazardous, because, unless they are sanctioned, *a posteriori*, by the recognition of the scientific community, the categories under which they operate remain invalid. In this exact sense, Piaget is the direct heir of Baldwin for two reasons. First the emphasis is put, in both epistemologies, on the social dimension in the control of facts. Second, in spite of his numerous and vigorous attacks on Darwinism, Piaget seems to accept, here, a form of natural selection of ideas (considered as analogous to mutations in biology) not unlike the natural selection of the fittest. In such a view, truth is conflated to life. In other words, errors ought to be lethal, if this viewpoint happened to be correct. Such an identification of truth with survival is indeed categorically mistaken.

Nevertheless, such a formulation has the important merit of changing global and unsolvable (philosophical) questions into specific, soluble ones. But one does not clearly see how genetic psychology of intelligence can contribute to such an effort. To do so, one must go back to the thrust of knowledge. No matter how far back one traces the origins of a sicentific concept, one always ends up with a construction of reality that has emerged from a more primitive one. From early constructs of reality into earlier ones, the starting point is always removed one step further and one is forced to recognize that there is no absolute zero in knowledge. One collective representation of knowledge hides always another one considered as more primitive. Each of these collective representations entails a specific form of ontology accepted as true by a given social group or propounded by some advanced thinkers. These successive ontologies presuppose correlative forms of understanding ("genetic modes" to use Baldwin's phrase) or systems of concepts that can be characterized and even logically formalized ultimately.

For Piaget, the safeguard to maintain in such an inquiry is the consideration that the *norms* of "scientific" inquiry of a given group at a given moment are *facts* for the modern researcher; no claim is made about the validity of these ancient norms. Hence, the phrase to designate them: "*normative facts*". They must be mentioned to account for the beliefs of those investigators who have used them as valid norms, but they cannot be used by subsequent researchers in a more advanced stage of knowledge.

This sort of historical and critical effort constitutes the *phylogeny* of minds and reality. Since, no one leaps from infancy into the adult views of one's own society, the phylogenetic research has to be complemented by ontogenetic investigations into the forms of understanding or ontologies of the infants, children, adolescents and youth who are, in the process of becoming full-fledged members of a social order.

In the same way as one could not reconstitute the world views of people in different places and times without active research, one should actually investigate the ways in which growing human beings understand themselves and their world. The form of this investigation is the same as that of the investigation of foreign cultures and past times: the aim is not the knowledge of the actual contents of thought but the architectural organization of thoughts and things.

Moreover, child psychology, à la Piaget, constitutes "a kind of mental embryology in that it describes the stages of the individual development, and particularly in that it studies the mechanism itself of this development" (Piaget 1952, p. 50).

"Once we admit comparisons of this kind, the history of the relations between embryology and other biological studies throws considerable light on the possible and, *to a certain extent*, the actual relationship between child psychology and epistemology." (Piaget 1952, p. 51 emphasis mine) This last statement shows exactly why and how Piaget believes genetic psychology to be relevant to genetic epistemology cosidered as a new scientific discipline.

6. HOW IS GENETIC PSYCHOLOGY RELEVANT TO GENETIC EPISTEMOLOGY?

The history of scientific thought inevitably poses the problem of changes of perspective and even of "revolutions" in "paradigms". As a student of the French philosopher Léon Brunschvig, Piaget is aware of this dynamic character of any epistemology that does not put forward either *a priorist* views or empiricist, where knowledge is subordinated to forms occuring in advance either in the subject (nativism) or in the object (empiricism). In addition, Baldwin's "genetic logic" had given Piaget some penetrating insights, to say the least, into the psychological construction of cognitive structures.

Piaget's concern with this question led him to a twofold task: 1) constituting a method capable of providing empirical tests, and 2) reaching back to the very origins of knowledge. But though this kind of analysis involves an essential element of psychological experimentation, it must not be confused with a study in pure psychology: the psychology dimension inherent in genetic epistemology is nothing but a by-product of the epistemological research.

There is here a strong belief that psychology or, more accurately, psychogenesis (or formation of knowledge) would give a positive answer to the problems of validity of knowledge, as if this answer were read directly in the empirical observation of reality. In a man who always claimed not to be an empiricist, such an attitude seems rather surprising, because it eliminates the side of the subjective construction in these findings, a side that should be essential to someone who claims to be a constructivist as does Piaget: "Knowledge arises neither from a self-conscious subject, nor from objects already constituted (from the point of view of the subject) which would impress themselves on him; it arises from interactions that take place midway between the two." (Piaget 1972, p. 19). Moreover, Piaget considers, as did Baldwin who coined for this the term "adualism", that "there exists at the start neither a subject in the epistemological sense of the word, nor objects conceived as such, nor invariant intermediaries." (Piaget 1972, p. 20). The initial problem of knowledge will therefore be the construction of such invariant intermediaries. This is the aspect of Piaget's work which is best known: the construction of invariants from the permanence of objects to the most elaborate forms of conservation. Starting from the point of contact between the body itself and external things, Piaget will construct, in the long series of his books on cognitive developmental psychology, the development of these intermediaries in two complementary directions: internal, leading to the construction of the self and its cognitive structures, and external, leading to the construction of reality and its properties.

7. THE STAGES OF DEVELOPMENT

Here again, there is a common misrepresentation of the importance of stages in Piaget's theory. This misrepresentation is threefold: 1) range of application of the concept of stage; 2) evidence for stages; and 3) function of the concept of stage in Piaget's theory.

Does the stage concept apply only to the three major "periods" of development (sensorimotor, concrete operations, formal operations) or does it apply also to the progressive steps or substages in the attainment of a cognitive structure within a stage? It seems self-evident that the idea of conservation of matter rests on and includes the idea of object permanence and, in turn, that the idea of conservation of volume rests on and includes the idea of conservation of matter. So it is clear that each major period of development prepares the next one logically and materially. But this is not so clear for each of the substages characterizing every major period. However, empirical research seems to agree with the general sequencing observed by Piaget and his collaborators in Geneva. This comparative flexibility of substages seems to be an argument in favor of constructivism, since, if both stages and substages were inflexible, development would be a mere unfolding of genetic competences. On the other hand, if it were possible to speed up cognitive development at the level of basic stages, the sequencing would be unnecessary.

Another question is raised by Piaget's stages: they do not seem to extend beyond the years of adolescence. What happens in the rest of life? Do adults cease to grow? Or do they grow in such idiosyncratic ways that it would be pointless to attempt to trace a coherence in their life? Neither alternative is very satisfying. That Piaget did not address himself to this question very clearly is the reflection of his position in the field of developmental psychology. Be it in his own auto-biographical works (Autobiography, his "novel" *Recherche* and other essays) or in his scientific papers, Piaget's categories of understanding human development have not gone beyond the categories of his time, just as Freud's theory of memory does not innovate on Ebbinghaus'. Piaget did not invent a new stage of development after adolescence and youth. In this sense, he was less creative than G. Stanley Hall who revived the notion of adolescence to explain what was happening to young men in America at the turn of the century.

If we, now, turn to the second misrepresentation of Piaget's ideas about stages, we see that some critics of Piaget believe that one could not possibly conclude to the existence of universal stages on the basis of homely observations of one's own three children, as did Piaget. After careful scrutiny, this is not exactly Piaget's claim. His contention really is that the sequence of stages is orderly and that this order is necessary because it reflects a "genetic logic" to use Baldwin's phrase.

To show that any sequence is orderly means to show that the observed sequence corresponds in some way to another sequence which is deemed orderly. To show that the logical structures underlying thought grow in an orderly way, we must have a method for ordering logical structures that is independent of any observations we may make of children's growth. This independent scale is found by Piaget in the scales of conservations. An example of this method can be given in the field of geometry by Felix Klein's Erlangen program. This program orders the different possible geometries according to the number of parameters that are conserved. At the lowest level, there is topological geometry which is strictly qualitative and conserves only the most general properties of space: inside, outside, and vicinity. One cut above one finds projective geometry which also conserves straight lines. Then comes Euclidian geometry which conserves, in addition, distances, angles and parallel.

The order of acquisition of these different geometries parallels their logical order of integration in terms of the number of invariants taken into consideration. But, showing that the thought of the child at a given age (or stage) corresponds to a given logical model is something less than showing orderly growth. As an empirical researcher, Piaget never attempted to go beyond this level of evidence. He never tried to test the reliability of his findings by the standard test-retest procedure.

But, he tried to demonstrate the limits of coherence of a stage by his studies of "decalages". Horizontal decalages are the evidence of the uneveness of stages, since they are the sign that a concept – conservation, for instance – appears in one form at first – conservation of matter, for instance – before being extended over its possible range: weight, volume and so on. This extension may take many months. For Piaget, this tension between the more advanced aspects and the less highly developed ones is the sign that there is an actual construction of any concept, because this concept does not spring fully formed from the child's mind, as Athena from Zeus' brow. This development of the same concept in its extension is considered by Piaget as evidence for a developmental mechanism that he called "cognitive conflict" among unevenly structured concepts. The dialectical relationship between less and more developed mental structures creates conflicts that foster further growth. But, how much conflict is productive and structuring in the formation of a given stage and how much is destructive and dangerous for further development? This is never clearly specified in Piaget's theory. Apparently, only small and local conflicts are acceptable as factors of cognitive growth.

This question leads to another question, more basic for a good understanding of Piaget's stage theory: the function of the notion within the general theory. Any careful review of Piaget's writings on the topic would show that he wrote very little about stages. This could be considered as evidence for a lack of interest in the subject as well as for the self-explanatory aspect of the concept of stage in Piaget's mind.

From the little writing by Piaget on stages, one gains the impression that two issues are important to him in this respect: the determination of a developmental level and the way in which cognitive development takes place. Determining developmental level is essentially structural in nature. It requires the recognition of the unity of a certain period of development when it is compared to another one. The order of succession of acquisition must be constant. The structures constructed at a given stage must be integrated in the structure of the following stage and the unity of the stage must be marked by a "structure d'ensemble". In order to maintain development on its optimal course, Piaget imagined a complex system of concepts: equilibrium, assimilation and accommodation, adaptation and so on. But, at this point in the discussion, the concept of "*homeorhesis*", borrowed from Waddington's work in theoretical embryology, is the relevant one. As its name indicates, homeorhesis is a system regulating the flow of different inputs in the developmental process of an organism. Another concept was also borrowed from Waddington's vocabulary, that of "*creode*" or due course. When the forces making for one choice of developmental pathway (creode) are not strong enough, the organism will develop in another way.

Contrary to the conclusions generally drawn in the literature about Piaget's notion of stage, we can conclude, from this brief summary of Piaget's writings on stages that the universality of stages is a claim made more often and more strongly by followers or detractors of the Genevan epistemologist than by himself. The very notion of creode indicates the polymorphic character of development for Piaget. So one has to take the following sobering view about Piaget's stages: stages are merely classificatory and hierarchizing devices imported by Piaget from his education in natural history. In addition, stages come in handy to fight empiricism, because they preserve the integrity of the organism in face of environmental pressure.

There are 4 main periods or levels of development in Piaget's theory: level 1: sensori-motor period (in Geneva: birth to age 2 roughly); level 2: pre-operational thought period (in Geneva: about age 2 to 7); level 3: concrete-operational period (about age 7 to 11); level 4: formal-operational period (about age 11 to 15).

The first level is composed of six substages which mark the infant's progress from only performing reflex actions in complete adualism (no dissociation of self and the rest of the world) to a gradual dissociation of the self from the objects and from others (especially the mother). The effects of such a dissociation are: perception of the effects of the infant's own actions upon objects, anticipation of the actions of people and objects and, at the very end of this period, the beginnings of a mental representation of things and thoughts, which existed only in actions so far.

This period is characterized by this passage, in the baby's mind, from action to thought. This means that during infancy, as for animals during all their life, thoughts and things exist only in the course of an action. They do not exist for and by themselves. Hence, no need to conserve them outside of action. There is a tool function, but tools are not kept from one action to the next (no memory). There is space, but only the space of action. Things are located in reference to specific actions that can be performed upon them: vicinity, in and out of the box, the room, the house, finding the cookie-jar when mother/father takes it away, and so on. There is time, but it is elastic. Depending on the actions performed during those minutes or hours, they pass by more or less quickly. There is causality, but at first there is one unique causal agent: the child. Then, when objects become permanent, the infant attributes to objects some physical and spatial connections independent of its actions on them.

The second level: pre-operational thought is characterized by the use of language. But language is only one of the signs of the child's mastery of the semiotic or symbolic function. Not only does the child speak, but he or she draws, imitates others intentionally, plays symbolic games (dolls, indians) becomes capable of minimal forms of cooperation with others (setting the table, carrying things for parents, helping friends) and moral actions (right/wrong).

But there are tremendous limitations to this new form of conduct. First of all, children still use language in an egocentric fashion; they talk *at* each other and not *to* each other. True dialogue is not yet possible. Second, they depend on perception in problem-solving situations: if a row of beads is longer than another containing the same number of beads, the child is convinced that the longer row has *more* beads, and so on, for other quantities.

Between ages 5 and 7, the child undergoes a certain change. The egocentric speech that accompanied the child's actions is becoming focused upon communicating something to others. He or she is more sensitive to contradictions brought about by the heavy reliance upon perception: that a row of beads could be, at the same time, longer than another and still contain the same number of beads becomes puzzling. Children, who used to focus exclusively on one dimension of a situation, begin to recognize that more than one factor at a time has to be taken into consideration to account for an event in a coordinated manner. This increasing coordination leads to the next level, that of concrete operations.

The term "operations", in Piaget's system, means actions that are *interiorized*, reversible, and coordinated into systems that have laws that apply to the entire system and not just to the single action itself. "Interiorization" means the eventual dissociation between the general form of a coordination and the particular content of an external action. Interiorization leads from "practical" (sensori-motor) to operational intelligence and is the precondition for objective knowledge as well as for symbolic representation, by opposition to "internalization" which is the eventual diminution of external movements that become covert and sketchy, illustrated in imitation and language. "Reversibility" is the possibility of performing a given action in a reversed direction. Its two chief forms are negation (not male = female) and reciprocity (not better = worse).

The stage of concrete operations is marked by the acquisition of conservation. Conservation is the maintenance of a structure as invariant during physical changes of some aspects. For Piaget, the stability of an objective attribute is never simply given, it is *constructed* by the living organization. Conservation therefore implies an internal system of regulations that can compensate internally for external changes. In the case of concrete operations, this system of regulations is made out of three main "*groupings*" (groupements): classification, seriation and number. Its applicability is limited to objects considered as *real* (concrete) by opposition to objects considered as *virtual* or possible (formal).

By the end of the period of concrete operations, children have markedly increased their abilities to account for the cause of physical events so that they are now ready to solve not only problems that involve real objects but also ones concerning hypotheses and propositions about relationships.

This readiness to think in hypothetico-deductive terms is the hallmark of the stage of formal operations. It is typically manifested in propositional thinking and a combinatory system that considers the real as one among other hypothetical possibilities. Formal operations are characteristic of the final stage of operational intelligence which "reflects" on concrete operations through the elaboration of formal "group" structures.

8. THE PSYCHOLOGICAL METHODS OF GENETIC EPISTEMOLOGY

The findings of Piaget in the domain of developmental psychology have been obtained by methods that are not standard in contemporary psychology. Piaget's main method for the study of intelligence has been labelled variously by him as "clinical" or "critical". Nothing could be more misleading than these two words for the reader. Clinical or critical investigations à la Piaget mean a free conversation between the child and the investigator when the child is older than four. It means also a free conversation dealing with a "matériel" ("experimental" device) that "does" something in front of the child who has to anticipate what will happen and, then, explain what has happened. This *matériel* can be two clay balls, matches, sticks, erector-sets, etc.... Here, Piaget is the true heir to the pedagogical method used, since the XVIIth century, in the little charity-schools for the poor and the needy by Protestant ministers in the Jura chain of mountains. Like these ministers and teachers, he used a *materiel* that was both inexpensive and simple: clay, water, sand, wood and so on. This *matériel* could be seen and manipulated by the child *ad libitum*. Sometimes, an imaginary child of the same age made counter-suggestions and the degree of resistance of the real child interviewed by the investigator is observed. Sometimes, the investigator in person makes counter-suggestions to the child. The general idea of this method is clinical in the sense that, like any competent clinician, the investigator probes carefully the extent and the nature of the thought process that produced a child's initial answer. It is critical to the extent that it confronts the child with the consequences of his or her assertions. It depends therefore very much on the argumentative talent of the child and of the investigator alike.

In fact, there are two rather prodoxical aspects in this method: 1) the reliance upon a method that is closer to juridicial cross-examinations than to standard experimental methodology; 2) the apparent internal contradiction between the idea that the ultimate stage of cognitive development is hypothetico-deductive thinking and the fact that this method is not used in his research by its extoller.

Nevertheless, one must recognize that Piaget used his clinical "method" with a tremendous flair. Those of us who had the privilege to work with him never forget how he could extract from an apparently pointless protocole, *the* significative statement and how reluctantly we would interview a few more kids, convinced as we were that the "boss" (le patron) was wrong, just to find out that he was "damned" right. In sum, Piaget's method was made by him and for him.

9. PIAGET'S THEORY OF INTELLECTUAL PROCESS

Each stage of intellectual development may be described as a set of organized structures of "schemes". For Piaget, a "scheme" is the internal general form of a specific knowing activity, be it sensori-motor or formal. Note that Piaget distinguishes "scheme" from "schema" which conveys a figurative, representational outline, by opposition to a scheme which is related to the operative side of knowledge.

As an individual encounters his world, he assimilates objects and events to his schemes. When this is not possible, because the existing schemes are inadequate, these mental structures modify themselves or accommodate. Assimilation and accommodation form the two poles of adaptation. Adaptation, in turn, is a balanced state of a biological organization within its environment. In behavior, it is the equilibrium between accommodation and assimilation which is reached only in intelligence by opposition to instincts and other cognitive modalities. Hence, Piaget's effort in this very domain.

Unlike Darwin, Marx or Freud who were also interested in the problem of dialectics between change and stability, Piaget had to discover and describe the very rudiments of the course of cognitive growth at the same time as he developed a theory to explain it. Darwin did not have to produce evidence for the existence of species or to construct a taxonomy *ex novo*, nor did Marx have to demonstrate the existence of different social forms and classes. Freud's case is somewhat different, since his stages of psychosexual development was a rather long-range inference about children, based on his work with adult patients and aimed mainly at developing his psychodynamic ideas. So much so that his stage theory can be easily separated from his psychodynamics.

Piaget had to transform each of the Kantian categories from the philosophical status of first principle into a subject of scientific investigation, in order to prove that there was a positive, experimental solution to epistemological dilemmas faced by philosophers of knowledge.

The strategy chosen by Piaget for the solving of this problem was neither the establishment of laws nor the search of elements constitutive of the universe of knowledge. Unlike the successful physical and chemical scientists, Piaget looked for a third strategy: the search for structures.

Piaget's understanding of structures is somewhat different from that of mathematicians, linguists and anthropologists. It is closest to Gestalt-psychologists, because, as he said once, he was "a Gestaltist with, in addition, the activity of the subject" (personal communication); which does not mean too much, because he also claimed to be a "subjective behaviorist". Indeed he was neither.

Piaget's use of structures and operations is at variance with what is usually intended by these concepts. The concept of group is a case in point. For Piaget, "A group is closed circle of operations that return to the point of departure through an operation of the group as a whole" (Piaget 1954 p. 105). Mathematically speaking, the group axioms imply that a sequence of two or more operations can result in a return to the point of departure, but they also imply that there are many more sequences that do not terminate in this way. It is also contrary to mathematical usage to speak of a series of operations as being a group, and still more to speak of two series, involving the same elements, as being two groups. Such a confusion of mathematical groups with the composition of interiorized actions allows Piaget to transform external acts into internal operations. Moreover, this confusion of the global properties of a group with the individual properties of operations pertinent to the group allows the fusion with modern mathematics of the Gestaltist idea that a part of a whole is determined by its relationship to the whole. Finally, the main difficulty of such a treatment of the transformation of object manipulations into mental operations is that there is no explanation for how object handlings see truth-values attached to themselves. It seems rather strange that a man who devoted an entire year to the study of negation did not perceive that implications and negations, bearing on propositions to which no action could ever correspond, could not be translated into a combination of acts, as Binet had shown at the time of his opposition to the Wurzburg School and to French empiricism.

Nevertheless, such a critique should not be taken too far. By its very structure, mathematics is a science for which an operation either exists fully formed or does not exist at all. Genetic epistemology and its main ancillary science, genetic psychology, are genetic and, as such, they displace the emphasis from structures to genesis or development. There is here a strife of systems that should be fully understood.

For structuralism, concepts do not develop; one just leaps from one form of understanding to another. The articulation between these two forms is never specified. In some forms of structuralism, this articulation cannot be specified in principle. For such a pure form of structuralism, it is obvious that either a concept is in operation or it is not. There is no solution of continuity. If the child, the "primitive" or the "patient" do not employ the valid criteria for the given concept, they are not manifesting a more primitive form of the concept, they are simply not manifesting the concept at all, period. From this point of view, all of developmental psychology of cognition would be dissolved and especially Piaget's theory of intelligence.

It should be clearly understood, on the other hand, that this general remark about the strife between structuralism and geneticism does not absolve Piaget of two mental confusions: 1) using logico-mathematical concepts in such a distorted way that their formalization becomes virtually impossible as for the notion of "*groupements*"; 2) trying to unite in one superordinate system structuralism and geneticism, thinking that, by so doing, he could solve the opposition between Gestaltism and behaviorism in psychology; Tarde and Durkheim in sociology; nativism and empiricism in philosophy. There is indeed a world of difference between attempting to solve such contradictions and the legitimate enterprise of a developmental psychology that would try to demonstrate the existence of attenuated forms of a given concept in the child, the "primitive" and the "patient" and would try to show the metamorphoses thereof. Too often, by a recourse of mere analogies taken for structural components, as in *Biology and Knowledge* (Piaget 1971b) Piaget has given his readers the impression of overcoming these contradictions too easily. All of the seven methods presented in the second chapter of this very large book are methods to generate analogies, not to explore their meaning.

This question leads us to the two other components of Piaget's epistemology: 1) the biogenesis of knowledge and 2) his reconsideration of classical epistemological problems.

10. BIOGENESIS OR THE PRELIMINARY ORGANIC CONDITIONS OF KNOWLEDGE

> Cognitive processes seem, then, to be at one and the same time the outcome of organic autoregulation reflecting its essential mechanisms, and the most highly differentiated organs of this regulation at the core of interactions with the environment so much so that, in the case of man, these processes are being extended to the universe itself. (Piaget 1971b, p. 26)

But for Piaget, there are some crucial differences between cognition and organic functioning: the completeness, stability, and flexibility of the structures of intelligence, the progressive dissociation of form and content, characteristic of cognitive organizations.

This convergence between cognition and organic regulations could be explained, according to Piaget, in three different ways: 1) behaviorism; 2) nativism; 3) interactionism.

Piaget's understanding of what he calls "Lamarckian empiricism" is rather different from Lamarck's ideas, since Lamarckism is, here, reduced to the heredity of acquired characters. This aspect of his transformism was secondary for Lamarck and pertinent only to what he called "circumstances". His main idea was that of a metaphysical scale of beings that was intangible.

Even limited to this form, Piaget's rejection of Lamarckism rests upon the absence of a regulatory mechanism between the stimulation of the environment and the responses of the organism. For him, Lamarckism is the biological version of behaviorism and thus falls under the same criticism.

At the opposite end of the spectrum of epistemological positions, Piaget locates nativism which corresponds to Lorenz's ideas at the level of the physiology of behavior and to neo-Darwinism at the more general level. Piaget's opposition to Lorenz is focused on Lorenz's attempt to show that Kantian categories are biologically preformed. If these categories are preformed, Piaget argues, they lose their necessity and their unity; which is unacceptable for logical structures.

In spite of its simplicism, this difference between Lorenz's and Piaget's geneticisms deserves some attention. Too often, rapid readings of Piaget's ideas have led some of his commentators to think that he was a nativist. His strong rejection of Lorenz's ideas shows indeed the contrary. Without arguing the case in detail, any claim for an invariant sequence or a necessary order of emergence must rest either on a *genetic* (nativistic) argument concerning a hierarchy of dispositions (evidence showing that putative later forms in the sequence cannot be manifested in the absence or suppression of earlier forms), or, on evidence of a *logical* order showing that the putative later forms presuppose the "earlier" forms. Clearly, the second alternative is chosen by Piaget.

This criticism of Lorenz is then doubled by a criticism of neo-Darwinism revolving around two key points: 1) random mutations and 2) natural selection. For him, mutations must, somehow, be related to the adaptive needs of the organism. In addition, chance is an improbable candidate to account for the evolution of complex organs and organisms. Natural selection is a process in which the organism is essentially passive. Consequently, Darwinism should be rejected.

Piaget wants a theory in which the initiative remains with the organism: the mutations that occur are part of the total process of organized self-regulation; and an internal process of what Baldwin called "organic selection" plays a major role in determining which variants will be presented to the environment for further natural selection. This is the "phenocopy hypothesis". This hypothesis proposes that there are both exogenous and endogenous variations in phenotypes. Once the exogenous form is established, the organism "re-invents" it by producing endogenously the same phenotypic result. This endogenous or genotypic form is, in some sense, a copy of the phenotypic one; hence "phenocopy". The sort of groping by which an organism senses that something is not working properly is similar to groping in intellectual problem-solving; hence the continuity from life to intelligence. Note that this continuity is based on an entirely speculative biology.

11. CLASSICAL EPISTEMOLOGICAL PROBLEMS RECONSIDERED

Once one accepts that "epistemology is the theory of valid knowledge" (Piaget 1971c) it follows that epistemology is by nature an interdisciplinary subject, since the process of knowing validly raises questions both of fact and of validity. If it were a question of validity alone, epistemology would merge with logic. If it were a question of facts, it would merge with the psychology of cognitive functions. Since it is both, developmental epistemology entails the co-operation (in the sense of Piaget's operations, of course) of psychologists who are studying development as such, of logicians who formalize the stages or states of temporary equilibrium in this development, of mathematicians who would ensure some liaison between logic and the field in question, and of cyberneticians who would connect logic and psychology.

The International Center for Genetic Epistemology in Geneva has been functioning according to this rule since 1955, and has worked on several classical epistemological problems. Some of them will be reviewed critically here.

1. Number and Space: Whitehead and Russell sought to make the ordinals equivalent to classes of classes by a one-to-one correspondence, whereas Poincaré considered that the concept of number relies upon an irreducible intuition of (n + 1). In a certain sense, the two Englishmen tried to reduce number to class, whereas the Frenchman attempted to reduce it to series or iteration. By looking at what happens in the child between the ages of four and seven, Piaget claimed to have observed the construction of three correlated systems of operations: classification, seriation and numeration. The concept of number appears then to be the result of the synthesis of classification and seriation, and genetic epistemology has put an end to the endless debate between seriationists and classificationists in an empirical, scientific manner.

As for the problem of space, the Italian mathematician Enriques attempted to reduce different geometric forms to separate sensory categories, whereas others, in the Kantian tradition, thought of it in terms of *a priori* categories preceeding any experience. Piaget shows how space is the result of the sensorimotor construction of a practical group of displacements analogous to Poincare's group and thus having all the properties of an Abelian group. In addition, he shows why the order of acquisition of different geometries is different in the child and in history. Historically, Euclidian metric preceeded projective geometry and topology, whereas developmentally the reversed order is observed. We have already seen that this last order is more logical in the sense of the number of parameters to be conserved. Developmental psychology discovers the logical (genetic) order of things rather than their historical (accidental) order which is linked, in this case, to the accident of perception that "sees", according to Piaget, Euclidian forms first. It is useless to point out here that perceptual space is not Euclidian. There are many perceptual spaces corresponding to different geometries.

2. *Time and Velocity*: In classical or Newtonian mechanics, time and space are both absolute, corresponding to simple intuitions (Newton's *Sensorium Dei*), while velocity is merely a relationship between them. In relativistic mechanics, by contrast, velocity becomes an absolute and time, like space, is relative to it. Piaget shows that there exists, developmentally, a basic intuition of speed, independent of any idea of duration and resulting from the primal concept of order: the intuition of kinematic "overtaking". Consequently, the very development of the ideas of space, time and velocity shows that "there is nothing inevitable about the intuition of universal and absolute time" (Piaget 1971c, p. 11).

3. *Object Permanence, Identity and Conservation*: The idea of permanence seemed to be both self-evident and necessary at the beginning of this century. Then, it had been cast into doubt by atomic physics. Once again, Piaget shows that his developmental researches give an empirical answer to this question by demonstrating the construction of object permanence in the course of the first year of life. Object permanence is, in addition, the first stage of qualitative identity of objects. Then this concept is transferred to quantities and will lead, in turn, to the various forms of conservation.

The fact that the concept of identity is attained before conservation of quantity seems interesting to Piaget from an epistemological point of view, because an epistemological school of thought, represented by Emile Meyerson, in the past, and by Jerome Bruner presently, holds that conservation derives from the concept of identity. In fact, Piaget's own studies show that there is a qualitative jump between identity and conservation. Identity has only qualitative significance; it is obtained by a simple dissociation of constant qualities from variable qualities. Conservation, on the other hand, presupposes operations of quantifying (compensations between the dimension which increases and the dimension which decreases and so on): the concept of quantity therefore requires an act of construction and is not given by simple perceptual verification as is the concept of quality, according to Piaget. Conservation embeds identity in the wider framework of reversibility, quantitative compensations, and the concepts of number and measure. Thus, for Piaget, developmental psychology is the royal road to the solution of all pending epistemological questions.

As we have already pointed out earlier in this paper, there is a touch of naivete in believing this, and one should question Piaget's sincerity on this central point of the very existence of a positive, empirical epistemology. Two sorts of questions are in order here: logical and methodological.

From a logical standpoint, it is clear that Piaget posits a "telos" (goal) to development such as to run the risk, in his historico-critical reconstruction of the various stages of a given science, of building up his concepts and beliefs in such a way as to comfort his own view, rather than as "immediate to God" and worthy of consideration independently of his own concerns. In my opinion, after having read the forthcoming (posthumous) book by Piaget and Garcia: "Histoire des sciences et psychogenèse", Piaget has not always resisted the temptation to write history in a way pleasing his theory.

From a methodological viewpoint, it is not less clear that sometimes Piaget's evidence is shaky or flimsy. The case in point is the genesis of the various geometries inventoried by Piaget in the child. This inventory rests upon two basic assumptions: 1) children's drawings have been completely analyzed by Luquet in the twenties and 2) drawings are the best means to use in order to assess the attainment of geometrical concepts in children. If we examine carefully these two assumptions, we are forced to recognize that they are a bit abusive. Luquet analyzed children's drawings with a strong prejudice: for him, the goal of children's drawings is academic realism. With that "telos" in mind, he proceeded to show that the stages of drawing develop from fortuitious realism (found in scribblings) to intellectual realism and then visual realism. This sort of analysis might have been possible in the late twenties in some uncultivated circles who wanted desperately to "defend" Western art against the invasion of "Negro Art" as did the Nazis and the Soviets who invented "socialist realism". Now, after the surrealists' revolution, the rupture with academism instituted by Mirò, Picasso, etc. ..., it is no longer possible to assess children's productions in the same way. This does not mean to say that children do not grow out of scribblings into more realistic drawings. It means that the artistic revolution of the beginning of this century has drawn our attention on the possibility that children's drawings are a language that should not be taken at face-value: they might contain some conventions to represent some reality and we might not read them correctly. An example of what is meant here can be given by the stage in which children, incapable of representing perspective according to the rules instituted by the Renaissance, draw very lacy roads to *signify* perspective. This stage is ignored by Luquet who called it "missed realism" demonstrating only that *he* missed the point.

Secondly, drawings are not necessarily the best means to represent the progresses in geometry made by children. If we accept the hypothesis that drawings are a sort of shorthand for children, that is to say a conventional means of communication, we must recognize the possibility for the child to neglect the uses that the adults who misunderstand this shorthand will make of it. In other words, it is not at all evident that children's drawings are the right way to assess geometrical knowledge in children. Other approaches should at least complement this one. Unfortunately, this has not always been the case. It is so much more regretful that this methodological sloppiness combined with bold theoretical conclusions has cast some doubt on the entire *oeuvre* of Piaget, whereas, when compared with what was done in the field by others at the same time, his work appears as so much superior. Think of language and moral development, for instance. At the time when Carmichael's *Handbook of Child Psychology* was happy to account for language by simple word counts at various ages, Piaget presented a complete theory of language. When Hartshorne and May were working on resistance to temptation, Piaget was studying moral judgment. These examples are only two among many possible.

University of Geneva

REFERENCES

Baldwin, J. M. (1902) *Development and Evolution*. London; New York: MacMillan.

Baldwin, J. M. (1906, 1908, 1911) *Thoughts and Things or Genetic Logic*, 3 vols. London: George Allen; New York: MacMillan.

Bosanquet, B. (1888) *On the Morphology of Knowledge*, 2 vols. London; New York: Oxford University Press.

Cassirer, E. (1903-1920) *Das Erkenntnisproblem*, 3 vols. Berlin: B. Cassirer.

Elkind, D. (1976) *Child Development and Education*. Oxford, New York.

Furth, H. G., Wachs, W. (1975) *Thinking Goes to School*. Oxford, New York.

Gruber, H. & Vonèche, J. (1977) *The Essential Piaget*. New York: Basic Books.

Hobhouse, L. T. (1888) *Theory of Knowledge*. London: MacMillan.

Hobhouse, L. T. (1912) *Mind in Evolution*. London: MacMillan.

James, W. (1880) Great Men, Great Thoughts and the Environment, *Atlantic Monthly*, Boston, *46*, 441-459.

James, W. (1890) *Principles of Psychology*, vol. II. New York: Henry Holt.

Piaget, J. (1947) Du rapport des sciences avec la philosophie. *Synthèse*, *6*, Amsterdam.

_________. (1952) Genetic Psychology and Epistemology. *Diogenes*, *1*, 49-63. Originally published in French in 1952 (De la psychologie génétique à l'épistémologie génétique).

_________. (1954) *The Construction of Reality in the Child*. London: Routledge and Kegan Paul. Originally published in French in 1937 (La construction du réel chez l'enfant).

________. (1957) Programme et méthodes de l'épistémologie génétique. In Beth, E. W.; Mays, W.; Piaget, J. (Eds.) *Epistemologie génétique et recherche psychologique*. Etudes d'épistémologie génétique, vol. 1, Paris: Presses Universitaires de France.

________. (1970a) *Genetic Epistemology*. New York; London: Columbia University Press.

________. (1970b) *Science of Education and Psychology of the Child*. New York: Orion. Originally published in French in 1969 (Psychologie et pédagogie).

________. (1971a) *Insights and Illusions of Philosophy*. New York, Cleveland: The World Publishing Co. Originally published in French in 1965 (Sagesse et illusions de la philosophie).

________. (1971b) *Biology and Knowledge. An essay on the relations between organic regulations and cognitive processes*. Chicago: University of Chicago Press. Originally published in French in 1967 (Biologie et connaissance).

________. (1971c) *Psychology and Epistemology*. Harmondsworth: Penguin. Originally published in French in 1957 (Psychologie et épistémologie).

________. (1972) *The Principles of Genetic Epistemology*. New York: Basic Books. Originally published in French in 1970 (L'épistémologie génétique).

Sigwart, C. (1894-1895) *Logic*, 2 vols. New York: MacMillan.

Wundt, W. (1880) *Logik*. Stuttgart: Enke.

PART II

CONCEPTUAL CONSIDERATIONS

ILKKA NIINILUOTO

TRUTHLIKENESS, REALISM, AND PROGRESSIVE THEORY-CHANGE

1. INTRODUCTION: SARTON'S DEFINITION OF SCIENCE

In his influential book *The Study of the History of Science* (1936), George Sarton - the first Professor of History of Science at Harvard University and the long-time editor of *Isis* - formulated a definition of science, dating from his earlier work in 1913:

> *Definition*. Science is systematized positive knowledge, or what has been taken as such at different ages and in different places.
>
> *Theorem*. The acquisition and systematization of positive knowledge are the only human activities which are truly cumulative and progressive.
>
> *Corollary*. The history of science is the only history which can illustrate the progress of mankind. In fact, progress has no definite and unquestionable meaning in other fields than the field of science. (Sarton 1957, p. 5)

In other words, Sarton claims here that scientific progress means the cumulation of systematized positive knowledge. Further, the notion of knowledge is associated by him with the notions of accuracy and truth: science is "search after truth", and

> our merit is in every case proportional to the ratio of the amount of truth which we discover to the amount discoverable. (Sarton, 1957 p. 18)

At first sight, Sarton seems to be a typical representative of the traditional empiricist school which views scientific development as piecemeal accumulation of truths.

J. C. Pitt (ed.), Change and Progress in Modern Science, 235–265.
© 1985 by D. Reidel Publishing Company.

However, his position is in fact more complicated – it contains tensions which also illuminate the reasons why the traditional conception of science was sharply criticized and re-evaluated in the 1960's and 70's. As a result of this recent discussion, it has been questioned whether the notion of progress has any "definite and unquestionable meaning" even in the field of science.[1]

As a historian of science, Sarton was fully aware that "knowledge is not something dead and static, but something fluid, alive, and moving" (Sarton 1957, p. 5.) He makes a contrast between historical knowledge, which "grows slowly and precariously" and tends asymptotically "towards completeness" without ever reaching its goal (Sarton 1957, p. 7), and scientific knowledge the growth of which is "unpredictable, luxuriant, and sometimes explosive in its intensity and destructiveness" (Sarton 1957, p. 8). Indeed, Sarton argues, any branch of natural science "may be completely revolutionized at any time by a discovery necessitating a radically new approach to the subject."[2] While we may "safely assume" that chemistry after five hundred years will be as unlike the present as the chemistry five hundred years ago, similar revolutions in the methods, scope, fundamental notions, and contents are not to be expected in "literary subjects", such as Latin grammar, Greek literature, and the history of the eighteenth century (Sarton 1957, p. 8).[3]

Sarton's conception of science thus contains three basic theses:

(1) Science, as systematic pursuit of knowledge, is *progressive*.
(2) The growth of science is *cumulative*.
(3) The development of science involves conceptual, theoretical, and methodological *revolutions*.

But how is it possible to combine the theses (1), (2), and (3)? If the progress of science is guaranteed by its cumulative nature, what does progress mean in connection with revolutions? How could science be cumulative and contain destructive explosions at the same time?

Many models of scientific change can be viewed as attempts to solve the conflict between theses (1), (2), and (3). In most caes, this is done by accepting two of them and by rejecting or reinterpreting the remaining third one. For example, naive empiricists and rationalists reject (3);

Feyerabend's (1975) thesis of permanent revolutions leads him to reject (1); Kuhn's (1962) compromise between cumulative normal science and revolutionary extraordinary science leads him to question (1); Popper (1975) tries to combine all the three theses by giving an essentially cumulative explication for the notion of verisimilitude.[4]

In this paper, I am primarily interested in the contrast between 'realist' and 'instrumentalist' views about scientific progress (cf. Niiniluoto 1980). The representatives of the various types of *scientific realism* accept thesis (1) in the strong form which takes science to be progressive as a *cognitive* (i.e., knowledge-seeking) enterprise. But when scientific theories are interpreted realistically, most revolutions in science have to be understood as changes where one false theory is replaced by another false theory. Therefore, critical realists - who reject the accumulation model and accept theses (1) and (3) - are bound to reject thesis (2) and replace it with a claim that the 'cognitive virtues' of scientific theories increase or may increase within the development of science.[5] The 'epistemic utility' which measures such virtues is usually taken to be *truthlikeness* (verisimilitude), essentially a combination of truth and information, so that scientific progress is viewed as 'approach to the truth' or towards better and more comprehensive representations of reality (e.g., Peirce, Engels, Lenin, Popper, Putnam, Nowak).[6]

For the scientific realists, theories are primarily attempts to give informative and lawlike descriptions of (some aspect of) reality. For the *instrumentalists*, theories are conceptual tools for performing certain tasks (prediction, control, systematization, problem-solving, etc.), so that scientific progress is definable by the achievements of theories relative to these tasks. Most instrumentalists are prepared to accept thesis (3), since there are revolutionary changes on the level of scientific theories. But since they refuse to give a realistic interpretation to theoretical terms, they are able to combine this view with modified theses (1) and (2): science is progressive and (possibly) cumulative on the level of the achievements of theories. A classical expression of this position was given by Duhem in 1908 (cf. Duhem 1954, pp. 38-39).[7] The characterizations of scientific progress in terms of problem-solving (cf. Kuhn 1962; Laudan 1977) or effectivity in prediction and control (cf. Rescher 1980) are variations of this Duhemian theme. This is the case in spite of the fact that Laudan does not subscribe to the instrumentalist interpretation of theories (see Laudan 1981c): he regards science as a problem-solving rather than a truth-seeking activity,[8] denies that science is cumulative as a procedure for problem-solving (cf. Laudan, 1977, pp. 149-150), but suggests that progress can be defined by the weighted number of the empirical problems solved by a theory minus the weighted number of the anomalies and the conceptual problems associated with it.

2. REMARKS ON LAUDAN'S PROBLEM-SOLVING MODEL

In a recent paper on scientific change, Larry Laudan makes the following statement:

> Given the notorious difficulties with notions of 'approximate truth' - at both the semantic and epistemic levels - it is implausible that characterizations of scientific progress which view evolution towards greater truthlikeness as the central aim of science will allow one to represent science as a rational activity. (Laudan 1981b, pp. 144-145)

Laudan is certainly justified in his criticism of those philosophers who have made their case for realism without adequate semantical and epistemological tools. But, while I agree that the notion of 'approximate truth' is difficult, I also happen to think that a satisfactory theory of degrees of truthlikeness and their estimation is already available. Before I defend this claim in the next sections, some comments on Laudan's own problem-solving model of scientific change are given in this section.

In my view, the recent debates about the nature of 'problems' their 'solutions' (Cf. Koertge 1981; Mellor 1981; Laudan 1981c; Sarkar 1981; Finocciaro 1981; Leplin 1981; Laudan 1982) indicate that at present the theory of truthlikeness is in fact better developed and understood than the problem-solving account of science. However, erotetic logic would give a valuable tool for analyzing the relevant notions – such as various kinds of questions, the criteria for a reply being a potential answer to a question, the distinction between potential and correct answers, the distinction between complete and partial answers, etc. Most of the work in this direction still remains to be done.[9] As most scientific questions are either truth-using or truth-seeking or both,[10] I am also convinced that the problem-solving account of scientific change ultimately needs – rather than makes unnecessary – the concepts of truth and truthlikeness.

Following Mellor and Cutting, Sarkar (1981) has argued that inconsistent theories create trouble for Laudan's model, since they entail every statement and, hence, are bound to have at least as high problem-solving ability as any other rival theory. Laudan (1982) correctly points out that Sarkar ignores here the fact that inconsistency is a serious conceptual problem which strongly reduces the problem-solving ability of a theory. But Laudan (1981c) and Finocciaro (1981) also claim that an inconsistent theory will "always be confronted by as many anomalies as solved problems". This is hard to combine with Laudan's conception of anomaly: according to him, a problem is anomalous for a theory T if it is not solved by T but is solved by another theory T' (Laudan 1977, p. 17). If problem-solving means that a "statement of the problem" is "deduced" from a theory (Laudan 1977, p. 25), then an inconsistent theory solves all problems. This observation has the following consequence: as soon as an inconsistent theory is introduced, there are no unsolved empirical problems any more, so that all the problems that a theory T does not solve are after all anomalies for it. Hence, Laudan's central distinction between unsolved and anomalous problems in the appraisal of theories disappears.

Laudan's other objection to inconsistent theories suggests a different view about problems. He says that the "*raison d'être* of science" is to put us "in a position to anticipate nature" – and this is precisely where inconsistent theories fail (Laudan 1982, p. 80). In this case, a problem is not represented by a statement which is believed to be true, but rather by a question 'Is P true or not?.' This predictive problem is solved by a theory T which either entails P or entails $\sim$P but not both. This account of problems would have the desirable consequence that there are problems unsolved by an inconsistent theory. However, it is unfortunately incompatible with Laudan's (1981c) claim that "to speak of those problems whose solutions can be credited to a theory is to refer to those *already observed* states of affairs which the theory entails." A further difficulty for Laudan is the following: if a theory gets credit for its predictions only after their correctness has been observed, then the aim of science is to make *true* predictions[11] – which would make science a truth-seeking enterprise. If we disregard issues about the truth value of predictions, then non-scientific predictive devices (horoscopes, chrystal balls, cards, etc.) could have a very high problem-solving ability.[12]

Laudan's restriction of problems to "already observed states of affairs" can be motivated by a wish to avoid such paradoxes which Grunbaum presented against Popper (cf. Sarkar 1981). Laudan also emphasizes that, in contrast to the always infinite class of the deductive consequences of a theory, the class of solved problems is always finite. This claim is hard to evaluate, however, until the individuation of problems is analyzed in more detail. For example, Newton's theory is able to solve the problem concerning the orbit of Mars around the sun. It follows that it can solve, precisely in the same way, the corresponding problem for any planet (real or imagined) relative to any other (real or imagined) star. Do we say that Newton's theory gets credit only for those planets whose orbits actually have been presented as problems? Or do we say that Newton's theory solves an infinite number of problems belonging to the same problem type? If we count this problem type as *one* solved problem, then it does not make sense to associate this problem to any "already observed state of affairs".

Similar troubles can be raised for problems like 'What is the position of Mars within interval (O,t)?.' If an answer is given as a function of time t, then the theory has given a (correct) description of the position of Mars in an *infinite* number of points. Did it solve one problem or an infinite number of problems? I don't think that these comments raise any unsolvable problems for the problem-solving approach. However, I don't see any reason to expect that it is easier to define a measure for the problem-solving ability of a theory than to succeed in the notoriously hard task of defining the degree of confirmation of a theory within inductive logic.

3. TRUTHLIKENESS AND ITS ESTIMATION

In this section, I shall outline a general theory of truthlikeness. In the next section, I develop some of its consequences for a realist view of scientific progress.

Let S be a set of alternative *states of nature*. Each element s in S represents a possible state in which the world (or rather some fragment of the world) may be. Let

L be a language which is sufficiently rich for describing the states s in S.[13] The statement in L which claims that the state s obtains in the world is denoted by the same symbol 's'. Thus,

(1)
$$\vdash \bigvee_{\underline{s} \in S} \underline{s}$$

$$\vdash \sim (\underline{s} \,\&\, \underline{s}') \text{ for all } \underline{s}, \underline{s}' \in S, \underline{s} \neq \underline{s}'.$$

Condition (1) guarantees that there is one and only one element of S which is *true* (in the ordinary correspondence-theoretical sense); let us denote it by s_*. In other words, statement s_* expresses what the state of the world actually is.

Set S defines a *cognitive problem* of the following form: Which of the states s in S is the true state s_*? The strongest *potential answers* to this proplem are just the elements s of S. All the other relevant answers can be expressed as disjunctions of elements from S. Let us define

(2)
$$D(S) = \left\{ \bigvee_{\underline{s} \in I} \underline{s} \;\middle|\; I \subseteq S \right\}$$

Hence, a statement h of L belongs to D(S) if and only if

(3)
$$\vdash h \equiv \bigvee_{\underline{s} \in I_h} \underline{s} \quad \text{for some } I_h \subseteq S.$$

The representation (3) is called the S-normal form of h. The elements h of D(S) are potential *partial answers* to the cognitive problem associated with S. A partial answer h is true if and only if $s_* \in I_h$, i.e., s_* belongs to its normal form. The information content of a partial answer h depends on the width $|I_h|$ of its normal form. If all elements s of S are regarded as equally informative, the *content* of h may be defined by

(4)
$$B(h) = \frac{1 - |I_h|/|S|}{1 - 1/|S|}$$

so that $B(s)=1$ for each $s \in S$ and $B(h)=0$ for a tautology h. Otherwise the content of h may be measured by the standard cont-measure of the semantic information theory (cf. e.g. Niiniluoto and Tuomela, 1973).

As examples of cognitive problems the following cases may be mentioned:

> (a) S consists of the *state descriptions* of a first-order language,
> (b) S consists of the *structure descriptions* of a first-order language,
> (c) S consists of the *constituents* of a first-order language,
> (d) S consists of the *constituents* of a higher-order language,
> (e) S consists of the *basic causal laws* of a language with the operation of nomic necessity,
> (f) S consists of *points* in a real-valued interval $[a,b] \subseteq \mathbb{R}$ (or in a region $C \subseteq \mathbb{R}^n$),
> (g) S consists of all real-valued *polynomials* of degree n ($n \in N$),
> (h) S consists of a class of continuous *functions* of the type $f: \mathbb{R}^n \rightarrow \mathbb{R}$
> (i) S consists of a class of *probability measures* on a given algebra of sets.

Of these situations, (a) is relevant for cases where we are interested in singular sentences, (b) for numerical statements about class occupancy, (c) and (d) for qualitative generalizations and theories,[14] (e) for nomic generalizations and theories,[15] (f) for point and interval estimation, (g) and (h) for quantitative laws and theories, and (i) for probabilistic laws.[16] It seems to me that all the interesting scientific problems can in principle be reduced to one of the cases (a)-(i).

Many scientific problems are *open* in the sense that we are not able to enumerate all of their potential answers. For example, usually we cannot give all the alternative theories that may be used in answering an explanatory question of the form 'Why p?'. However, the basic method of approaching open questions is to transform them to *closed* questions with a known set of potential answers, i.e., to cognitive problems in our sense above. (See Niiniluoto, 1976.) In particular, a question can be 'closed' by fixing the language L in which its answers are supposed to be formulated. If the formulation of the problem turns out to be insufficient or inadequate, we have to proceed to consider *deeper* cognitive problems by extending the language L with new vocabulary. It seems to me that, in this sense, all the interesting scientific problems can in principle be reduced to cognitive problems of the types (a) - (i).

The next move in the theory of truthlikeness is to introduce a metric d in the set S. If S does not already have a natural topological structure, this can be done by utilizing the syntactic structure of the elements s of S. I shall illustrate the basic idea only in cases (c), (f), and (h).

Case (c). If L is a monadic first-order language with a finite vocabulary, its constituents s are statements for the form

(5) $$\bigwedge_{j \in CT_i} (\exists x)\, Ct_j(x) \;\&\; (x)\Big[\bigvee_{j \in CT_i} Ct_j(x)\Big]$$

where $Ct_j(x)$ are Carnapian Q-predicates. This means that the predicates $Ct_j(x)$, $j=1,\ldots,K$, constitute a complete classification system with K mutually exclusive cells. Constituent (5) claims that cells $Ct_j \in CT_i$ are non-empty and other cells are empty. The simplest measure for the distance $d(s_i, s_k)$ between two constituents s_i and s_k of form (5) is due to W. K. Clifford (1877):

(6) $$d(\underline{s}_i, \underline{s}_k) = |CT_i \,\Delta\, CT_k| / K.$$

i.e., the normalized cardinality of the symmetric difference of sets CT_i and CT_k. Definition (6) can be modified by taking into account the distances between Q-predicates. These ideas can also be generalized to arbitrary first-order languages with polyadic predicates.[17]

Case (f). The distance between two points $x, y \in \mathbb{R}$ is the absolute value of their difference or the square of their difference:

(7) $$d(x,y) = |x - y|$$

$$d(x,y) = (x-y)^2.$$

If $x, y \in \mathbb{R}^n$, where $x = (x_1, \ldots, x_n)$ and $y = (y_1, \ldots, y_n)$, their distance may be defined by

(8)

$$d^k(x, y) = \left(\sum_{i=1}^{n} |x_i - y_i|^k\right)^{\frac{1}{k}}$$

Here the choice k=1 gives the *city-block distance* and k=2 the *Euclidean distance* between x and y.

Case (h). The distance between two n-place functions $f: C \rightarrow \mathbb{R}$ and $g: C \rightarrow \mathbb{R}$, where $C \equiv \mathbb{R}^n$, can be defined by

(9)
$$d^k(f, g) = \left(c\int |f(x) - g(x)|^k dx\right)^{\frac{1}{k}} \quad (K = 1, 2)$$

or by

(10)
$$d(f, g) = \sup_{x \in C} |f(x) - g(x)|.$$

When a metric $d: S \rightarrow \mathbb{R}$ has been introduced to set S, we may say that the value $d(s, s_*)$ expresses the *distance* of s *from the truth* s_*. The *degree of truthlikeness* of s (relative to s_*) can then be defined by one of the following functions:

(11)
$$M(S, S_*) = 1 - d(\underline{S}, S_*)$$
$$M(S, S_*) = 1/d(S, S_*).$$

The next step is to extend the function $d: S \rightarrow \mathbb{R}$ to a function $d: D(S) \rightarrow \mathbb{R}$. Let h D(S). The simplest strategy is to define the distance of h from s by some function G of the distances d(s',s) for states s' in the normal form of h:

(12)
$$d(h,\underline{s}) = G(\{d(\underline{s}',\underline{s}) \mid \underline{s}' \in I_n\}).$$

Then $d(h,s_*)$ is the distance of h from the truth, and the degree $M(h,s_*)$ of truthlikeness of h is defined as in (11). Further, h is *closer to the truth* than h' if and only if $M(h,s_*)>M(h',s_*)$, and h is *approximately true* if its degree of truthlikeness is sufficiently high.

The form of function G in (12) is a matter of dispute. One possibility is to choose G so that the degree of truthlikeness of h is given by

(13)
$$M(h,\underline{s}_*) = \gamma M_{min}(h,s_*) + (1-\gamma) B(h) M_{max}(h,s_*)$$

where 0< <1, B(h) is defined by (4), and

$$M_{min}(h,s_*) = 1 - \min_{\underline{S} \in I_h} d(S,S_*)$$

$$M_{max}(h,s_*) = 1 - \max_{\underline{S} \in I_h} d(S,S_*)$$

(cf. Niiniluoto 1978). Hence, $M(s,s_*)=1-d(s,s_*)$ for $s \in S$, and $M(h,s_*)=$ for a tautology h. Thus, s is closer to the truth than a tautology if and only if $d(s,s_*)<1-\gamma$.

In the cognitive problem associated with S the true state s_* is unknown. Let e be a statement of the available (observational and theoretical) *evidence* which is relevant to the task of finding s . Assume that a probability measure P, given e, is defined on set S. In other words, P(s/e) is the probability given e that s is the true state s . (If S is an infinite subset of $\mathbb{R}^n$, P is a n-dimensional probability density on S.) Here P is an epistemic probability measure, and the values P(s/e) are obtained by Bayesian conditionalization from prior probabilities, specified by some system of inductive logic or simply expressing the initial epistemic state of the investigator. To *estimate* the degree of truthlikeness of a hypothesis h in D(S) we calculate the expected value of M(h,s), where the possible values of s are weighted with the probabilities P(s/e):

(14) $$\text{ver}(h/e) = \sum_{\underline{s} \in S} P(s/e)\, M(h,s).$$

If $S \subseteq \mathbb{R}^n$, (14) is replaced by the integral

$$\int_S P(\underline{s}/e)\, M(h,s)\, d\underline{s}.$$

By (4), (13), and (14), ver(h/e)=γ for a tautology h. If $e \vdash s_0$ for some s_0 in S, then (14) reduces to $\text{ver}(h/e)=M(h,s_0)$. Similarly, if $P(s_0/e) \to 1$ for some s_0 in S, when the size of the evidence e increases, then

$$\text{ver}(h/e) \to M(h,\underline{s}_0).$$

In the same situation,

(15) $$\text{ver}(h/e) \to 1 \text{ iff } \vdash h \equiv s_0$$ [18]

It follows that

(16) $$\text{ver}(h/e) \to M(h,s_*) \text{ iff } s_0 = s_*.$$

For problems of type (c), the condition $s_0=s_*$ holds if and only if evidence e is both true (i.e., does not eliminate s_*)

and non-misleading (i.e., exhibits all the qualitative variety relative to L there is in the world).[19] More generally, if $P(s_i/e) \to p_i$ for some $i=1,\ldots,m$, then

$$\text{(17)} \qquad ver\,(h/e) \to \sum_{i=1}^{m} P_i\, M(h,s_i).$$

In this case, the disjunction $s_1 \vee \ldots \vee s_m$ has a relative high expected verisimilitude.[20]

4. PROGRESS AS INCREASE OF TRUTHLIKENESS

Assume now that h_1 and h_2 are two theories in D(S). Then the step h_1/h_2 (i.e., the step from h_1 to h_2) is *progressive* if and only if $M(h_1,s_*)<M(h_2,s_*)$. Definition (12) entails then the following theorem (cf. Niiniluoto 1979b, p. 251).

*Thm.*1. (i) If $h \neq s$, then h/s_* is progressive.
(ii) If h is false, then $h/(h \vee s\ )$ is progressive.
(iii) If h and h are true and if h_2/h_1 and $h_1 \vdash h_2$, then h /h is progressive.
(iv) If h_1 and h_2 are both false and $h_2 \vdash h_1$, then h_1/h_2 may be either progressive or non-progressive.
(v) If h_1 is true and h_2 is false, then h_1/h_2 may be progressive.

Here (i) says that the step to s_* from any other theory h in D(S) is progressive. According to (ii), the step from a false theory h to the strongest true theory entailed by h is progressive. Among true theories, but not among false theories, increase in logical strength entails progress (by (iii) and (iv)). Finally, (v) follows from the fact that sufficiently informative false theories may have a higher degree of truthlikeness than a less informative true theory.

Let us say that step h_1/h_2 *seems progressive on evidence* e if and only if $ver(h_1/e)<ver(h_2/e)$. Then, corresponding to Theorem 1, we have:

*Thm.*2. Assume that $P(s_0/e)$ is sufficiently close to one.
(i) If $h \neq s_0$, then h/s_0 seems progressive on e.

(ii) If $s_0 \notin I_n$, then $h/(h \vee s_0)$ seems progressive on e.

(iii) If $s_0 \in I_{n_1}$ and $s_0 \in I_{n_2}$ and if $h_2 \vdash h_1$ and $h_1 \nvdash h_2$, then h_1/h_2 seems progressive on e.

(iv) If $s_0 \in I_{n_1}$ and $s_0 \notin I_{n_2}$, then h_1/h_2 may seem progressive on e.

As a corollary from Thm.2(iii), we obtain the following result:

Thm.3. Assume that $P(s_0/e\&h_1)$ is sufficiently close to one. If $s_0 \in I_{n_1}$ and $s_0 \in I_{n_2}$ and if $h_2 \vdash h_1$ and $h_1 \nvdash h_2$, then h_1/h_2 seems progressive on $e\&h_1$.

Thm. 3 shows that, in the case where we have accepted a theory h_1 in D(S) as part of the evidence, a step from h_1 to a logically stronger theory h_2 may seem progressive on the same evidential situation. In some cases, such a theory h_2 may be even incompatible with h_1 (see Niiniluoto 1979b, p. 253). On the other hand, if we have accepted an element s_0 of S as our evidence, then no step from s_0 to any other theory h in D(S) seems progressive on s , since:

$$\text{ver}\ (h/\underline{S}_0) = M\ (h\ ,\underline{S}_0) \leq 1 = M(\underline{S}_0,\underline{S}_0) = \text{ver}\ (S_0,S_0).$$

The only way of making further progress is by *conceptual enrichment*, i.e., by introducing new descriptive vocabulary to the language – and thereby replacing the original problem S with a deeper problem S' such that s D(S') for each s S (cf. Niiniluoto 1979b, p. 254).

If we have accepted a theory as evidence, it is possible to make *retrospective* comparative judgments of verisimilitude concerning historically earlier theories. This idea is illustrated by the following examples.

Example 1. When a light ray passes from one medium to a denser medium, it bends towards the perpendicular drawn at the point of incidence. Let α be the angle of incidence and β the angle of refraction (see Fig. 1). Since Ptolemy's *Optica*, it was known that, for two given media, β depends in some regular way on α. The problem of refraction was discussed by many medieval scientists (Alhazen, Grosseteste, Roger Bacon, Witelo). Ptolemy's data was sometimes taken to suggest that α/β = constant. For light passing from air to water, we have roughly:

(18) $$\beta = \frac{2}{3} d$$

In fact, a more accurate formula which fits Ptolemy's data is:

(19) $$\beta = \frac{33}{45} \alpha - \frac{1}{400} \alpha^2$$

(cf. Cohen and Drabkin 1948, pp. 27[illegible]-281). Grosseteste suggested that the refracted ray divides into two equal parts the angle between the perpendicular and the imagined straight continuation of the incident ray. His law of refraction thus takes the form

(20) $$\beta = \frac{1}{2} \alpha$$

The sine law of Snell (1620) has later been accepted as giving the correct relation between α and β. In Newton's *Opticks*, it is presented as Axiom V: "The Sine of Incidence is either accurately or very nearly in a given Ratio to the Sine of Refraction." In particular, for the case of air and water Newton gives the formula

$$\frac{\sin \alpha}{\sin \beta} = \frac{4}{3}$$

which can be written in the form:

(21) $$\beta = \arcsin \left(\frac{3}{4} \sin \alpha \right)$$

If we have accepted law (21) as true, then the metrics (9) and (10) give the result that the false laws (20), (18), and (19) - in this order - are increasingly better approximations to the truth.

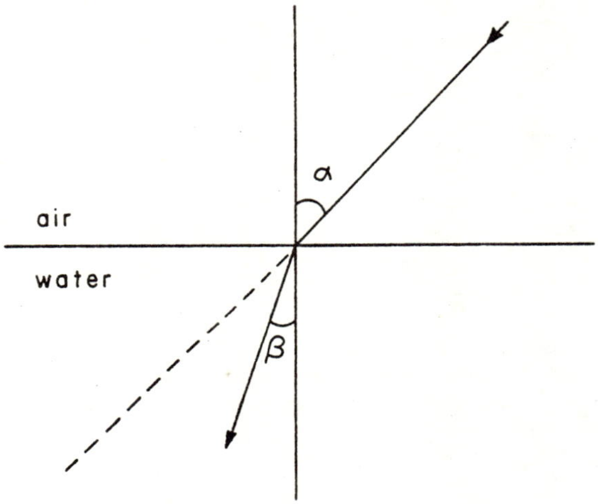

Example 2. According to van der Waals's law,

$$(p + \frac{a}{V^2})\ (V-b) = RT,$$

where p is the pressure of a gas, V is its volume, T its absolute temperature, R is the gas constant, a expresses the intermolecular attractive forces, and b the finite size of the gas molecules. When this equation is solved with respect to p, we obtain

$$(22) \quad p = \frac{RT}{V-b} - \frac{a}{V^2}$$

If we assume that (22) is true, then the following claims can be seen to be increasingly truthlike attempts at formulating the gas law:

(23) $P \; {}^{1}/_{V}$ (Boyle)

(24) $P = RT/V$ (Boyle-Mariotte)

(25) $P = \frac{RT}{V} - \frac{a}{V2}$

Here the "ideal gas law" (24) is in the relation of *correspondence* to the laws (25) and (22); similarly (25) is in the relation of correspondence to (22). Namely, from equation (22) we may deduce the following *idealizational laws*:

(26) If $b=0$, then $p = \frac{RT}{V} - \frac{a}{V2}$.

(27) If $a=0$ and $b=0$, then $p = RT/V$

It can also be seen from (22) that, when V is sufficiently large in relation to a and b, the law (24) is approximately true (in the sense of Niiniluoto 1982b).

We have assumed above that the rival theories h_1 and h_2 belong to the same set D(S), so that they can be expressed as disjunctive statements in the same language L associated with S. This assumption may seem problematic in cases where h_1 and h_2 are *incommensurable* theories which have been formulated in different conceptual frameworks. To see how such situations can be handled, let us note first that the comparison of the truthlikeness of two theories h_1 and h_2 is interesting only if h_1 and h_2 are *rival* theories. It may be that statements of the form 'Newton's celestial mechanics is more truthlike than Darwin's theory of evolution' could be shown to be meaningful, but I doubt that. For all purposes that I can imagine, it is sufficient to define the relation

h_1 is more truthlike than h_2

for theories which are rival solutions to a cognitive problem. But this condition means that there has to be a set S, and a language L associated with S, such that h_1 and h_2 belong to D(S). If h was originally a potential solution to problem S_i (i=1,2), then the common problem S for h_1 and h_2 will be deeper than S_1 and S_2, and the language L associated with S will be a common extension of the languages L_1 and L_2 (for S_1 and S_2, respectively). Further, if languages L_1 and L_2 are 'incommensurable' in the sense that they presuppose different, or even conflicting, principles as 'analytic truths' or 'meaning postulates' (say, MP and MP , respectively), then in language L the relevant comparison has to be made between h_1&MP_1 and h_2&MP_2 (cf. Niiniluoto 1980, pp. 445-446).

To define scientific progress in terms of increasing truthlikeness amounts to the claim that science is primarily a cognitive enterprise which aims at finding highly truthlike information about reality. This definition gives us a methodological norm for advancing science: prefer those theories which can be claimed to be more truthlike than their rivals. It does not entail that all the steps in the actual historical development of science have been and will be progressive.[21] It is also compatible with different views about the end of scientific progress: some realists think that a 'final theory' will eventually be reached, while some regard such a final theory only as an ideal towards which we shall indefinitely approach without ever reaching it.[22]

Some realists have tried to prove that science approaches to the truth by a sort of natural necessity.[23] The strongest results in this direction show that, on certain conditions, our empirical procedures lead *with probability one* to approximately true results (cf. Niiniluoto 1980). This is sufficient to confute cognitive skepticism which views truth and truthlikeness "utopian" as the aims of science (cf. Laudan 1981a,c). On the other hand, these results do not guarantee that all truths are accessible to human beings and will be revealed within the history of science. Whether this is the case or not cannot be settled *a priori* by philosophers.[24] In particular, there are no convincing proofs that the predicates 'true' and 'limit of scientific inquiry' when applied to scientific theories, are coextensive.[25] We may therefore claim that truth cannot be defined by epistemic terms, just as the realist correspondence theory of truth assumes, but at the same time note that this position is compatible with the existence of many kinds of *factual dependencies between truth and our knowledge-seeking* procedures. In other words, while truth is non-epistemic, our scientific procedures may have the property that they yield results which approach to the truth at least in the long run.[26]

5. APPROXIMATE TRUTH AND SUCCESS

In his recent criticism of "convergent realism", Laudan includes among "urgent items on the realists' agenda" the following two tasks: first, to articulate "a coherent account of approximate truth which entails that approximately true theories will, across the range where we can test them, be successful predictors"; secondly, to give a "criterion that would *epistemically* warrant the ascription of approximate truth to a theory" (Laudan 1981a, p. 32). It seems to me that the account of truthlikeness, outlined in Section 3, gives a satisfactory answer to Laudan's challenge.

Laudan's basic strategy is to point out difficulties with two principles:

(T1) If a theory is approximately true, then it will be empirically successful.

(T2) If a theory is empirically successful, then it is probably approximately true.

He assumes that a realist accepts, besides (T1) and (T2), the view that a theory is not approximately true if its central theoretical terms fail to refer (Laudan 1981a, p. 33), i.e.,

(T3) If a theory is approximately true, then it is genuinely referential.

Then Laudan argues that there is a "plethora of theories" (such as the aether theory) which were empirically successful and "by present lights" non-referential (Laudan 1981a, p. 33). By (T3), such theories are not approximately true; hence, they violate (T2) and block the "upward path" from empirical success to approximate truth. Against the "downward path" from approximate truth to empirical success Laudan does not present any counterexamples. He notes that there are genuinely referential theories which were not empirically successful (such as 19th century atomism) (Laudan 1981a, p. 24). Further, he argues that, while the principle (T1') is "self-evident" (Laudan 1981a, p. 30), the realists have so far failed to give a notion of approximate truth which would justify the principle (T1).

To answer Laudan's criticism, let us first note that our definition of function ver gives us an epistemic criterion for the ascription of degrees of truthlikeness and approximate truth to a theory - and thus solves the problems concerning the 'upward path'. As all our knowledge about matters of fact is fallible, this criterion is not infallible. Thus, it is possible that ver(h/e) is high but ver(h/e), where e' is new evidence, is not high, i.e., it may be rational to claim on evidence e that h is approximately true but not rational any more on evidence e & e'. There is nothing strange in J. C. Maxwell's remark that the aether *was* "better confirmed than any other theoretical entity in natural philosophy" (Laudan 1981a, p. 27). By our present lights, the status of the aether may quite well seem different than in the nineteenth century. To ask for an infallible criterion which would prove the approximate truth of a scientific theory is to demand too much of the realists.

But – the critic of realism may continue – if a theory may be empirically successful without really being approximately true, how could the approximate truth of a theory *explain* its empirical success? And how could one claim that the success of science would be a "miracle" (cf. Putnam 1975) unless its theories were approximately true? These questions can be answered only by analyzing in more detail the notion of 'empirical success'.

It is easy to see that condition (T1), as formulated above, cannot be generally valid. The same holds even of condition (T1') – which Laudan finds "self-evident". Assume that h is a theory which is formulated in a purely theoretical language L_T. Then, depending on the factual structure of the world relative to L, theory h may be highly truthlike or true even though it does not have any empirical consequences! Assume then that h is formulated in a language L which contains an empirical sublanguage L_0. Then the following holds:

(T4) If h is true, then all its deductive consequences in L are true.

On the other hand, the following principle is not valid:

(T4) If h is true, then all its deductive consequences in L_0 are true.

On the other hand, the following principle is not valid:

(T5) If h is true, then all its deductive consequences in L_0 are highly truthlike,

since some of the true consequences of h are uninformative. For the same reason, the following principle fails:

(T6) If h is highly truthlike, then all its deductive consequences in L are highly truthlike.

The failure of (T5) and (T6) does not block the 'downward path' from approximate truth to empirical success, however. It only shows that this success typically concerns *the most informative empirical predictions* derivable from a theory: it need not be the case that most or many of these predictions are true, since in fact they all may be false, but nevertheless many of them cannot be at a great distance from the truth. Typical illustrations of this situation are the following (cf Niiniluoto 1982a). If a quantitative law $y=f(x)$ is approximately true within range $C \subseteq \mathbb{R}$, given the metric (10), then *all* of its predictions of the form $y_a=f(x_a)$, when $x_a \in C$, are close to the true value $f_*(x_a)$. (Here y_a and x_a are the values of quantities y and x for an object a.) If a law $y=g(x)$ is approximately true within $C \subseteq \mathbb{R}$, given the metric (9), then its predictions of the form $y_a=g(x_a)$ will *in the average* be close to the true values $f(x_a)$, for objects a such that $x_a \in C$. (see Fig. 3.)

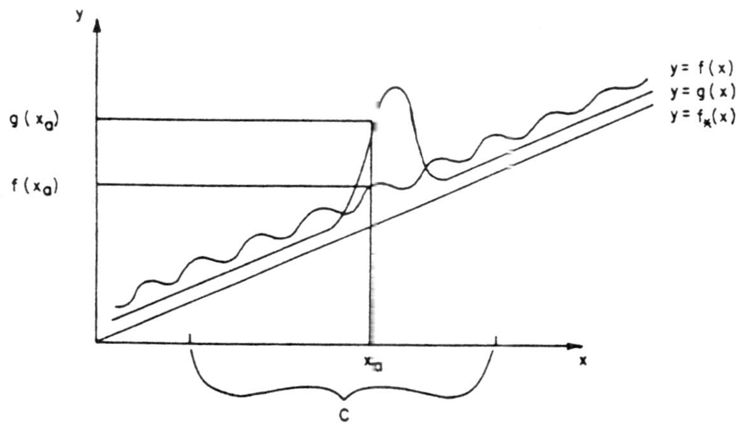

These observations can be extended to theories which contain theoretical terms. Laudan demands the realists to give an argument to the effect that "greater correspondence at the level of unobservable claims is more likely than not to reveal itself in greater accuracy at the experimental level" (Laudan 1981a, p. 35), but again this is too strong a requirement. If h is a theory in language $L=L_T \cup L_O$, and h_T is the purely theoretical part of h in L_T, then the question whether an increase in the truthlikeness of h_T has any effects upon the empirical consequences of h in L_O depends essentially upon the connections between the theoretical and empirical terms in h. If these connections are completely wrong, it does not 'help' theory h even if its theoretical part h_T would be maximally truthlike in L_T. However, in this case the theory h would not be highly truthlike in the whole language L, since its empirical part would be far from the truth. Similarly, if h in L does not have empirical consequences, it is empirically uninformative - and cannot be highly truthlike in the whole language L. It follows that a highly truthlike theory h in $L=L_T \cup L_O$ has to be at least to some extent empirically successful.

A similar remark can be made about purely observational 'theories': they cannot be highly truthlike in the whole language L. For example, Hesse thinks that the Bayesian theory is partial "toward empirical rather than theoretical content" (Hesse 1981, p. 17). It is true that an informative theory h in L is always less *probable* than its observational consequences e in L_O, but the estimated *truthlikeness* of h may be larger than that of e.

Laudan (1981a) presents also a number of objections to the view that later theories should retain appropriate portions of the "laws and mechanisms" of earlier theories. I don't want to make any historical claims about the frequency in which this 'retentionist' strategy is in fact followed by scientists. It is clear that this strategy is not necessary for scientific progress. However, our function ver helps us to understand why it is often very reasonable. Krajewski claims that the great advantage of the Correspondence Principle is the following: if the new theory contains the old theory as a counterfactual special case, it

> has its empirical support not only in the results of new, specific ... experiments but also in the results of the whole experience confirming the old theory. (Krajewski 1977, p. 1)

It is difficult to justify this idea on the basis of the traditional concept of confirmation. Assume that the new theory h_2 contains a special case h_0 which is close to the old theory h_1 but inconsistent with h_1. Then if evidence e highly confirms h_1, it cannot highly confirm h . If we claim that e confirms h_0 (and thereby h_2), it has to be admitted that e did not confirm h in the first place. The notion of estimated verisimilitude behaves in a nicer way in this situation, since it generally satisfies the following principle:

(28) If $ver(h_1/e)$ is high and h_1 is close to h_0, then $ver(h_0/e)$ is high.[27]

And if h_2 entails h_0, which seems highly truthlike on evidence e, there is a good chance that h_2 itself is highly truthlike.

8. CONCLUDING REMARKS

Dilworth (1981) has recently presented an account of scientific progress which starts from Kuhn's idea of theory change as analogous to a Gestalt switch. He ends up with a definition which can readily be understood on the realist basis - even though Dilworth himself mistakenly thinks that his view is in sharp contradiction with "empiricism" and "Popperianism". He assumes that there is something common to which rival theories refer, and compares theories with respect to their accuracy, simplicity, and scope. It is obvious that our definition of progress in terms of the function ver takes into account such factors as accuracy (between theory and observations) and scope. It may also reflect considerations of simplicity through the probability measure P: in some systems of epistemic probabilities, there is a built-in preference in the long run for the simplest hypothesis compatible with the evidence.

It remains to be studied more precisely how the concept of truthlikeness is related to other epistemic and methodological notions which might be taken to be relevant to the aims of science.[28] The definition of scientific progress in terms of increasing truthlikeness expresses the realist conviction that information about the reality is the *basic* aim of science. This is compatible with the introduction of other 'utilities' as independent but secondary goals of scientific inquiry. Moreover, there may be special fields of science (e.g. the engineering or technological sciences[29]) where a particularly strong weight on some of these secondary utilities (e.g. practical applicability and computability) can be motivated by the intended use of their results.

University of Helsinki

NOTES

[1] For a survey of problems concerning the notion of scientific progress, see Niiniluoto (1980). See also Niiniluoto (1983a).

[2] According to Cohen (1976), the notion of 'revolution' was first applied to science in the 18th century by Turgot and Bailly.

[3] It seems that Sarton underestimates here the possibilities of revolutions within the humanities: think, for example, about the 'Chomskyan revolution' in linguistics or the radical re-evaluation of mediaeval science in this century! For attempts to apply Kuhn's notions of 'normal science' and 'revolution' to the human and social sciences, see the articles and bibliography in Gutting (1980).

[4] Popper (1975) argues that scientific revolutions are "conservative" in the sense that the new theory "must always be able to explain fully the success of its predecessor". It is precisely this cumulativist trend that has lead his account of scientific progress to troubles (cf. Miller, (1974), Grunbaum (1976) Popper (1979, Appendix 2)).

[5] On the other hand, I do not include among these realist those pragmatist philosophers who define truth in epistemic terms - and thereby make the thesis about the convergence of knowledge "towards the truth" an analytic truth. (Cf. Putnam (1980) (1981) and Shapere (1980) for epistemic conceptions of truth.)

[6] Cf. Niiniluoto (1980). Also Sarton can be included in this group of realists, but his characterization of our "merit" (quoted in Section 1) illustrates the difficulties with presystematic notions of truthlikeness: his "ratio" is not well-defined, since the amount of truth which is "discoverable" is infinite. For Sellars, the basic epistemic utility is explanatory coherence (cf. Pitt's (1981) exposition).

[7] Duhem's position is complex, because he tries to combine an instrumentalist view of theories with the idea that one task of theories is to classify experimental laws. He

further thinks that there is a "natural classification" of these laws towards which scientific theories converge. (I am grateful to Mr. Veli Verronen for challenging my earlier account of Duhem.) As Duhem explicitly speaks of the "correspondence" between the classification given by a theory and the transcendental natural classification (Duhem 1954, p. 335), his view may be taken to represent a peculiar type of convergent realism. Duhem's position can be criticized on the following grounds. It seems reasonable to say that a theory of electricity has the function of grouping together all laws concerning electric phenomena. But what other reason for this grouping there might exist except the fact that *there is electricity* which these laws talk about? But this existential claim is not accepted in Duhem's instrumentalism.

[8] The idea that science is a problem-solving activity is central to Popper as well. In contrast with Popper's realist view, Laudan wants to separate questions about truth from problem-solving.

[9] For attempts to apply some ideas from the logic of questions to science, see Niiniluoto (1976) and Hintikka (1981). Cf. also Rescher (1979).

[10] See my discussion in Niiniluoto (1980) and in the article "Paradigms and Problem-Solving in Operations Research" (1983b) (forthcoming in Niiniluoto (1983a) as well).

[11] Laudan (1982) gives, as an example of a truth-independent characterization of the aims of science, the requirement that theories should entail 'novel facts". However, a *fact* is a state of affairs described by a *true* sentence. Note also that, at least in some cases, Laudan (1982) speaks of situations where theory T entails P and theory T' entails P, but only the former gets credit for the solution of a problem.

[12] Such devices solve all empirical problems, have no anomalies, and at least are not associated with the same conceptual problems as inconsistent theories.

[13] In some cases, the relevant states s cannot be expressed by any single sentences of language L. For example, the structures for a first-order language with relations are limits of structures of depth d, when d , where a structure of depth d is described by a constituent with the quantificational depth d. For details, see Niiniluoto (1979a).

[14] See Niiniluoto (1978) and Oddie (1979) (1982).

[15] See Niiniluoto (1982d).

[16] For the treatment of (g) and (h), see Niiniluoto (1982b). For (i), see Rosenkrantz (1980).

[17] For discussion about alternative definitions, see. Niiniluoto (1978) (1980) (1982c) and Oddie (1981).

[18] Hintikka's inductive logic which is designed for situations of type (c) satisfies the principle that $P(s_0/e) \to 1$ for the constituent s_0 of L which is qualitatively similar to the evidence e. Cf. Niiniluoto and Tuomela (1973).

[19] This condition has an interesting connection to Peirce's notion of a 'likely' or 'verisimilar' theory (see Niiniluoto 1978b, p. 250).

[20] This situation will obtain, if constituents s_i are expressed in a richer language than evidence e. It is thus related to the Quinean problem concerning the "underdetermination of theories by observations". (See note 41 below.)

[21] For example, we have seen above that Grosseteste's law of refraction is farther from the truth than Ptolemy's.

[22] Cf. Niiniluoto (1980). Pitt interprets Sellars's account of science to require that "we finally reach a completed science" or "final theory" FT, but he finds this aim "ill-conceived" (Pitt 1981, pp. 134-137). He argues against Sellars that FT cannot be a theory at all, since it does not have any "external subject matter" (Pitt 1981, p. 134). I don't find this convincing: if the final theory FT will be reached in the development of science, then FT will be, for some n, the *nth* member T_n of a sequence of theories

$$(*) \quad F_{cs}, T_1, T_2, \ldots, T_{n-1}, T_n.$$

where F_{cs} is the framework of common sense and each theory T_i externally explains its predecessor T_{n-1}. Therefore, the 'external subject matter' of FT is simply its predecessor T_{n-1}. If FT is *not* reached except as an ideal limit, so that the sequence (*) is infinite, FT would not have a predecessor any more. But even in this case one might perhaps say that the external subject matter of FT is given by all theories preceding it in (*).

[23] This seems to be a common assumption of Marxism-Leninism and Sellarsian realism. Jardine (1980) has investigated general conditions which would guarantee the convergence of science towards "absolute truth".

[24] This point can be illustrated by Quine's thesis of the "underdetermination" of theories by observational data. Boyd (1973) has argued that underdetermination does not exist, while Newton-Smith (1980) defends the possibility that there may be incompatible theories which "fit all actual and possible observations". In such cases, Newton-Smith favors the "arrogance response": if P is "empirically undecidable", then the world is indeterminate relative to P. Arguments of this type ignore the fact that our capability for observation is influenced by new instruments and theories - and therefore it increases with the progress of science. Even if something is empirically undecidable *today*, we cannot know that it will still be empirically undecidable tomorrow. Hence, we are never in a position to give the "arrogance response". Hesse (1981) defends a form of the underdetermination thesis which is so weak that, in my view, it does not challenge realism at all. If "uniquely true theories cannot be known to be attained by science in any finite time", it does not follow that "realism is in practice an epistemologically and scientifically empty doctrine" (Hesse 1981, p. 9): realism is "non-empty" as soon as one can support non-tautologous disjunctions of theoretical hypotheses.

[25] Putnam has proposed a model-theoretic argument to show that an "epistemically ideal" theory could not be false: if a theory T_I satisfies all desirable epistemic conditions and all "theoretical and operational constraints", the claim that it is false is unintelligible (see 'Realism and Reason' in Putnam 1978, and Putnam 1980). This argument fails for several reasons, since Putnam makes a number of questionable assumptions. (For criticism, see Merrill 1980; Tuomela 1980; Pearce and Rantala 1982.) Thus, Putnam presupposes that the ideal theory T_I is a *first-order* theory which entails *singular atomic sentences* about macroscopic objects; that T_I is *uniquely* determined by the constraints; that T_I may be semantically closed in the sense of containing a theory of reference for its own terms; and that the intended interpretation(s) of T_I have to be chosen among the models of T_I.

[26] Horwich (1982) fails to note this distinction in his - otherwise convincing - criticism of Putnam and Dummett.

[27] The distance between theories can be defined as a generalization of the notion of truthlikeness. See Tuomela (1979) and Niiniluoto (1980).

[28] The connection between the notions of truthlikeness and explanatory power can be illustrated by the following result: if theory T is maximally truthlike in language L and if C is the class of constituents of the observational sublanguage L_0 of L, then the explanatory power of T with respect to C is also maximally great - assuming that explanatory power is measured by $syst_i(T,C)$, i = 1,2,3, defined in Niiniluoto and Tuomela (1973), pp. 91-92.

[29] See my essay 'Technological Progress', forthcoming in Niiniluoto (1983a).

REFERENCES

Boyd, R., (1973), 'Realism, Underdetermination, and a Causal Theory of Evidence', *Nous* 7, 1-12.

Cohen, I. B., (1976), 'The Eighteen-Century Origins of the Concept of Scientific Revolution', *Journal of the History of Ideas* 37, 257-88.

Cohen, M. R. and Drabkin, I. E. (eds.), (1948), *Source Books in Greek Science*, Cambridge: Harvard University Press.

Dilworth, C., (1981), *Scientific Progress: A Study Concerning the Nature of the Relation Between Successive Scientific Theories*, Dordrecht: D. Reidel.

Duhem, P., (1962), *The Aim and Structure of Physical Theory*. Princeton: Princeton University Press.

Feyerabend, P., (1975), *Against Method*, London: New Left Books.

Finocciaro, M. A., (1981), 'Remarks on Truth, Problem-Solving, and Methodology', *Studies in History and Philosophy of Science* 12, 261-268.

Grunbaum, A., (1976), 'Can a Theory Answer More Questions than one of its Rivals?', *The British Journal for the Philosophy of Science* 27, 1-23.

Gutting, G. (ed.), (1980), *Paradigms and Revolutions*: Applications and *Appraisals of Thomas Kuhn's Philosophy of Science*, University of Notre Dame: University of Notre Dame Press.

Hesse, M., (1981), 'The Hunt for Scientific Reason', in P. D. Asquith and R. Giere (eds.), *PSA 1980*, vol. 2, Philosophy of Science Association, East Lansing, pp. 3-22.

Hintikka, J., (1981), 'On the Logic of an Interrogative Model of Scientific Inquiry', *Synthese* 47, 69-84.

Horwich, P., (1982), 'Three Forms of Realism', *Synthese* 51, 181-201.

Jardine, N., (1980), 'The Possibility of Absolutism', in D. H. Mellor (ed.), *Science, Belief and Behaviour: Essays in Honour of R. B. Braithwaite*. Cambridge: Cambridge University Press, pp. 23-42.

Krajewski, W., (1977), *Correspondence Principle and the Growth of Knowledge*. Dordrecht: D. Reidel.

Kuhn, T. S., (1962), *The Structure of Scientific Revolutions*. Chicago: University of Chicago Press. (2nd ed. 1970.)

Laudan, L., (1977), *Progress and Its Problems*, London: Routledge and Kegan Paul.

__________, (1981a), 'A Confutation of Convergent Realisim', *Philosophy of Science* 48, 19-49.

__________, (1981b), 'A Problem-Solving Approach to Scientific Progress' in Hacking, pp. 144-155.

__________, (1981c), 'The Philosophy of *Progress* ...', in P. Asquith and I. Hacking (eds.), *PSA 1978*, vol. 2, Philosophy of Science Association, East Lansing.

__________, (1982), 'Problems, Truth and Consistency', *Studies in History and Philosophy of Science* 13, 73-80.

Leplin, J., (1981), 'Truth and Scientific Progress', *Studies in History and Philosophy of Science* 12, 269-292.

Merrill, G. H., (1980), 'The Model-Theoretical Argument Against Realism', *Philosophy of Science* 47, 69-81.

Miller, D., (1974), 'Popper's Qualitative Theory of Verisimilitude', *The British Journal for the Philosophy of Science* 25, 166-177.

Newton-Smith, W., (1980), 'The Underdetermination of Theory by Data', in R. Hilpinen (ed.), *Rationality in Science*. Dordrecht: D. Reidel, pp. 91-110.

Niiniluoto, I., (1976), 'Inquiries, Problems, and Questions: Remarks on Local Induction', in R. J. Bogdan (ed.), *Local Induction*. Dordrecht: D. Reidel, pp. 263-296.

__________, (1978), 'Truthlikeness: Comments on Recent Discussion', *Synthese* 38, 281-330.

__________, (1979a), 'Truthlikeness in First-Order Languages', in J. Hintikka, I. Niiniluoto, and E. Saarinen (eds.), *Essays on Mathematical and Philosophical Logic*. Dordrecht: D. Reidel, pp. 437-458.

_________, (1979b), 'Verisimilitude, Theory-Change, and Scientific Progress', in Niiniluoto and Tuomela, pp. 243-264.

_________, (1980), 'Scientific Progress', *Synthese* 45, 427-462.

_________, (1982a), 'What Shall We Do With Verisimilitude?', *Philosophy of Science* 49, 181-197.

_________, (1982b), 'Truthlikeness for Quantitative Statements', in *PSA 1982*, vol. 1, Philosophy of Science Association, East Lansing, 208-216.

_________, (1982c), 'On Explicating Verisimilitude: A Reply to Oddie', *The British Journal for the Philosophy of Science* 33, 290-296.

_________, (1982d), 'Verisimilitude vs. Legisimilitude', in *Bulletin of the Section of Logic*, Polish Academy of Science, Wroclaw.

_________, (1983), *Aspects of Scientific Progress*, forthcoming.

Niiniluoto, I. and Tuomela, R., (1973), *Theoretical Concepts and Hypothetico-Inductive Inference*, Dordrecht: D. Reidel.

Niiniluoto, I. and Tuomela, R. (eds.), (1979), *The Logic and Epistemology of Scientific Change*, Acta Philosophica Fennica 30:2-4, North-Holland, Amsterdam.

Oddie, G., (1979), 'Verisimilitude and Distance in Logical Space', in Niiniluoto and Tuomela, pp. 243-264.

_________, (1981), 'Verisimilitude Reviewed', *The British Journal for the Philosophy of Science* 32, 237-265.

_________, (1982), 'Cohen on Verisimilitude and Natural Necessity', *Synthese* 51, 355-379.

Pearce, D. and Rantala, V., (1982), 'Realism and Formal Semantics', *Synthese* 52, 39-53.

Pitt, J. C., (1981), *Pictures, Images, and Conceptual Change: An Analysis of Wilfrid Sellars' Philosophy of Science*. Dordrecht: D. Reidel.

Popper, K., (1972), *Objective Knowledge*, Oxford University Press, Oxford, Revised edition 1979.

Popper, K., (1975), 'The Rationality of Scientific Revolutions', in R. Harre (ed.), *Problems of Scientific Revolution*. Oxford: Oxford University Press. pp. 72-101. (Also in Hacking, 1981, pp. 80-106.)

Putnam, H., (1975), *Mind, Language and Reality: Philosophical Papers*, vol. 2. Cambridge: Cambridge University Press.

__________, (1978), *Meaning and the Moral Sciences*. London: Routledge and Kegan Paul.

__________, (1980), 'Models and Reality', *The Journal of Symbolic Logic* 45, 464-482.

__________, (1981), *Reason, Truth, and History*. Cambridge: Cambridge University Press.

Rescher, N., (1979), 'Some Issues Regarding the Completeness of Science and the Limits of Scientific Knowledge', in G. Radnitzky and G. Andersson (eds.), *The Structure and Development of Science*. Dordrecht: D. Reidel, pp. 19-40.

__________, (1980), 'Conceptual Schemes', in P. A. French, T. E. Vehling, Jr. and H. E. Wettstein (eds.), *Midwest Studies in Philosophy V: Studies in Epistemology*. Minneapolis: University of Minnesota Press, pp. 323-345.

Rosenkrantz, R., (1980), 'Measuring Truthlikeness', *Synthese* 45, 463-488.

Sarkar, H., (1981), 'Truth, Problem-Solving and Methodology', *Studies in History and Philosophy of Science* 12, 61-74.

Sarton, G., (1936), *The Study of the History of Science*. Harvard: Harvard University Press. (Also Dover, New York, 1957.)

Shapere, D., (1980), 'The Character of Theory-Change', in T. Nickles (ed.), *Scientific Discovery, Logic, and Rationality*. Dordrecht: D. Reidel, pp. 61-102.

Tuomela, R., (1979), 'Scientific Change and Approximation', in Niiniluoto and Tuomela, pp. 265-297.

__________, (1980), 'Metaphysical versus Internal Realism', *Reports from the Department of Philosophy, University of Helsinki* N:o 8/1980.

MARCELLO PERA

IN PRAISE OF CUMULATIVE PROGRESS*

> The progress of science is generally regarded as a kind of clean, rational advance along a straight ascending line; in fact it has followed a zigzag course, at times more bewildering than the evolution of political thought.
>
> A. Koestler

1. THREE PROBLEMS CONCERNING CUMULATIVE PROGRESS

According to a view which is so widespread and reasonable as to seem a truism and which we may call the classic conception, science is a form of knowledge that grows continuously. Certain authors have even gone so far as to maintain that science is the only intellectual activity which is progressive and that progress may be used as a demarcation criterion for scientific knowledge. Kant was perhaps the first to explicitly voice this point of view. He maintained that a form of knowledge is only on the "secure path of a science" when, amongst other things, the various thinkers add their result to those of previous thinkers and are not obliged, each time, to begin right from the beginning, as happens in metaphysics (see Kant 1787, B VII).

According to the classic conception, the progress of science is cumulative. The common explication of cumulative progress (*CP*) is that a later theory *T*2 is progressive in relation to a previous theory *T*1 when *T*2 explains all the facts explained by *T*1 and something else as well. Whewell would seem to have been the first to give systematic expression to this cumulativist and retentionist idea of scientific progress when he upheld the view – which he *also* attempted to demonstrate figuratively in his "Inductive Tables" – that, in theory change, "the earlier truths are not expelled but absorbed, not contradicted but extended",

J. C. Pitt (ed.), Change and Progress in Modern Science, 267–282.
© 1985 by D. Reidel Publishing Company.

and that "the history of each science, which may thus appear like a succession of revolutions, is, in reality, a series of developments" (Whewell 1857, p. 8).

As is well known, the concept of cumulative progress has recently come under various forms of attack. It has been called a "dogma" (Laudan 1976), or it has been seen as a myth or an unreasonable demand. To be more precise, the attack is threefold. Certain authors (Grunbaum 1976, Feyerabend 1975) claim that *CP* – when referred to theory change – is, in the first place, impossible; others (Kuhn 1962; Feyerabend 1975; Laudan 1976, 1977, 1981) maintain it is as impossible logically as it is inadequate historically; others yet again (Feyerabend 1965) are of the opinion it is not even desirable.

In actual fact, *CP* sets three quite separate problems:

– is it possible?
– is it adequate?
– is it desirable?

The first problem is whether the (syntactical and semantical) conditions on which *CP* is based can be satisfied; the second is whether the history of science effectively demonstrates that scientific theories follow one another and grow in conformity with the notion of *CP*; lastly, the third problem is whether there are good reasons for taking *CP* as a selection criterion between rival theories. To say "no" to the first question obviously implies saying "no" to the other two. Whereas saying "yes" to the first problem is compatible with saying either "yes" or "no" to the other two.

In this paper I shall maintain 1) that *CP* cannot be logically confuted on principle; 2) that *CP* is inadequate from a historical point of view, but any other model is also inadequate; 3) that, lastly, *CP* is still a good criterion for the heuristics of theory construction.

Even though it is not my intention to rehabilitate the optimistic classic conception – which surely needs revision – I am forced to recognize that my ideas on the subject of scientific progress may seem conservative. They very probably are. I have had the opportunity once before, concerning the conditions of scientific progress, to say that, to a certain extent, I am advocating an old solution to an old problem. (Pera 1983) However, in these times of anarchy, I should like to remark that often the best wine is to be found in old barrels.

2. THE CONDITIONS OF CUMULATIVE PROGRESS

The idea of *CP* requires that in theory change certain true propositions – usually empirical laws or low level empirical regularities – be retained. When some truth or other is reached, it is "preserved for ever" (Post 1971, p. 238); at the very most, its range may be narrowed so that it will appear in the new theory as a limiting case. In this way, a necessary condition of *CP* is that there be permanent facts and concepts in theory change, that is, observation invariance and meaning invariance.

Both these presuppositions have come under heavy attack from the "new philosophy of science". To be more precise, it has been objected to the first that "each theory will possess its own experience, and there will be no overlap between these experiences" (Feyerabend 1965, p. 214); to the second, it has been objected that "the meaning of observational propositions depends upon the theory they are connected with" (Feyerabend 1965). This point of view clearly undermines the possibility of *CP*; it is doubtful whether the said point of view also upholds the concept of progress in a cognitive sense.

Now I believe observation variance and meaning variance, such as they are upheld by the new philosophy of science, to be unacceptable. They are, in my opinion, a perverse and unnecessary consequence of the point of view according to which observation is theory-laden. I, therefore, consider that this point of view must be maintained but on the sole condition that a logical, functional and pragmatic distinction be made between various kinds of theories. Before generalizing on this point, I should like to briefly examine an elementary case taken from the history of electricity.

In 1747 Franklin discovered "the wonderful effect of pointed bodies, both in *drawing off* and *throwing off* the electrical fire" (Franklin 1941 p. 171); that is, in more modern terms, he discovered the fact

(1) Pointed conductors repel and attract electric charges.

This is not merely an observational fact; it is certainly theory-laden. This fact presupposes, *inter alia*, – in Franklin's formulation – that electricity is a fluid, that this fluid is mobile, that quantities of it in excess spread out along the surfaces of bodies in the form of "electrical atmosphere", and so on. These are all theoretical presuppositions and it can be demonstrated that when they change completely, the fact quoted above also changes completely or at least considerably.

Take, for instance, Nollet's theory of affluents and effluents. From Nollet's point of view, Franklin's discovery was not a genuine fact. Must the conclusion be drawn that Nollet and Franklin were seeing different things, that they were living in two different worlds, that there was no overlap between their experiences? In a certain way, this was really so. Nollet's and Franklin's theories were frameworks with relevant ontological consequences: what was a real fact within Franklin's framework (for instance positive electrification) seemed to be an illusion or a groundless interpretation when judged within Nollet's. This does not mean, however, that the two theories were incommensurable. Even though Nollet confuted the fact discovered by Franklin, he nevertheless recognized another fact, in his own words;

(2) "A piece of electrifiable matter takes electricity from a body quicker if the point, rather than the rounded part, is presented". (Nollet 1774, p. 127)

Now, this too can be said to be a theory-laden fact. It is, however, a more elementary fact than the previous one, since it does not depend upon Nollet's particular theory. The theories it is connected with – the assumption, for example, that electricity is a very subtle fluid which can pass from one body to another – are of a deeper kind because they express convictions – shared by Franklin and Nollet – about the fundamental entities of the world.

But it must be recognized that facts exist which are even more elementary than the previous ones; take for instance:

(3) A glass tube, when rubbed, quickly stops attracting small bodies when it is near a pointed metal.

This, too, to a certain extent, is a theory-laden fact and yet the theory involved here is of an even more profound and firm kind, because it concerns our very manner of perceiving objects. Therefore, expressed schematically and without going into details, a distinction may possibly be made between three different types of theories, according to their logical level and their function; that is:

– *explanatory theories or hypotheses,* which explain facts or solve problems; for instance, the theories of Nollet and Franklin;

– *interpretative theories or assumptions,* which express general or local world views; mechanism and behaviorism, for example;

– *categorial theories or categories,* which determine our way of perceiving objects and of expressing judgments; for instance, the geometry of our perception[1].

We may therefore distinguish three different kinds of facts corresponding to and laden with different kind of theories; that is:

– *theoretical facts,* whose existence depends upon a true explanatory theory; for example, the fact (1) recognized by Franklin concerning the power of points, or the fact that the earth revolves around the sun;

– *general facts,* the existence of which depends upon the ontological commitments of some conception or other of natural phenomena or upon a certain metaphysical assumption; for example the fact (2), recognized by both Nollet and Franklin, concerning the behavior of points, or the fact, shared by Ptolomy and Copernicus, that the orbits of planets are circular;

–*perceptive facts,* the existence of which depends upon categories incorporated in our own physiology; for example the fact (3) concerning the attraction of a small body by a glass tube which has been rubbed, or the fact that, at dawn, the distance between the sun and the horizon increases.

Now, since interpretative and categorial theories do not change, or do not necessarily change, when explanatory theories change, facts may be said to exist – general and, above all, perceptive facts – which are invariant in theory change. Those facts can be used to test rival theories and so they can also be used to establish whether the later theory preserves the truth content of the earlier one or not. In this manner, the necessary condition of *CP* can be satisfied even without upholding that there exist purely observational facts and concepts totally distinct from theoretical ones, as was maintained by the old empiricist doctrine.

I should like to point out incidentally that the invariance of at least certain facts, laws or empirical regularities, is necessary, and in actual fact assumed, even in non-cumulative but cognitive conceptions of scientific progress.

Take, for example, Laudan's approach according to which progress is to be explained and measured in terms of problems solved. Here we are faced with the following alternative. If two theories have not at least a certain number of empirical problems in common, then Laudan's criterion is totally paradoxical and useless, because it would make sense of a comparison even between theories with completely different ranges and aims[2]; on the other hand, if – as Laudan himself recognizes – "permanent" empirical problems exist which are neutral and common to two or more theories or research traditions (Laudan 1977, p. 140), then such problems can be said to be permanent only in so far as they refer to facts which remain invariant in theory change. And since these permanent facts, too, are theory-laden, it follows even from Laudan's approach that, over and above explanatory theories, which change, and research traditions which also change, but less rapidly, there exist categorial theories – or whatever you like to call them – which are virtually invariant.

So we can answer the first of our questions in the affirmative: *CP* can be said to be possible; we must now go on to see if it is an adequate model from the point of view of the history of science.

3. THE HISTORY OF SCIENCE AND CUMULATIVE PROGRESS

One of the most widespread objections to *CP* is that a "wholesale discrepancy" exists between it and the "realities of history" (Laudan 1976, p. 591). Before going on to make more general remarks, it may be worthwhile examining this objection with reference again to the theories of Nollet and Franklin. Let us assume that this theory succession, on the basis of our so-called "pre-analytical intuitions", is progressive and let us try to understand, historically, the relationship between the two theories at the time when most physicists preferred Franklin's.

Some help is to be obtained from the *Comparison between the theories of Franklin and Nollet* which the first French translator of Franklin, Barbeu Dubourg, added as an Appendix to his translation. According to Barbeu Dubourg, Franklin's theory had two advantages. First of all, it explains "all the known facts and . . . sets them in such good order that not only are the relationships between them easy to understand, but it is also possible, almost at first sight, to see what may still be missing". Whereas

Nollet's theory "does not teach us to make any distinctions and, even less, to foresee anything". Secondly, Franklin's theory "points out the road that must be followed in order to make new discoveries, to add these to previous ones, to widen the horizons of science and to make science not only pleasant but also useful" (Franklin 1773, vol. I, p. 337). Nollet's theory, on the contrary, is like a "sterile fig".

The conclusion to be drawn from this comparison would seem to be that Barbeu Dubourg had in mind a model of progress identical or very similar to *CP*. He certainly thought Franklin's theory explained all the facts explained by Nollet's theory and that, besides doing this, it also made new confirmed predictions. This was also the opinion of many physicists of the time and is also the way certain historians of science have reconstructed the episode[3].

It must, however, be recognized that, strictly speaking, this theory change did not come about in conformity with the *CP* model[4]. To the contrary, the change in question was an open violation of that model. For Franklin's theory did not explain *all* the known facts as Barbeu Dubourg claimed; what is more and what counts more here, it did not even explain all the facts explained by Nollet's theory.

In the first place, Franklin's theory met with great difficulties when it came to explaining the power of the points. What it did give was an explanation of the repulsive power, on the basis of the theoretical assumptions that "particles of electrical matter repel each other [while] they are strongly attracted by all other matter" (Franklin 1941, p. 213) and that "the form of the electrical atmosphere is that of the body it surrounds" (Franklin 1941 p. 216). Franklin's theory, however, could not explain the attractive power in the same way. Franklin himself was well aware of this difficulty and, to overcome it, he resorted to a positivistic attitude. "I must own", he wrote, "I have some doubts about [these explanations]; yet as I have at present nothing better to offer in their stead I do not cross them out"; he then added: "nor is it of much importance to us to know the manner in which nature executes her laws; it is enough if we know the laws themselves" (Franklin 1941, p. 219).

Secondly, Franklin's theory did not explain the repulsion between negatively charged bodies or between a neutral body and a negatively charged one. The principle or assumption of the electrical atmospheres surrounding bodies did not work in this case, because no excess of fluid could be imagined around a negative body. Nollet's theory, to the contrary, did offer an explanation of this phenomenon, even though it was a very complicated one. By means of a system of interplay between affluents and effluents, Nollet was able to demonstrate why bodies with the same kinds of electricity – either "vitrous" or "resinous" – had to repel one another (Nollet 1976, part III, "Second Fact").

We must therefore conclude that the relationship between Franklin's theory and that of Nollet was not cumulative; and so, if we do not wish to abandon our "pre-analytical intuitions", we must find models of progress which are more suitable for the case in hand.

Perhaps the best way to find these alternative models is to draw up a list of the main criteria on the basis of which one theory may be said to be "better" than another. These criteria can be divided up into two groups.

The criteria of the first group are *empirical* and refer to the quantity and quality of the facts explained or of the empirical problems solved or to be solved. Such criteria are, for instance:

*P*1: *T*2 explains all the facts of *T*1, plus a few others;
*P*2: The facts explained by *T*2 are more numerous than those explained by *T*1, whereas the facts not explained (the "anomalies") are fewer or in the same numer;
*P*3: *T*2 solves more empirical problems than *T*1;
*P*4: *T*2 suggests more empirical problems than *T*1, or makes more predictions in different fields.

The criteria of the second group are *theoretical*; they refer to the manner in which facts are explained or problems are solved and depend, therefore, upon the conditions governing the explanations or the solutions to problems. The following, for instance, are examples of such criteria:

*P*5: *T*2 is simpler than *T*1;
*P*6: *T*2 is more unitary than *T*1;
*P*7: *T*2 is more accurate than *T*1;

*P*8: *T*2 fits in better than *T*1 with the prevailing world view.

Alone or combined, these criteria may be chosen to form a model of progress and the model chosen can then be compared with historical realities. Two remarks must be made, however, in this connection.

The first is that, whereas the criteria of the first group are totally or partially cumulative, because in this case progress is assessed in terms of common or new facts, the criteria of the second group may also be non-cumulative since they refer to theoretical requisites alone. In this way, a simpler theory may have more content than a less simple one, but the history of science shows – and, in any case, it is not impossible – that the opposite may happen, too. What can be said, in general terms, is that the criteria of empirical progress do not necessarily conflict with those of theoretical progress but they are not even necessarily compatible. Whether they are one way or the other depends upon the particular theories being compared. From a logical point of view, the two kinds of requisites can be satisfied separately.

But there is another, more important, remark to be made. Let us suppose we want to introduce a complex criterion of choice, that is a criterion formulated in terms of *both* empirical and theoretical progress. Then it must be noted that the constituent criteria can be combined in different ways, and an historical situation may be found for which one combination is suitable, while another combination holds good for a different situation. We may say, for instance, that the substitution of Nollet's theory for Franklin's theory was progressive according to the combination (*P*2, *P*6, *P*8). It would be easy to show that things were different yet again in other cases

Now, in all these historical situations, we can reasonably speak of progress because we may, in fact, observe that, in each of them, one theory was replaced by another which was "better" when judged on the basis of some cognitive value or group of values. Nevertheless – *and this is the moral to be drawn* – the history of science is there to show that every theory change has been progressive according to this or that empirical or theoretical criterion, but not always according to the *same* criterion or set of criteria.

So saying, it is not my intention to reach the conclusion that "anything goes". I am not maintaining there are not *common* favorite criteria used by individual scholars or even by whole scientific communities at different times.

Indeed there are. They are those criteria that are based upon cognitive values such as empirical content, simplicity, accuracy, completeness of the explanation, etc. These may be called *regulative criteria or principles*, since they determine the different directions of scientific research and the various images of science. In actual fact, these principles are extremely stable and do not usually change during theory change.

Nor am I maintaining that *non-invariant* criteria are to be found in the history of science. The criterion of agreement with the facts, based on the cognitive value of truth, is invariant. It may be called a *constitutive criterion or principle*, because it is the very condition of scientific research. In every period, however different the prevailing images of science may have been, scientists have always aimed at explanations and/or descriptions of facts which were, first of all, *true* and then, as a secondary requisite, simple, unitary, complete, beautiful, pleasant, etc. Even those who do not wish to recognize that the search for truth is the driving force of science, and prefer to talk about a process of problem solution, must admit that scientists attempt to solve problems not with *any theory whatsoever* but with theories which do at least agree with the main known facts. Not even the most radical instrumentalist may entirely give up the value of truth from scientific research. He will deny that the predicate "true" may be properly applied to laws and, above all, to theories, but he will surely agree it may be used with reference to descriptions and predictions of singular facts.

If we look at the history of science with a view to obtaining a model of progress suitable for all episodes of theory change or even for the most important alone, we are bound to be disappointed every time. No *single* (non generic) criterion of progress can be obtained from the history of science nor can it be shown that science grows constantly according to the model based on such a criterion. So, even if not anything goes, many different things, in actual fact, go.

I feel this is hard to deny. In any case, it is something which *may* happen, for at least two reasons. The first reason is that regulative criteria – even though they are stable – may have different distributions of weight, and there is no reason for thinking that only one sort of distribution has always prevailed or will always prevail in the history of science. In Ptolomaic astronomy, for instance, accuracy of predictions was considered to be more important than simplicity; whereas in Copernican astronomy simplicity carried greater weight. The second reason is that, even if we suppose that the constitutive criterion is invariant, we must admit that it can have various explications, because what "agreement with the facts" means largely depends upon epistemological and metaphysical views, and in any case it can be combined in various ways with the regulative criteria. Once again, there is no reason for thinking there is always, or always will be, a single favorite combination. Both these reasons lead me to believe that a model of progress obtained from the history of science is not like an L.C.M. which contains the given numbers exactly so many times, or like the G.C.D., which is contained so many times in them. I see it rather as a tailor-made garment for the Average Man: it may fit anybody but it is usually tight or loose for the majority.

It has been maintained by Post that the Correspondence Principle, and accordingly *CP*, is valid for any theory change in the history of science (Post 1971); Krajewski holds the more cautious view that it is valid only in mature science. (Krajewski 1977, p. 88) If it is not a tautology, this is a disputable question of fact. Anyway, supposing *CP* is a characteristic of past history, there is no warrant it will also be preserved in the future.

In conclusion, the situation with *CP* is that it is really at variance with the history of science; I maintain, however, that this is equally true of any other criterion that is not generic or empty. Science may be said to grow and perhaps it may even be said – in the long run – to grow cumulatively. What cannot be said is that every scientific change is an example of cumulative progress.

But, if *CP* is not always a suitable model of growth, what reasons can there be for advising it to be adopted as a choice criterion between rival theories? This is the last point I have still to examine.

4. A FEW REASONS IN SUPPORT OF CUMULATIVE PROGRESS

It has been said that "all methodologies, even the most obvious ones, have their limits" (Feyerabend 1975, p. 32). I agree with this. I have even attempted to show that, in the field of methodology, there exist something akin to an "uncertainty theorem" (Pera 1978, II p. 4-5 and 1982,

Chapter I.): the conceptual accuracy and historical suitability of a system of methodological rules are properties which cannot be enjoyed at one and the same time. If the rules are accurate, then they are not suitable in all situations, not even in the most important of them; on the other hand, should they be suitable, then they are either generic or incomplete. This means that, in the field of methodology, a choice must be made and, as always happens when a choice has to be made, something must be given up. We must even be ready to give up our most favorite "pre-analytical intuitions".

The fact all non-generic methodologies are unsuitable does not mean, however, they are all on the same level. In my opinion, a methodological code that, for instance, contains the rule that a theory must be accepted even if it does not agree with the main facts it is meant to explain, would not be acceptable. But the reason for this opinion cannot be found in the history of science where cases of dogmatism and refusal to accept even the obvious also exist. The reason, rather, is that such a rule is contrary to the constituitive criterion of scientific research, and sidetracks us from the basic aim of science. This seems to me the most satisfactory justification that can be found for that rule: *if* we wish to attain such an end, *then* it is rational to choose those means.

A similar justification can also be adduced to support the criterion of cumulative progress. If we attribute to science the aim of producing theories which are in agreement with experience and which reveal constantly new orders of reality, then a necessary condition for the attainment of this aim, or advancement towards it, is that the new theories widen the range of the previous ones and preserve their truth-content as a limiting case. Even if the history of science showed there are many cases in which the Correspondence Principle has not worked, there would still be good reason for advising it to be adopted as a rule for the heuristics of theory construction.

I call this kind of justification for the rules an "axiological vindication". This can be obtained on two conditions: that a value and a set of rules are indicated and that the rules are shown to follow from the value. Given its means-ends structure, this kind of argument is, in principle, analytical; in actual fact, however, if, as happens in our case, the end – that is, truth – is not known and may be explicated in various ways, then it is not always possible to tell precisely which rules or set of rules follow from the ends. Nevertheless, the axiological vindication does not allow us to do anything we like; it is not anarchism in disguise but it seems precise enough to exclude certain policies.

A number of objections may be raised at this point. The first is that an axiological vindication is weak. But this objection may be countered by saying that every justification must stop at some point if it wants to avoid endless regression, and no stopping point seems better than the unswerving decision to choose values.

The second objection is that to defend cumulative progress and to ask for content increase is the expression of an inductivist and conservative policy. This is true in my opinion. But I also answer that such a policy is not incompatible with great scientific revolutions or advancements. In any case, I am not sure any better alternative exists. To the contrary, the anti-inductivist and anarchist policies (if they are policies) of introducing theories which are contrary to the facts and incommensurable, seem to me to be impossible to uphold unless one is ready to do away with truth as the aim of science.

Finally, there is a third and still stronger objection concerning precisely this aim. It might be said that the axiological vindication of cumulative progress proposed here is circular, because it takes for granted that truth is the real aim of science and that the choice of this aim is advisable[5]. But I do not propose to justify the criterion of cumulative progress on the basis of the aim and then the aim on the basis of the science of today or of the "last two centuries". As is emphasized by the "if-then" structure of the vindicatory argument. I am simply expressing and proposing a personal choice (even though I believe it is not mine alone), and I do not care much about the history of science. Indeed, in making my choice, I am assuming, at least in part, that the science of today or of the "last two centuries" is a standard of rationality. And if I am asked to give reasons for this assumption, then I have no choice but to adduce further axiological vindications based on further values.

Lastly, there is one point which would require lengthy treatment, but which I can only mention here; it is about the weight of the history of science in this kind of discussion.

The introduction and justification of a criterion of progress are one of the tasks of the methodology of science. Now I do not believe the methodology of science must be assessed on the basis of history or of the present-day manner of practicing science. Nor do I believe that, in order to avoid the onus of the historical unsuitability of *all* methodologies, we ought to conclude that it is not their task to establish what scientists should do.[6] On the contrary, I believe that a methodology is a code of conduct and I maintain that this code cannot be assessed on a historical basis. In the first place, because this would go against "Hume's law", which there is no reason not to abide by even if it is not a dogma; secondly, because history has a nasty habit of going along with those who want to take lessons from it, whatever lessons these might be.

Perhaps a fair suggestion is to respect the division of labor. As far as methodology is concerned, philosophers might deal with questions of relationships (of consistency, contradiction, etc.) between the ends scientists claim to pursue and the means they use. In this way, they would give vent to their normative passions without doing too much harm. Scientists, on the other hand, should be allowed to choose their aims and research programmes freely. So doing, they would satisfy their need for independence. Besides, methodology seems too serious a thing to be left to methodologists.

University of Pisa

NOTES

(*) I am indebted to several participants at the Blacksburg Conference, especially to Larry Laudan, for their comments. I am also indebted to Gerd Buchdahl and Paul Feyerabend who read and commented on an early version of this paper. Their suggestions and criticisms considerably helped me in clarifying my ideas even when, obstinately and perhaps mistakenly, I did not change them.

[1] For a more analytical examination of these different kinds of theories, see Pera (1978) and (1982); see also (1981).
[2] See Kordig (1980).
[3] See Bernard Cohen in Franklin (1941, p. 73). However the acceptance of Franklin's theory was a complicated affair. It was surely connected with the explanation of the Leyden jar, but it was immediately clear that the theory met with serious difficulties. Home (1972), (1975) has well stressed these shortcomings and the attempts of the electricians to overcome them. In actual fact, the first result was a combination of parts from both effluvial theories and Franklin's.
[4] On this point I completely agree with Laudan (1976, p. 590) and (1977, p. 149), even if I draw different conclusions concerning the role that the idea of cumulative progress plays in scientific research.
[5] A similar objection may be drawn from Feyerabend (1978).
[6] This is John Worrall's way of getting out of troubles. See Worrall (1976), Section 5.

REFERENCES

Feyerabend, P. (1965) "Problems of Empiricism", *Beyond the Edge of Certainty*. ed. R. Colodny. Englewood Cliffs: Prentice-Hall, pp. 145-260.

__________ (1975) *Against Method*. London: New Left Books.

__________ (1978) "In Defense of Aristotle: Comments on the Condition of Content Increase", *Progress and Rationality in Science*. eds. G. Radnitzky and G. Andersson. Dordrecht: Reidel, pp 143-180.

Franklin, B. (1773) *Oeuvres de M. Franklin*. traduites de l'Anglais par M. Barbeu Dubourg, , II, Paris.

__________ (1941) *Benjamin Franklin's Experiments*. ed. I. Bernard Cohen. Cambridge, Mass.: Harvard University Press.

Grunbaum, A. (1976) "Can a Theory Answer more Questions than One of Its Rivals?", *British Journal for the Philosophy of Science*, 27, pp. 1-23.

Home, R. (1972) "Franklin's Electrical Atmospheres", *British Journal for the History of Science*, 6, pp. 131-151.

_______ (1975) "Electricity in France in the Post-Franklin Era", Congress Int. Hist. Sciences (1974), *Actes*, II, pp. 269-272.

Kant, I. (1787) *Critique of Pure Reason*. 2nd Edition; trans. N. Kemp Smith. London: Macmillan, 1978.

Kordig, C. (1980) "Progress requires Invariance", *Philosophy of Science*, 47, p. 141.

Krajewski, W. (1977) *Correspondence Principle and Growth of Knowledge*, Dordrecht: Reidel.

Kuhn, T. (1962) *The Structure of Scientific Revolutions*. Chicago: Chicago University Press.

Laudan, L. (1976) "Two Dogmas of Methodology", in *Philosophy of Science*, 43, pp. 585-597.

_________ (1977) *Progress and its Problems*. Berkeley: University of California Press.

_________ (1981) "A Confutation of Convergent Realism", *Philosophy of Science*, 48, pp. 19-49.

Nollet, J. (1746) *Essai sur l'electricite des corps*. Paris.

_________. (1774) *Lettres sur l'electricite*. Premiere Partie, Nouvelle edition, Paris.

Pera, M. (1978) *Induzione e metodo scientifico*. Pisa: Edizioni Tecnico-Scientifiche.

_______ (1981) "Inductive Method and Scientific Discovery", *On Scientific Discovery*, eds. M. D. Grmek, R. Cohen, G. Cimino. Dordrecht: Reidel.

_______ (1982) *Apologia del metodo*. Roma-Bari: Laterza.

_______ (1983) "The Justification of Scientific Progress" (forthcoming).

Post, H. (1971) "Correspondence, Invariance, and Heuristics; In praise of Conservative Induction", *Studies in History and Philosophy of Science*, 2, pp. 213-255.

Whewell, W. (1857) *History of the Inductive Sciences*. 3rd edition. London.

Worrall, J. (1976) "Thomas Young and the 'Refutation' of Newtonian Optics", *Method and Appraisal in the Physical Sciences*. ed. C. Howson. Cambridge: Cambridge University Press, pp. 102-179.

LARRY LAUDAN

KUHN'S CRITIQUE OF METHODOLOGY

1. INTRODUCTION

For centuries, philosophers of science have regarded their primary role as that of identifying and justifying the methodological rules which inform and shape the learning and testing techniques of the sciences. A science for which no rules could be given, a science without a methodology of inquiry and testing, seemed unthinkable to many of our forebears. But not so to many of our contemporaries. However plausible the identification of rules of assessment with scientific rationality may once have seemed (and, for some of us, still seems), several recent writers have criticized – and in some cases thoroughly repudiated – the enterprise of identifying and clarifying the methodological procedures or rules utilized in scientific assessment. A growing chorus of voices insists that the identification and analysis of the rules of 'scientific method' is not the route to a philosophical understanding of science.

Paul Feyerabend, for instance, preaches that scientific methodology is dead; that *all* methods of inquiry are flawed; that "anything goes". Coming to the same conclusion from a different direction, numerous followers of Wittgenstein insist that communal life generally, and the life of the scientific community in particular, is mis-characterised if one imagines it to be grounded in clear rules of assessment or evaluation. This repudiation of methodology is not restricted to philosophers. New-wave sociologists of knowledge, rebelling against Merton and his focus on the 'norms' of scientific behavior, see the rules of scientific rationality proposed by philosophers as little more than *post hoc* rationalizations for beliefs and actions which are grounded, not in an abstract, objective and disinterested 'scientific method', but rather in the subjective and professional self-interest of individual researchers. Michael Polanyi has argued that the 'rules' of science typically cannot even be articulated, and that much of science consists of 'tacit knowledge' which defies explicit formulation.

J. C. Pitt (ed.), Change and Progress in Modern Science, 283–299.
© 1985 by D. Reidel Publishing Company.

I think that this whole approach is fundamentally wrong-headed. But it is much too ambitious a task to attempt to take on in omnibus fashion the entire range of arguments which have been directed against the possibility of methodology as a normative and descriptive enterprize. I shall focus instead on one of the most influential critics of classical conceptions of scientific methodology, namely, Thomas Kuhn. I shall seek to show in this paper that his specific arguments about the unavailingness of appeals to scientific methods and standards will not stand up to sustained analysis. (I shall leave it to another occasion to examine the arguments of several of Kuhn's fellow travellers!)

I have chosen to deal with Kuhn's critique of methodology because it strikes me as possessing a great deal more prima facie plausibility, and to be more closely based on how science actually works, than the discussions of most of the other authors in this tradition. In brief, Kuhn's view is this: if we examine situations where scientists are required to make a choice between the handful of paradigms which confront them at any time, we discover that the relevant evidence and appropriate methodological standards fail to pick out one contender as unequivocally superior to its extant rival(s). I shall call such situations cases of 'local' underdetermination, by way of contrasting them with the more global and more familiar forms of underdetermination associated with Quine and Duhem (which say, in effect, that the rules are insufficient to pick out *any* theory as being uniquely supported by *any* data). Kuhn offers four distinct arguments for local underdetermination. Each is designed to show that, although methodological rules and standards do constrain and delimit to some degree a scientist's choices or options, those rules and standards are never sufficient to compel or unequivocally to warrant the choice of one paradigm over another.

2. 'THE AMBIGUITY OF SHARED STANDARDS' ARGUMENT

Kuhn's first argument for local underdetermination rests on the purported ambiguity of the methodological rules or standards which are *shared* by advocates of rival paradigms. The argument first appeared in the *Structure of Scientific Revolutions* (1962) and has been expanded considerably in *The Essential Tension* (1972). As he put it in the earlier work: "lifelong resistance [to a new theory]...is not a violation of scientific standards...though the historian can always find men – Priestley, for instance – who were unreasonable to resist for as long as they did, he will not find a point at which resistance becomes illogical or unscientific." (Kuhn 1962, p. 159) Many of Kuhn's readers were perplexed by the juxtaposition of claims in such passages as these. On the one hand, we are told that Priestley's continued refusal to accept the theory of Lavoisier was 'unreasonable'; but we are also told that Priestley's refusal was neither 'illogical' nor 'unscientific'. To those of us accustomed to thinking that being 'scientific' (at least in the usual sense of that term) required one to be 'reasonable' about shaping one's beliefs, Kuhn seemed to be talking jibberish. On a more sympathetic construal, Kuhn seemed to be saying that a scientist could always interpret the applicable standards of appraisal, *whatever they might be*, so as to 'rationalize' his own paradigmatic preferences, whatever *they* might be. This amounts to saying that the methodological rules or standards of science never make a difference to the outcome of a decision-making process; for, if any set of rules can be used to justify any theory whatever, then methodology would seem to amount to just so much window-dressing. But this cynical and minimalist construal of this passage turns out to be a far cry from what Kuhn intended; as he has made clear in later writings, he wants to allow a positive, if (compared to the traditional view) much curtailed, role to methodological standards in scientific choice.

What Kuhn apparently had in mind was this: the shared criteria, standards, and rules to which scientists explicitly and publicly refer in justifying their choices of theory and paradigm are typically 'ambiguous' and 'imprecise' (Kuhn 1977, p. 322), so much so that "individuals [who share the same standards] may legitimately differ about their application to concrete cases". (Kuhn 1972, p. 322) Kuhn holds that, although scientists share certain cognitive values "and must do so if science is to survive, they do not all apply them in the same way. Simplicity, scope, fruitfulness, and even accuracy can be judged differently (which is not to say they may be judged arbitrarily) by different people." (Kuhn 1972, p. 262) Because then, the shared standards are invariably ambiguous, two scientists may subscribe to 'exactly the same standard' (say, the rule of simplicity) and yet come to endorse opposing viewpoints.

Kuhn draws some quite large inferences from the presumed ambiguity of the shared standards or criteria. Specifically, he concludes that *every* case of theory choice must involve an admixture of 'objective' and 'subjective' factors, since (on Kuhn's view) the shared – and presumably 'objective' – criteria are too amorphous and ambiguous to warrant a particular preference. He puts the point this way: "I continue to hold that the algorithms of individuals are all ultimately different by virtue of the subjective considerations with which each [scientist] *must* complete the objective criteria before any computations can be done." (Kuhn 1977, p. 329; my italics) As this passage makes clear, Kuhn believes that, because the shared criteria are too imprecise to justify a choice, and because – despite that imprecision – scientists do manage to make choices, those choices 'must' be grounded in 'individual' and 'subjective' preferences. As he says, "every individual choice between competing theories depends on a mixture of objective and subjective factors, or of shared and individual criteria." (Kuhn 1977, p. 325; cf. also *ibid.*, p. 324) Again, he writes: the shared criteria "are not by themselves sufficient to determine the decisions of individual scientists." (Kuhn 1977, p. 325)

This is a very ambitious claim; if true, it would force us drastically to re-think our views of scientific rationality. Among other things, it would drive us to the conclusion that *every* scientist has *different* reasons for his theory preferences from those of his co-workers. The view entails, among other things, that it is a category mistake to ask (say) why physicists think Einstein's theories are better than Newton's; for, on Kuhn's view, there must be as many different answers to that question as there are physicists. It evidently turns out to be a cosmic coincidence that scientists, each working with his own subjective criteria, are so often able quickly to achieve a consensus about which theories to accept and to reject. We might note in passing that this is quite an ironic conclusion for Kuhn to reach. Far more than most writers on these subjects, he has tended to stress the importance of community and socialization processes in understanding the scientific enterprize. Yet the logic of his own analysis drives him to the radically *individualistic* position that every scientist has his own set of reasons for theory preferences and thus that there is no real consensus whatever with respect to the *grounds* for theory preference – not even among the advocates of the same paradigm! Seen from this perspective, Kuhn tackles what I have elsewhere called the problem of consensus formation in science (Laudan 1982, 1984) by a maneuvre which trivializes the problem: for, if we must give a separate and discrete explanation for the theory preferences of each member of the scientific community – which is what Kuhn's view entails – then we are confronted with a gigantic mystery at the *collective* level, to wit, why the scientists in a given discipline – each supposedly operating with his own individualistic and idiosyncratic criteria, each giving a different 'gloss' to the criteria which are shared – are so often able to agree about which theories to bet on. But we can leave to Kuhn to sort out how he reconciles his commitment to the social psychology of science with his views about the individual vagaries of theory preference. What must concern us is the question whether Kuhn has made a plausible case for thinking that the shared or collective criteria 'must' be supplemented by individual and subjective criteria.

The first point to stress is that Kuhn's thesis purports to apply to *all* scientific rules or values which are shared by the partisans of rival paradigms, not just to a selected few, notoriously ambiguous ones. We can grant straightaway that some of the rules, standards and values used by scientists ('simplicity' would be an obvious candidate) exhibit precisely that high degree of ambiguity which Kuhn ascribes to them. But Kuhn's *general* argument for the impotence of shared rules to settle disagreements between scientists working in different paradigms cannot be established by citing the occasional example. Kuhn must show us, for he claims as much, that there is something in the very nature of those methodological rules which come to be shared among scientists that makes the application of those rules or standards invariably inconclusive. He has not established this result, and there is a good reason why he has not: it is false. To see that it is, one need only produce a methodological rule widely accepted by scientists which can be applied to concrete cases without substantial imprecision or ambiguity. Consider, for instance, one of Kuhn's own examples of a widely shared scientific standard, namely, the requirement that an acceptable theory must be internally consistent and logically consistent with accepted theories in other fields. (One may or may not favor this methodological rule. I refer to it here only because it is commonly regarded, including by Kuhn, as a methodological rule which frequently plays a role in theory evaluation.)

I submit that we have a very clear notion of what it is for a theory to be internally consistent, just as we understand perfectly well what it means for a theory to be consistent with accepted beliefs. Moreover, there are at least *some* occasions when we can tell whether a particular theory has violated the standard of (internal or external) consistency. Kuhn himself, in a revealing passage, grants as much; for instance, when comparing the relative merits of geocentric and heliocentric astronomy, Kuhn says that "the consistency criterion, by itself, therefore, spoke unequivocally for the geocentric tradition." (Kuhn 1977, p. 323) (What he has in mind is the fact that heliocentric astronomy, when introduced, was inconsistent with the then-reigning terrestrial physics; whereas the assumptions of geocentric astronomy were consistent with that physics.) Notice that in this case we have a scientific rule or criterion "speaking unequivocally" in favor of one theory and against its rival. Where is the inevitable 'imprecision' and 'ambiguity' which is supposed by Kuhn to afflict all the shared values of the scientific community? What is "ambiguous" about the notion of internal consistency? The point of these rhetorical questions is to drive home the fact that, even by Kuhn's lights, some of the rules or criteria widely accepted in the scientific community fail to exhibit that multiplicity of meanings which Kuhn has described as being entirely characteristic of methodological standards.

One could, incidentally, cite several other examples of reasonably clear and unambiguous methodological rules. For instance, the requirements that theories should be deductively closed or that theories should be subjected to 'controlled experiments' have not generated a great deal of confusion or disagreement among scientists about what does and does not constitute deductive closure or a control group. Or, consider the rule that theories should lead successfully to the prediction of results unknown to their discoverer; so far as I am aware, scientists have not differed widely in their construal of this rule. The significance of the non-ambiguity of many methodological concepts and rules is to be found in the fact that such non-ambiguity refutes one of Kuhn's central arguments for the incomparability of paradigms and for its corollary, the impotence of methodology as a guide to scientific rationality. There are at least some rules which are sufficiently determinate that one can show that many theories clearly fail to satisfy them. We need not supplement the shared content of these 'objective' concepts with any private notions of our own in order to decide whether our theories satisfy them.

3. 'THE COLLECTIVE INCONSISTENCY OF RULES' ARGUMENT

As if the ambiguity of particular standards was not bad enough, Kuhn goes on to argue that the *shared* rules and standards, when taken as a collective, "repeatedly prove to conflict with one another". (Kuhn 1977, p. 322) For instance, two scientists may each believe that empirical accuracy and generality are desirable traits in a theory. But, when confronted with a pair of rival (and thus incompatible) theories, one of which is more accurate and the other more general, the judgments of those scientists may well differ about which theory to accept. One scientist may opt for the more general theory; the other, for the more accurate. They evidently share the 'same' standards, says Kuhn, but they end up with conflicting appraisals. Kuhn puts it this way: "...in many concrete situations, different values, though all constitutive of good reasons, dictate different conclusions, different choices. In such cases of value-conflict (e.g. one theory is simpler but the other is more accurate) the relative weight placed on different values by different individuals can play a decisive role in individual choice." (Kuhn 1972, p. 262)

Because many methodological standards do pull in different directions, Kuhn thinks that the scientist can pretty well go whichever way he likes. Well, not quite *any* direction he likes since – even by Kuhn's very liberal rules – it would be unreasonable for a scientist to prefer a theory (or paradigm) which failed to satisfy *any* of the constraints. On Kuhn's view, we should expect scientific disagreement or dissensus to emerge specifically in those cases where (a) no available theory satisfied all the constraints and (b) every extant theory satisfied some constraints not satisfied by its rivals. That scientists sometimes find themselves subscribing to contrary standards, I would be the first to grant. Indeed, the discovery of that fact about oneself is often the first prod toward readjusting one's cognitive values. But Kuhn is not merely saying that this happens occasionally; he is asserting that such is the nature of any set of rules or standards which any group of reasonable scientists might accept. As before, I think our verdict has to be that Kuhn's highly ambitious claim is just that; he never shows us why families of methodological rules should always or even usually be internally inconsistent. He apparently expects us to take his word for it that he is just telling it like it is. ["What I have said so far is primarily simply descriptive of what goes on in the sciences at times of theory choice." (Kuhn 1977, p. 325.)] I see no reason why we should follow Kuhn in his global extrapolations from the tiny handful of cases he describes. On the contrary, there are good grounds for resisting, since there are plenty of sets of methodological standards which are entirely internally consistent. Consider, for instance, one of the most influential documents of 19th-century scientific methodology, John Stuart Mill's *System of Logic*. Mill offered there a set of rules or 'canons' for assessing the soundness of causal hypotheses. Nowadays those rules are still called 'Mill's methods', and much research in the natural and social sciences utilizes them, often referring to them as the methods of agreement, difference and concomitant variations. To the best of my knowledge, no one has ever shown that Mill's methods exhibit a latent tendency toward contradiction or conflict of the sort which Kuhn regards as typical of systems of methodological rules. To go back further in history, no one has ever shown that Bacon's or Descartes' or Newton's or Herschel's famous canons of reasoning are internally inconsistent.

The fact that numerous methodologies of science can be cited which have never been shown to be inconsistent raises acute doubts about Kuhn's claim that any methodological standards apt to be shared by rival scientists will tend to exhibit internal inconsistencies.

Kuhn could have considerably strengthened his argument if, instead of focussing on the purported contradictions in sets of methodological rules, he had noted, rather, that whenever we have more than one standard in operation, it is conceivable that we will be torn in several directions. And this claim is true, *regardless* of whether the standards are inconsistent with one another (just so long as there is not a complete co-variance between their exemplifying instances). Thus, if two scientists agree to judge theories by standards a and b, then it is trivially true that, depending upon how much weight each gives to a and b respectively, their judgements about theories may differ. Before we can make sense of how to work with several concurrent standards, we have to ask (as Kuhn never did) about the way in which these standards do (or should) control the selection of a 'preferred' theory. Until we know the answer to that question, we will inevitably find that the standards are of little use in explaining scientific preferences. Kuhn simply assumes that all possible preference structures (i.e., all possible differential weightings of the applicable standards) are equally viable or equally likely to be exemplified in a working scientist's selection procedures.

To sum up the argument to this point: I have shown that Kuhn is wrong in claiming that all methodological rules are inevitably ambiguous and in claming that scientific methodologies consisting of whole groups of rules always or even usually exhibit a high degree of internal 'tension'. Since these two claims were the lynchpins in Kuhn's argument to the effect that shared criteria "are not by themselves sufficient to determine the decisions of individual scientists" (Kuhn 1977, p. 325), we are entitled to say that Kuhn's effort to establish a general form of methodological underdetermination falls flat.

4. 'THE SHIFTING STANDARDS' ARGUMENT

Equally important to Kuhn's critique of methodology is a set of arguments having to do with the manner in which *standards* are supposed to vary from one scientist to another. In treating Kuhn's views on this matter, I will be following Gerald Doppelt's excellent and sympathetic explication of Kuhn's position. (Doppelt 1978) (Where Kuhn's own discussion of these questions in the *Structure of Scientific Revolutions* rambled considerably, Doppelt has offered a succinct and perspicacious formulation of what is, or at least what should have been, Kuhn's argument. Although I quarrel with Doppelt's analysis at several important points, my own thoughts about these issues owe a great deal to his writings.)

In general, Kuhn's model of science envisages two quite distinct ways in which disagreements about standards might render scientific debate indeterminate or inconclusive. In the first place, the advocates of different paradigms may subscribe to different methodological rules or evaluative criteria. Indeed, 'may' is too weak a term here, for Kuhn evidently believes that associated with each paradigm is a set of methodological orientations which are (at least partially) at odds with the methodologies of all rival paradigms. Thus, he insists that whenever a 'paradigm shift' occurs, this process produces "changes in the standards governing permissible problems, concepts and explanations." (Kuhn 1962, p. 104) This is quite a strong claim. It implies, among other things, that the advocates of different paradigms invariably have different views about what constitutes a scientific explanation and even about what constitutes the relevant facts to be explained (the "permissible problems"). If Kuhn is right about these matters, then debate between the proponents of two rival paradigms will involve appeal to different sets of rules and standards associated respectively with the two paradigms. One party to the dispute may be able to show that his theory is best by his standards, while his opponent may be able to claim superiority by his.

Kuhn is surely correct to say that scientists sometimes subscribe to different methodologies (including different standards for adequacy of explanation and of facticity). But he has never shown, and I believe him to be chronically wrong in claiming, that disagreements about matters of standards and rules neatly *coincide* with disagreements about substantive matters of scientific ontology. Rival scientists advocating fundamentally different theories or paradigms often espouse the same standards of assessment (and interpret them identically); on the other hand, adherents to the same paradigm will frequently espouse different standards. In short, methodological disagreements and factual disagreements about basic theories show no striking co-variances of the kind required to sustain Kuhn's argument about the intrinsic irresolveability of inter-paradigmatic debate.

But, of course, a serious issue raised by Kuhn still remains before us. If different scientists sometimes subscribe to different standards of appraisal (and that much is surely correct), then how is it possible for us to speak of the resolution of such disagreements as anything other than an arbitrary closure? To raise that question presupposes a picture of science which I have elsewhere sought to demolish. (Laudan 1984, Chs. 2,3) Provided that there are mechanisms for rationally resolving disagreements about methodological rules and cognitive values, the fact that scientists often disagree about such rules and values need not, indeed should not, be taken as showing that there must be anything arbitrary about the resolution of such disagreements.

5. 'THE PROBLEM-WEIGHTING' ARGUMENT

As I said earlier, Kuhn has another argument up his sleeve which he and others think is germane to the issue of the rationality of comparative theory assessment. Specifically, he insists that *the advocates of rival paradigms assign differential degrees of importance to the solution of different sorts of problems*. Because they do, he says that they will often disagree about which theory is better supported, since one side will argue that it is most important to solve this problem, while the other will insist on the centrality of solving a very different problem. Kuhn poses the difficulty in these terms:

> if there were but one set of scientific problems, one world within which to work on them, and one set of standards for their solution, paradigm competition might be settled more or less routinely by some process like counting the number of problems solved by each. But, in fact, these conditions are never met completely. The proponents of competing paradigms are always at least slightly at cross purposes...the proponents will often disagree about the list of problems that any candidate for paradigm must resolve. (Kuhn 1962, pp. 147-148)

In this passage, Kuhn runs together two issues which it is well to separate: one concerns the question (just addressed in the previous section) whether scientists have different standards of explanation or problem solution; the other (and the one which will concern us here) is the claim that scientists working in different paradigms want to solve different problems and that, because they do, their appraisals of the merits of theories will typically differ. So we must here consider the case where scientists have the same standards for what counts as solving a problem, but what they disagree about is which problems are the most important to solve. As Kuhn puts it, scientific controversies between the advocates of rival paradigms:

> involve the question: which problems is it more significant to have solved? Like the issue of competing standards, that question of values can be answered only in terms of criteria that

> lie outside of normal science altogether... (Kuhn 1962, p. 110)

Kuhn is surely right to insist that partisans of different global theories or paradigms often disagree about which problems it is most important to solve. But the existence of such disagreement does *not* establish that inter-paradigmatic debate about the epistemic support of rival paradigms is inevitably inconclusive or that it must be resolved by factors which lie outside the 'normal' resources of scientific inquiry.

At first glance, Kuhn's argument seems very plausible: the differing weights assigned to the solution of specific problems by the advocates of rival paradigms may apparently lead to a situation in which the advocates of rival paradigms can *each* assert that their respective paradigms are the best because they solve precisely those problems which they respectively believe to be the most important. No form of reasoning, insists Kuhn, could convince either side of the merits of the opposition nor of the weakness of its own in such circumstances.

To see where Kuhn's argument goes astray in this particular case, we need to dissect it at a more basic level. Specifically, we need to distinguish at least two quite distinct senses in which solving a problem may be said to be 'important'. A problem may be important to a scientist just in the sense that he is particularly curious about it. Equally, it may be important because there is some urgent social or economic reason for solving it. Both sorts of considerations may explain why a scientist regards it as urgent to solve the problem. Such concerns are clearly relevant to explaining the *motivation* of scientists. But these senses of problem importance have no particular epistemic or probative significance. When we are assessing the evidential support for a theory, when we are asking how well-supported or well-tested that theory is by the available data, we are not asking whether the theory solves problems which are socially or personally important. Importance, in the sense chiefly relevant to this discussion, is what we might call epistemic or probative importance. *One problem is of greater epistemic or probative significance than another just in case the former constitutes a more telling test of our theories than does the latter*.

So, if Kuhn's point is to be of any significance for the epistemology of science (or, what amounts to the same thing, if Kuhn's point is thought to be relevant to ascertaining how belief-worthy a theory is), then we must imagine a situation in which the advocates of different paradigms assign conflicting degrees of *epistemic import* to the solution of certain problems. Kuhn's thesis about such situations would be, I presume, that there is no rational machinery for deciding who is right about the assignment of epistemic weight to such problems. But that seems wrong-headed, or at least unargued, for philosophers of science have long and plausibly maintained that the primary function of scientific epistemology is precisely to ascertain the (epistemic) importance of any piece of confirming or disconfirming evidence. It is not open to a scientist simply to say that solving an arbitrarily selected problem (however great its subjective significance for that scientist) is of great probative value. Indeed, it is often the case that the epistemically most salient problems are ones with little or no prior practical or even heuristic significance. (Consider that Brownian motion was of decisive *epistemic* significance in discrediting classical thermodynamics, even though it had little intrinsic interest prior to Einstein's showing that it was 'anomalous' for thermodynamics.) The whole object of the theory of evidence is to *de-subjectify* the assignment of evidential significance by indicating the sorts of reasons that can be legitimately given for attaching a particular degree of epistemic importance to a confirming or refuting instance. Thus, if I maintain that the ability of a theory to solve a certain problem is much more significant epistemically than its ability to solve another, I must be able to give reasons for that preference. Put differently, I have to be able to show that the probative significance of the one problem for testing theories of a certain sort is indeed greater than that of the other. I might do so by showing that the former outcome was much more surprizing than the latter or more general than it. I may thus be able to motivate a claim for the greater importance of the first problem over the second by invoking relevant epistemic and methodological criteria. But if none of these options is open to me; if I can only answer the question "why is solving this problem more important probatively than solving that one?" by replying, in effect, "because I am interested in solving this rather than that", then I have surrendered any claim to be shaping my beliefs rationally in light of the evidence.

We can put the point more generally: the rational assignment of any particular degree of probative significance to a problem must rest on one's being able to show that there are viable methodological and epistemic grounds for assigning that degree of importance rather than another. Once we see this, it becomes clear that it is in principle possible to ascertain the empirical support for each theory in a manner which does not require a prior commitment to one paradigm or the other.

Let me expand on this point by using an example which has been cited extensively by both Kuhn and Doppelt: the Daltonian 'revolution' in chemistry. As Doppelt summarizes the Kuhnian position:

> ... the pre-Daltonian chemistry of the phlogiston theory and the theory of elective affinity achieved reasonable answers to a whole set of questions effectively abandoned by Dalton's new chemistry. (Doppelt 1978, p. 42)

Because Dalton's chemistry failed to address many of the questions answered by the older chemical paradigm, Kuhn thinks that the acceptance of Dalton's approach deprived "chemistry of some actual and much potential explanatory power" (Kuhn 1962, p. 107). Indeed, for most of the 19th century, Daltonian chemists were unable to explain many things which the older chemical traditions could make sense of. On the other hand, as Kuhn stresses, Daltonian chemistry could explain a great deal which had eluded earlier chemical theories. In short, "the two paradigms seek to explain different kinds of observational data, in response to different agendas of problems..." (Kuhn 1962, p. 43).

This 'loss' of solved problems during transitions from one major theory to another is an important insight of Kuhn's; I have elsewhere tried to trace out some of the implications of non-cumulative theory change for scientific epistemology (Laudan 1984, ch 5.). But this loss of problem-solving ability through paradigm change, although real enough, does not entail – as Kuhn claims – that proponents of old and new paradigms will necessarily be unable to make congruent assessments of how well-tested or well-supported their respective paradigms are.

What leads Kuhn and Doppelt to make this mistake is their assumption that the centrality of a problem on one's explanatory agenda necessarily entails one's assigning a high degree of epistemic or probative weight to that problem when it comes to determining how well-supported a certain theory or paradigm is. In general the observations to which a reasonable scientist attaches the greatest probative or epistemic weight are those instances which test a theory especially 'severely' (to use Popper's marvellous phrase). The instances of greatest probative weight in the history of science (e.g., the oblate shape of the spherical earth, the Arago disk experiment, the bending of light near the sun, the recession of Mercury's perihelion, the re-constitution of white light from the spectrum) have generally *not* been instances which came high on the list of problems which scientists developed their theories to solve. A test instance acquires probative weight because (for instance) it involves testing one of a theory's surprizing or counter-intuitive instances or because it represents a kind of 'crucial experiment' between rival theories. The point is that a problem or instance does *not* acquire great probative strength simply because scientists would like to be able to solve it. (Quite the reverse, many scientists and philosophers would say. After all, it is conventional wisdom that a theory is not very acutely tested if its primary empirical support is drawn from the very sort of situations it was designed to explain. Most logics of experimental design urge – in sharp contrast to Kuhn – that theories are not to be given high marks simply because they can solve the problems which they were invented to solve.) In arguing that the explanatory agenda which a scientist sets himself automatically dictates that scientist's reasoned judgments about well-testedness, Kuhn and Doppelt seem to have profoundly misconstrued the logic of theory appraisal.

Let us return for a moment to Kuhn's Dalton example. If I am right, Dalton might readily have conceded that pre-Daltonian chemistry solved a number of problems which his theory failed to address. Judged as theories about the qualitative properties of chemical reagents, those theories could even be acknowledged as well-supported *of their type*. But Dalton's primary interests lay elsewhere, for he presumably regarded those earlier theories as failing to address what he considered to be the central problems of chemistry. But this is not an epistemic judgment: it is a pragmatic one. It amounts to saying "these older theories are well-tested and reliable theories for explaining certain features of chemical change; but those features happen not to interest me very much." In sum, Kuhn and Doppelt have failed to offer us any grounds for thinking that a scientist's judgment about the degree of evidential support for a paradigm should or does reflect his personal views about the problems he finds most interesting. That, in turn, means that one need not share an enthusiasm for a certain paradigm's explanatory agenda in order to decide whether the theories which make up that paradigm are well- or ill-tested. It appears to me that what the Kuhn-Doppelt point really amounts to is the truism that scientists will tend to invest their efforts exploring paradigms which address problems those scientists find interesting. But that is a subjective and pragmatic matter which can, and should, be sharply distinguished from the question whether one paradigm or theory is better tested or better supported than its rivals. Neither Kuhn nor Doppelt has made plausible the claim that, because two scientists have different degrees of interest in solving different sorts of problems, it follows that their *epistemic* judgments of which theories are well-tested and which are not will necessarily differ.

We are thus in a position to conclude that the existence of conflicting views among scientists about which problems are interesting apparently entails nothing about the *incompatibility* or *incommensurability* of the epistemic appraisals which those scientists will make. That in turn means that these real differences of problem-solving emphasis do nothing to undermine the viability of a methodology of comparative theory assessment – insofar as that methodology is epistemically rather than pragmatically oriented. It seems likely that Kuhn and Doppelt have fallen into this confusion because of their failure to see that acknowledged differences in the motivational appeal of various problems to various scientists constitute no rationale for asserting the existence of correlative differences in the probative weights properly assigned to those problems by those same scientists.

The appropriate conclusion to draw from the features of scientific life to which Kuhn and Doppelt properly direct our attention is that the *pursuit* of (and doubtless the recruitment of scientists into) rival traditions is – and doubtless should be – influenced by pragmatic as well as by epistemic considerations. That is an interesting thesis, and surely a sound one; but it does nothing to undermine the core premise of scientific epistemology, viz., that there are principles of empirical or evidential support which are neither paradigm specific, hopelessly vague nor individually idiosyncratic. Not only may such principles exist, but they are sometimes sufficient to guide our preferences unambiguously.

Virginia Tech

REFERENCES

Doppelt, G. (1978) "Kuhn's Epistemological Relativism," *Inquiry*, 21, pp. 33-86.

Kuhn, T. (1962) *The Structure of Scientific Revolutions*. Chicago: University of Chicago Press.

_______ (1970) "Reflections on My Critics, *Criticism and the Growth of Knowledge*. eds. I. Lakatos and A. Musgrave. Cambridge: Cambridge University Press.

Kuhn, T. (1977) *The Essential Tension*. Cambridge: Cambridge University Press.

Laudan, L. (1982) "Two Puzzles about Science," *Minerva*, 20, pp. 253-268.

_________ (1984) *Science and Values*, Berkeley: University of California Press.

JOHN WORRALL

SCIENTIFIC DISCOVERY AND THEORY-CONFIRMATION*

1. INTRODUCTION

Although I find most recent challenges to older 'positivistic' views in philosophy of science either unchallenging or unconvincing, there is one respect in which the new 'post-positivists' are, I believe, definitely right and the older 'positivists' definitely wrong. Reichenbach, Carnap, Popper and others all agreed that philosophy of science is exclusively concerned with the logical analysis of the merits of theories already 'on the table'. Of course, these thinkers were ready to allow that the question of how a theory arrived on the table could be a fascinating one, but they held that it was a question of no interest to a philosopher as such. In particular, to hold that the origins of a theory have any relevance for the appraisal of its scientific merits was, according to these philosophers, to commit one form of the 'genetic fallacy'.

While I agree with the 'post-postivists' that this thesis is wrong, I also agree with Laudan, who, in a recent and hard-hitting article (Laudan 1980), claimed essentially that, while the 'post-positivists' have produced a good deal of bluster, they have produced no really solid argument for just why considerations of theory-discovery should have any relevance for any specifically philosophical issue. I shall try here to supply some solid argument for this claim.

In fact I shall argue the following specific *thesis*: in order to decide whether a particular empirical result supports or confirms or corroborates a particular theory the way in which that theory was developed or constructed needs to be known – more especially, it has to be checked whether or not that empirical result was itself involved in the construction of the theory.[1] If this thesis is correct then it follows, of course, that heuristic questions are indeed of central philosophical significance - for the question of the rationale for scientific change is surely a philosophical one, and any adequate rationale for scientific change is surely bound to give a substantial role to consid-

J. C. Pitt (ed.), Change and Progress in Modern Science, 301–331.
© 1985 by D. Reidel Publishing Company.

erations of empirical support. So, if indeed empirical support, probably *the central* notion from the 'context of justification', is discovery-dependent, then justification and discovery are not separate but rather interdependent contexts.

2. THE 'HEURISTIC VIEW' OF EMPIRICAL SUPPORT

When does a piece of empirical evidence support a scientific theory? This clearly has a lot to do with the logical relation between the theory and the evidence. Ideally, the evidence statement will assert that certain initial conditions held and that the experiment or observation had a certain outcome *and* the theory together with the statement of initial conditions will logically entail the outcome statement. Equally clearly, support or confirmation cannot be *simply* a question of whether or not the theory entails the evidence (I shall, from now on, talk about the theory entailing the evidence as shorthand for the above more complicated statement concerning initial conditions and outcome). For one thing, a theory which entails a given piece of evidence *e* could be created quite trivially by taking any theory *T* of which *e* is independent and forming a new theory *T′* by simply conjoining *T* and *e*. *T′* obviously entails *e* but surely no one would want to say that it is confirmed by it – at least not if the judgment is meant to carry with it any consequences, no matter how tentative or hedged, for the rationality of believing *T′* or using it in technological applications. One reason why such a 'cobbled up' *T′* is not regarded as thus confirmed by *e* despite entailing it is, of course, that exactly the same trick could be pulled starting from a theory *S* with radically different observational consequences than *T*: we could just as well 'tack' *e* onto this different theory *S*, and the resultant *S′* would guide applications quite differently than does *T′*. This is, of course, the venerable 'tacking paradox', so often raised, only to be swept under the carpet.[2]

Scientists and historians tend to ignore the 'tacking paradox' as a philosophers' plaything: no scientist would ever cheat so blatantly as by tacking. There are, however, real historical cases of procedures not so very different from tacking. Take the case of the early nineteenth century revolution in optics. The story usually given is that this revolution occurred because the wave theory of light turned out to make many predictions (for example about diffraction and polarisation effects) which proved correct observationally and which the rival, longer established corpuscular theory either contradicted, or, more often, simply could not match. Such accounts make it easy to see why scientists would rationally prefer the wave theory; unfortunately a little historical research shows them to be factually quite inaccurate. Far from diffraction and interference effects being new phenomena which surprised and defeated the early nineteenth century corpuscularists, the phenomena had been known since at least the mid-17th century and corpuscularists had been in the business of explaining them since Newton. They attributed the effects to close-range forces of 'inflexion' which ordinary, gross matter exercises on the minute corpuscles of light, thus bending them away from their naturally rectilinear paths. Since the 'inflecting' forces are alternately attractive and repulsive at different distances from the body, light and dark *fringes* are produced. As for polarisation effects, again many of these (those connected with the two rays produced by birefringent crystals) had been known since before Newton, who suggested that they could be explained by endowing the light-corpuscles with 'sides' or 'poles'. Newton's suggestion was turned into a highly elaborate theory by the French scientist J. B. Biot. Biot's theory was of polarised light-particles revolving around 'axes of polarisation', axes which themselves could perform complicated movements.[3]

I do not claim that a corpuscular theory was ever actually produced which was fully observationally equivalent to the wave theory as, say, developed by Fresnel in the 1820s. I *do* claim that many of the so-called crucial effects could in fact be explained within the corpuscular approach and that many more could have been brought within the corpuscularists' net by straightforwardly developing the ideas of Newton and Biot. The reason this exercise was not completed was not its impossibility but the fact that, as corpuscularists eventually realised, it was not scientifically worthwhile. Corpuscularists had managed, by various contrivances, to account within their theory for certain effects which are entailed quite naturally by Fresnel's wave theory. Yet neither the scientific community of the time nor subsequent generations of scientists took this achievement of the corpuscularists at all seriously. The great majority of scientists took it that these effects which the corpuscularists had forced into their framework still provided no genuine support for their theories. In other words, the wave theory was regarded as well ahead of its rival in terms of empirical support even before the corpuscularists finally surrendered; the wave theory's superiority was taken as established even by some phenomena, a correct description of which fully-fledged versions of the corpuscular theory entailed.

I labor this point because it does seem to me that we must capture this sort of intuitive judgment above all others if we are to have any hope of explaining scientific change as rational. Clearly if our formal account of empirical support *is* to capture this intuition then it must make derivability of the evidence at most a necessary and *not* a sufficient condition for empirical confirmation.

Someone who realised this rather clearly long ago is William Whewell. Writing about this same optical revolution he said:

> When we look at the history of the emission theory of light, we see exactly what we may consider as the natural course of things in the career of a false theory. Such a theory may, to a certain extent, explain the phenomena which it was contrived to meet; but every new class of facts requires a new supposition – an addition to the machinery; and as observation goes on, these incoherent appendages accumulate, till they overwhelm and upset the original framework. Such was the history of the hypothesis of solid epicycles; such has been the history of the hypothesis of the material emission of light. In its simple form, it explained reflection and refraction; but the colours of thin plates added to it the hypothesis of fits of easy transmission and reflection; the phenomena of diffraction further invested the particles with complex hypothetical laws of attraction and repulsion; polarisation gave them sides; double refraction subjected them to peculiar forces emanating from the axes of crystals; finally, dipolarisation loaded them with the complex and unconnected contrivance of moveable polarisation; and even when all this had been assumed, additional mechanism was still wanting. There

> is here no unexpected success, no happy coincidence, no convergence of principles from remote quarters; the philosopher builds the machine, but its parts do not fit; they hold together only while he presses them; this is not the character of truth. (Whewell 1837, II, 340)

The reason why Biot's theory of fixed and moveable polarisation is not genuinely confirmed by the polarisation effects seems obvious: the theory was actually constructed using those facts, it was rigged precisely to yield the known phenomena, or, in Whewell's phrase, it was 'contrived to meet' the known results. It had to look for real support to facts other than these, but no such support was forthcoming. To take a more recent example, the reason why classical physics as modified by the incorporation of the Lorentz-Fitzgerald contraction hypothesis is not generally taken to be supported by the result of the Michelson-Morley experiment is surely this. It is generally assumed that the amount of contraction postulated by Lorentz and by Fitzgerald was "read off" the null experimental result.[4] In other words, *the experimental outcome was used in the construction of the theory and hence cannot then be used again in its support.*

This seems the obvious lesson to draw from such cases and several philosophers have at first drawn it. Here, for example, is a passage from Moritz Schlick, which is also quoted with approval by Karl Popper in an early work:[5]

> the confirmation of a prediction means nothing else but the corroboration of a formula for those data which were not used in setting up the formula. Whether these data had already been observed or whether they were subsequently ascertained makes no difference at all.

There is, however, no trace of any heuristic element in the formal accounts of confirmation developed by members of Schlick's Vienna Circle and its later followers. And, as for Popper, he and later Popperians quickly switched to the 'historical' or 'temporal' account of confirmation.[6] According to this temporal account, the actual time-order of theory and evidence is a crucial factor in support. If a theory entails some empirical result but the result was already known to hold when the theory was first formulated, then the theory receives much less confirmation than if the empirical effect was found to occur only as a result of the theory's predicting it. (Indeed on Lakatos's extreme version of the 'temporal view' a theory is confirmed *only* by correct predictions and *not at all* by any evidence which it yields but which was already known.) (Lakatos 1978, I, p. 36)

The temptation to embrace this temporal view is strong: it captures some of the intuitive confirmation judgments which I have highlighted and makes them depend on a factor which is reasonably clear-cut and objective. We know that the Michelson-Morley result was obtained in 1887, while the Lorentz-Fitzgerald contraction hypothesis was formulated only in 1892. On the other hand, the existence of the planet Neptune was definitely *pre*dicted by Newtonian theory, the existence of the star-shift by general relativity theory, and so on. The important factor in determining the extent of confirmation is, at any rate usually, unambiguous. But if we have to look at the way that a theory was constructed in order to decide on confirmation, then things seem to become very messy. Wouldn't we need to have Einstein's psyche available for inspection in order to know how he arrived at his theory of relativity? Surely the philosopher should shun such subjective matters. And if we did know how theories were arrived at, mightn't it sometimes be true that two scientists arrived at the same theory in two quite different ways, and wouldn't this mean that a theory as arrived at by scientist *A* might be confirmed by evidence *e* on this heuristic view, while the *selfsame* theory as arrived at by scientist *B* is not confirmed by the self-same evidence? Surely this is a *reductio ad absurdum* if ever there was one?[7]

But the 'new predictions count more' view, whatever its appeal, is wrong. First, it seems altogether mysterious just why the time-order of theory and evidence *should* matter. My intuitions here are all on the side of John Stuart Mill who, in an often quoted passage wrote:[8]

> it seems to be thought that an hypothesis ... is entitled to a more favourable reception, if, besides accounting for all the facts previously known, it has led to the anticipation and prediction of others which experiment afterwards verified ... Such predictions and their fulfillment are, indeed, well calculated to impress the

> ignorant vulgar, whose faith in science rests solely on similar coincidences between its prophecies and what comes to pass. But it is strange that any considerable stress should be laid upon such a coincidence by persons of scientific attainments.

Mill's sentiments were equally famously echoed in this century by John Maynard Keynes, who wrote:

> [the] peculiar value of prediction ... is altogether imaginary ... The question of whether a particular hypothesis happens to be propounded before or after examination of [its experimental consequences] is quite irrelevant. (Keynes 1921, p. 305)

Mill and Keynes were surely right that the time-order suggestion lacks both inherent plausibility and intuitive justification. There is however another strong argument against it: it fails to capture all of scientists' intuitions about particular cases. It is certainly true that *some* results which a theory can account for but which were already known before the theory was developed are not regarded as lending the theory support. But this is far from *always* being the case. The facts about Mercury's perihelion advance, for instance, had long been known and been at the center of astronomical concern before the development of general relativity. Yet that theory is regarded as receiving enormous support from the facts about Mercury's orbit – quite as much support as it receives from the newly predicted facts about star-shift. Similarly, Newton's theory's precise account of the already known facts of the moon's orbit and of the earth's oblateness seem to have been taken as supporting his theory quite as strongly and dramatically as any novel prediction.[9]

It seems that while all novel, correct empirical predictions count significantly in favor of the theory which made them, *not* all empirical results which count significantly in favor of a theory are novel. The time-order of theory and evidence cannot *in itself* be the important factor in distinguishing genuine and spurious confirmations.

Where else might the missing factor be found? Let's next try the mainstream tradition in confirmation theory, namely Bayesianism. One usual Bayesian account is that the important quantity so far as confirmation is concerned is the ratio $p(h/e)/p(h)$: confirmation is the greater the higher is this ratio. One problem is that this account is principally designed for the case of statistical hypotheses and even some of its best friends would admit that it leaves something to be desired in the case of deterministic theories (the only ones I am concerned with in this paper). Nevertheless, let's see what Bayesianism can tell us about this case. Since $p(h/e)=p(e/h).p(h)$, the above ratio can also be expressed as $p(e/h)/p(e)$. I am interested only in the case in which empirical results are entailed by the relevant theories, that is in which $p(e/h)=1$; and so the Bayesian confirmation ratio is always going to be just $1/p(e)$ – one over the "prior probability" of the evidence e. The question, as so often with Bayesianism, is 'prior to what'?

If $p(e)$ is taken as an 'absolute probability', as some measure simply of the logical strength of e then, while such intuitive judgments might be delivered as that several pieces of independent evidence confirm h more strongly than a single piece, there seems no hope of making the sort of discrimination which I have argued is necessary among the confirming effects of 'single', 'atomic' pieces of empirical evidence.

If we interpret $p(e)$ as the probability of e, given everything that was accepted at the time the relevant h was proposed, then the Bayesian approach seems simply to incorporate a formalised version of the temporal account of confirmation – an account which, I have argued, is unacceptable.

The only hope that I see of successfully utilising the Bayesian system is to identify $p(e)$ with the probability of e, relative not to the whole of 'background knowledge' at the time h was proposed, but relative only to that part of 'background knowledge' which was used in the construction of h. We could then say that h is not confirmed by e unless $p(h/e)/p(h)>1$. Assuming that h entails e, this ratio would be greater than one if $p(e)$ is less than one, and would be equal to one only if $p(e)$ equals one; and $p(e)$ would equal one on this construal precisely if e were already entailed by some results used in the construction of h. While this is the right result, it clearly amounts to nothing more than a reformulation of the heuristic view, and giving it this Bayesian formulation has in itself taken us no further forward. We are still under an obligation to explain precisely what it means for empirical results to be used in the construction of a theory and to face certain difficulties with the view. Once these problems have been solved, that is, once the hard work has been done, then we certainly can, if we choose, express the result in Bayesian terms.

The Bayesian might resort to other functions of probabilities to represent confirmation values. The only other factor at his disposal, however, in building these functions seems to be $p(h)$ – the 'prior probability' of the hypothesis. The confirmation function might be so constructed that, if the prior probability of *h* is sufficiently low, even a large body of positive evidence leaves *h* without appreciable confirmation. If assignments of prior probabilities to hypotheses are not to be simply reflections of the quirks of individual scientists, then the suggestion amounts, in less formal terms, to the proposal that account should be taken not only of whether a theory entails the right results, but also of whether the theory is sufficiently plausible or simple or unified.

The suggestion that plausibility or simplicity or some kindred notion is the key to this problem is an attractive one, and one which several philosophers, both Bayesian and non-Bayesian, have adopted. The general idea would be that once a theory becomes sufficiently complex, it is not confirmed even by empirical results which it correctly entails. In the particular case of early 19th Century optics, the suggestion would be that the fully-fledged corpuscular theory of light, complete with alternately attractive and repulsive forces of inflexion and particles with axes of polarisation which perform jerky and irregular movements, is just too implausible and complicated for the fact that it entails correct results about diffraction and polarisation effects to count in its favor.

Biot's theory is undeniably enormously implausible and complex, intuitively speaking. Moreover, there is, as I shall indicate later, a close connection between the heuristic view of confirmation and the intuitive idea of simplicity. Still simplicity is *not* the obvious (and old) solution to the problem that I have been focussing on.

First, it is by no means obvious that we would do ourselves any favors by invoking plausibility or simplicity or the like. Allowing that heuristics play a role does indeed threaten to make confirmation a dangerously unclear and subjectivist notion. But does invoking plausibility or simplicity instead really improve matters? After all, centuries of effort aimed at producing clear and objective notions of plausibility and especially simplicity have failed notoriously to bear fruit.

Secondly, if something like simplicity/complexity *were* the key to this problem, then there would have to be no cases of the following sort. A theory is proposed which is (*i*) intuitively simple, (*ii*) logically entails an accepted evidence statement *e*, and yet (*iii*) is intuitively not supported by *e*. But there are such cases. The following is an artificially simple, yet instructive, abstract example. There are two theories T and T' which appear equally simple; and in the area of the two theories only one relevant piece of evidence is known. This evidence, *e*, is that the two variables x and y take on the values 2 and 10 respectively in a certain circumstance. One of the theories, T say, yields *e* directly and without artifice. The relevant equation yielded by T', however, is $y = ax$, where a is, at this stage, a free parameter. We can now use *e* to fix the value of a at 5 – thus creating a new theory T'' which (in the sense is which I am using the word in this connection) entails *e*. There are now, therefore, two theories, T and T'' each of which entails the known relevant evidence. I claim we should nonetheless say that – provided other things are equal, as we are assuming they are – *e* favors T above T'': T was genuinely tested by *e* (or so we are assuming), but T'' on the contrary was *not* genuinely tested by *e* – T'' *had* to get *e* right because of the way that it was constructed. If the invocation of simplicity did indeed solve this problem by yielding this judgment, it would have to be clear that T'' is more complex than its predecessor T', that is, simply filling in the value of an initially free parameter would have to increase automatically the complexity of a theory. This is surely not right.

I have argued, then, that the attempts to avoid drawing what I claim is the obvious lesson of cases like Biot's theory all fail. This 'obvious' lesson, remember, was that even if a theory entails an accepted empirical result, the theory is not supported by that result, if the result was *used in the construction of* the theory. I claim that we must return to this obvious lesson (all else having failed) and face up squarely to any difficulties it brings in its wake.

Perhaps the most obvious difficulty with this heuristic view of empirical support is its vagueness: what exactly does it mean for an empirical result to be used in the construction of a theory? I shall try to show that this is at any rate not a hopeless problem – by showing that there are some especially clear-cut and clearly describable cases. I shall then outline and try to rebut various more specific criticisms that have been made of this heuristic criterion.

3. SOME CLEAR-CUT CASES OF THE USE OF EMPIRICAL RESULTS IN CONSTRUCTING THEORIES

(a) "*Exception-Incorporation*"

Thomas Young did not produce one version of his celebrated principle of interference, but several quite different versions. The principle began life in 1802 in the quite general form that *any two* near-parallel beams of light which affect the same area will there produce alternating light and dark interference fringes.[10] Several critics, notably Young's arch enemy Henry Brougham, pointed out that this general proposition is quite readily refuted by a host of everyday facts – for example, if the proposition were correct, two closely contiguous candles casting their light on some nearby wall ought to produce fringes. Young had accounted for external straight-edge diffraction fringes by assuming that they are caused by the interference of two 'portions' of light – a 'portion' of direct light and another 'portion' 'inflected' by the diffracting object.[11] The obvious difference between this case and the two candles case is that the two candles are two *separate* sources of light, whereas the two 'portions' in the diffraction case originate in the same source. Young therefore responded to Brougham's objection by switching to the claim that interference occurs when, *and only when*, the two interfering 'portions' of light originate in the same source. (It is important to remark here that Young had no clear ideas about what came to be called the coherence of different light beams.) The modified principle entails, of course, the experimental result that the two contiguous candles will *not* produce interference fringes. While this is precisely what is observed, no one (least of all Young himself) would claim that it constitutes empirical support for his modified theory.

The logical structure of this episode is clear. It involves a rather stronger version of what Lakatos in his studies of mathematical discovery called 'exception barring'

(Lakatos 1976), – an appropriate name might be 'exception-incorporation'. An exception having been found to some general principle, some characteristic is sought which the exception does not possess, but which known positive instances do, and the original principle is then claimed to hold *when and only when* this characteristic is satisfied.

To avoid misunderstanding, let me say immediately that there is of course no question but that Young was right to take the empirical facts into account. His modified principle is consistent with experiment and hence is clearly preferrable to the original principle which is inconsistent with experiment. I claim only, first, that this is a clearly articulable way in which an experimental result may be used in the construction of a theory, and, second, that the intuition is clear that, although the modified principle entails the candles result which refuted its predecessor, the modified principle is not supported by the result. Moreover, should a different theory be produced which, directly and without artifice, entails the same results, then this new theory should, I claim, be judged superior to Young's modified theory *on this very ground* – that is, without necessarily requiring that the new theory be confirmed independently of these results.[12]

(b) *Parameter-adjustment*

A second type of case has already been touched upon and is, perhaps, clearest of all. In cases of parameter-adjustment, a theory is first proposed which has some free parameters – these may be parameters in the usual specific sense (constants in some mathematical equation) or in a more general vaguer sense. For example, the basic idea behind the mainstream corpuscular-theoretic account of diffraction – that the phenomenon was due to alternately attractive and repulsive forces of 'inflexion' – gave corpuscularists not a few degrees of freedom: the distances at which the force turned from attractive to repulsive and back again, the intensity of the force at given distances, the mass of the corpuscles themselves could all be juggled with in an attempt to accommodate the known experimental results. It is clear that there will be whole sets of results which the theory can be made to yield by judiciously reading off suitable values of the parameters from those very results.

It is easy to see the methodological defects of the original corpuscular theory with all its free parameters. It has a low, perhaps zero, degree of testability. But what of the theory with all the free parameters filled in? It certainly is testable in the logical sense – it has consequences which are directly comparable with, and indeed which compare favorably with, experiment. But this is no wonder since these consequences were all 'written into' the theory *via* the parameter-fixing. Many of Popper's most perspicacious remarks are, I believe, based on an *intuitive* notion of testability which readily applies here. The theory with the fixed parameters is certainly not intuitively testable against the observations used to fix the parameters; since, *because of the way it was constructed,* it was never at risk of refutation from these observations. *This fact cannot be discovered by simply inspecting the logical form of the theory.* Hence Popper's formal account of testability, which is in logical terms (set of 'potential falsifiers'), fails to do full justice to the notion; and Popper has never, I think, fully and clearly realised that his intuitive remarks about testability rest on heuristic considerations. It is easy to make a theory testable in the logical sense of having it entail empirical results which are already known. It will then, of course, be true that *were* these results not correct, the theory would be refuted. But this is no genuine *test,* since the results were already known to hold before being incorporated into the theory. Perhaps the clearest way to state my thesis is, then, this: Popper is right that a theory is genuinely supported only by passing real tests, but in order to decide whether some empirical result constituted a real test of some theory, we have to look at how the theory was constructed.[13]

Once again I should emphasize that I am *not* claiming that there is anything unscientific about parameter-adjustment – many of the best scientific theories contain parameters whose values had to be 'read off' the facts. (An example is the classical wave theory of light and the precise wavelengths of light of different colors.) These theories then went on to make *further* empirical predictions which did indeed lend them support. I *do* claim however that the results which were used to fix the parameter values provide no such support; and, especially, that when one theory has accounted for a set of facts by parameter-adjustment, while a rival accounts for the same facts directly and without contrivance, then the rival does, but the first does not, derive support from those facts.[14]

Many cases in which an empirical result was used in constructing a theory are cases of parameter-adjustment in either the more specific or broader sense. The reason why the Lorentz-Fitzgerald hypothesis, for example, is not regarded as supported by the Michelson-Morley result is that Lorentz and Fitzgerald seemed to provide themselves with a length contraction parameter, only to adjust this parameter nicely to yield the known null result. (But see footnote 4 above.)

(c) *The 'Correspondence Principle' as a Heuristic Device*

Perhaps the most important way in which empirical results may be used in the construction of a theory is rather more indirect than the two considered so far. Some theories in science have been developed *via* the use of the 'correspondence principle' as a heuristic device. It has proved difficult to give a precise general formulation of this principle. But it does seem clear that the idea that a new theory should, in some way or other, explain the *empirical* success of its predecessor has operated in the history of science – and it has operated, not just as an *adequacy requirement*, but sometimes as a heuristic device in the actual development of that new theory. Here empirical results play an indirect role – the correspondence principle guarantees that the new theory will 'go over' to the older one in those domains where empirical results have shown that the older theory is correct. I shall try to make this rather vague talk more precise by considering a famous historical example.

No logical claim in the history of physics has excited more comment than Newton's claim to have deduced his theory of gravitation 'from the phenomena' supplied by Kepler. It was for a long time widely believed that Newton had done precisely what he claimed to have done and that his deductive method should serve as a paradigm for other scientific innovations. On the other hand, it is clear that Newton's theory and Kepler's laws are logically inconsistent with one another. Although Newton was clearly aware of this plain fact, it was only pointed out plainly by Duhem in 1904 and was subsequently re-emphasized by Popper and others. (Duhem 1904 and Popper 1973) It is, of course, impossible for a valid deduction to lead from consistent premises to conclusions which contradict those premises.

Scientists, though, have either ignored Duhem's point or regarded it as an over-fastidious piece of logic-chopping. Their chief spokesman was Max Born who insisted that, whatever Duhem and formal logic might say, Newton did indeed infer his theory from Kepler's laws.

This confused situation can be clarified in a way which should satisfy both logician and scientist and which, at the same time, throws a great deal of light on the process of scientific discovery.

First, there is no doubt, of course, that Duhem was correct that Newton's theory logically contradicts Kepler's laws – the path which a planet ought to follow according to Newton's theory (and given uncontested initial conditions) is *not* an ellipse, for example. Newton could not have validly derived his theory from Kepler's laws. However, the deduction of various results from Kepler's laws did indeed play an important role in Newton's discovery of his theory of universal gravitation – as Newton himself explained clearly in the *Principia*. He for instance proved that if the sun is considered either at rest or in uniform rectilinear motion, and if Kepler's laws are strictly true of a planet orbiting the sun, then the planet is subject to a net force directed towards the sun, a force which is inversely proportional to the square of the distance between the planet and the sun, and directly proportional to the mass of the planet. The deduction involves Newton's own second law of motion but no other synthetic assumption. This result does not of course imply the principle of universal gravitation – indeed the assumption on which the result is based, that the sun is at rest or in uniform rectilinear motion, is logically inconsistent with the gravitational principle. Nonetheless, instead of some enormous and inexplicable leap being needed for Newton to arrive at his theory, the gap between this result (arrived at deductively remember) and the gravitational theory is intuitively quite small.[15] The gap is bridged by the third law of motion. Again Newton did not simply assert this law, but instead argued for it. He had in fact two main arguments: one largely observational and the other theoretical. The latter is especially simple. It starts from the Cartesian principle of the conservation of total momentum, applied to a two body system. Newton gave this principle the vectorial form:

$$m_1 \vec{v_1} + m_2 \vec{v_2} \quad \text{constant.}$$

Differentiating with respect to time gives

$$m_1 d\vec{v_1} / dt = -m_2 d\vec{v_2} / dt;$$

that is, given the second law of motion,

$$f_{12} = -f_{21},$$

which is the third law for the two body case.

It follows, of course, that if a planet is acted on by a force directed towards the sun, the sun must be acted on by a force directed towards the planet and hence will not be unaccelerated. Kepler's laws were predicated on the assumption that the sun is motionless. Hence again we have inconsistency between Newton and Kepler. This inconsistency is, however, surely unimportant when compared to the following facts. First, Newton knew that since planetary masses are very small compared to that of the sun, the latter will be *almost* unaccelerated according to his theory. Second, Newton showed that it follows from the assumption that a planet describes a strict ellipse relative to the sun and with the sun at one focus that (*a*) the center of gravity *C* of the system consisting of the sun and planet is either fixed or moves uniformly, (*b*) the planet and sun each describe ellipses relatively to the other and these ellipses have a common focus at *C*, (*c*) the force acting on the planet is constantly directed towards *C* and therefore towards the sun, and (*d*) this force is inversely proportional to the square of the distance between the planet and *C* and therefore to the square of the distance between planet and sun.

Newton still needed to assume that this inverse square force operates between *any two* bodies; and this assumption leads to a contradiction with the assumption of perfectly elliptical planetary orbits. Nonetheless, this extension from the two body case is the obvious one to make: Newton's achievement does not of course lie in his having made this step, but instead in the earlier and purely mathematical, deductive steps to the single planet results.[16]

This example points to many methodological lessons, not all of which have yet been drawn clearly – despite intensive study of the case. Some lessons are, however, clear. The case shows for example that the requirement that a new theory match its predecessor's success may *automatically* be satisfied in certain cases ('by construction'). It hence explains what Lakatos may have had in mind – if only in a vague and confused way – in making the seemingly extreme claim that in science the *only* empirical consequences of a new theory which count are those which go beyond its predecessor. (Lakatos 1978, I, 36) *Of course* it matters that the new theory get right what its predecessor already did, but this may sometimes be *guaranteed* by the method of construction of the new theory. In that case the sole question of interest may well indeed be whether the new theory is also empirically correct where it transcends the old.

The central points which this example illustrates are, however, these. First that it is indeed a fact, *and a plain one,* that at any rate weak versions of Kepler's laws were used by Newton in the development of his theory. Second that this fact needs to be taken into account in assessing correctly that theory's empirical support. The first of these facts underscores the point that the heuristic path leading to a theory may well be open to public inspection and not be hidden in the recesses of some scientist's psyche.[17] Other equally clear examples can be cited – for example, it is a matter of public record that Planck used the correspondence principle in his development of the relativistic version of the second law of motion. The requirement that the new theory go over to the classical law as v/c tends to zero figured as an explicit premise in Planck's development of the new theory (Zahar 1973).

The second point concerning empirical support is not quite as clear in this case as it is in some others. First of all, there is a measure of *overdetermination* here – Newton's theory can be derived without making use of the full power of all three of Kepler's laws. So the fact that Newton's theory yields all three laws as approximately correct should clearly be regarded as lending *some* support to the theory. Moreover, at least after the demise of Cartesian vortex theory, Newton had no real competitors and so not much hangs in this case on whether or not we judge those parts of Kepler's laws which Newton did use as supportive.[18] Given that Newton's theory made extra predictions which proved correct – notably about the corrections of Kepler's laws, so-called perturbations, but also about the oblateness of the earth, movements of the moon's axis and so on – it won't matter much if we say that this independent support licenses counting even those facts used in the construction of the theory as supportive. The principles of support-accounting advocated by the heuristic view really come into their own only when two rivals are vying – one really predicting results, the other accommodating itself to those results. Nonetheless, it is surely better even in this case to make a distinction between the results *used* in the construction of the theory and those predicted by the theory once constructed: the extra predictions (perturbations, earth's oblateness, *etc.*) surely count as stronger support for Newton's theory than do the results already taken to support Kepler's laws. I think this reflects the attitude of most scientists. Here, for example, is a very revealing passage from a recent textbook on Newtonian Mechanics:

> If universal gravitation had done no more than to relate planetary periods and distances, it would still have been a splendid theory. But like every other good theory in physics, it had predictive value; that is, it could be applied to situations besides the ones from which it was deduced. Investigating the predictions of a theory may involve looking for hitherto unsuspected phenomena, or it may involve recognizing that an already existing phenomenon must fit into the new framework. In either case the theory is subjected to searching tests, by which it must stand or fall. With Newton's theory of gravitation, the initial tests resided almost entirely in the analysis of known effects – but what a list! (French 1971, pp. 5-6)

And the author goes on to give the list which includes the bulging of the earth, the variation of the gravitational acceleration with latitude and the change in direction of the earth's axis of rotation. The passage conforms precisely to the heuristic account: the theory 'stands or falls' by the results of 'searching tests' on it; a test can concern *either* 'hitherto unsuspected phenomena' *or* 'already familiar phenomena' (both stand entirely on a par), the only empirical consequences which do *not* supply tests are those from which the theory was 'deduced'.

3. SOME DIFFICULTIES FACING THE HEURISTIC ACCOUNT OF EMPIRICAL SUPPORT

(a) *Is the Account Subjectivist and/or Relativist?*

The heuristic account of confirmation is far from new. Hints of it, usually mixed in with elements of other views, are to be found in several earlier philosophers. The account's present lease of life perhaps began in 1973 – fathered by some remarks in a paper by Elie Zahar (1973). The account has already met a fair amount of criticism (Musgrave 1974 and 1978, Gardner 1981).

Underlying several of these criticisms is the fear that the heuristic path which led to a theory may not be fully articulable – who knows what really went on in Einstein's or Newton's or Fresnel's minds before they produced their great theories? I hope that what I have said already about the clear-cut cases will have allayed this fear and indicated that we don't in fact need as much information about scientists' psyches as might appear. Scientists usually *argue* for their theories, trying to show that they follow 'naturally' from a combination of certain general metaphysical principles and certain empirical results. They *design* their theories using certain materials and we can always ask whether or not certain particular experimental results were among those materials. Moreover, this question is not usually so difficult to answer. It is clear that versions of Kepler's laws played a role in the genesis of Newton's theory, while the earth's equatorial bulge, say, did not. It is clear that general relativity theory arose from Einstein's extension of his relativity programme to the case of accelerated frames and that Einstein needed no results about the orbit of Mercury, for example, to fix any aspect of his theory. Hence the fact that general relativity yields the (more or less) correct result about Mercury's orbit is surely a great feather in its cap – independently of the fact that the facts about Mercury's orbit were already well-known *and* independently of the historical fact that Einstein was hoping that his theory would indeed deal with the Mercury anomaly. (For it is no part of the heuristic view that it should matter what Einstein was worrying about at the time he produced his theory, what matters is only whether he needed to use some result about Mercury in order to tie down some part of his theory.)

Even when we have no account from the author himself of how the theory was constructed it may not be so difficult to see whether or not some particular empirical result was involved in the process. The case of Thomas Young and his modified principle of interference is again instructive. The modified principle says, remember, that interference occurs beween two 'portions' of light *when, and only when*, certain special conditions apply – in particular the condition that the two 'portions' of light were derived from the same source. Young was not saying that interference fringes are *observable* only under these conditions, but rather that interference *occurs* only under these conditions. Why are these restrictions necessary? This seems a 'natural' question to ask in view of the fact that the basic idea behind the wave theory would seem to make any such restriction superfluous. If light consists of waves in the ether then *any* two light beams affecting the same area ought to interfere in the sense that the resultant disturbance should be the vector sum of the disturbances which would be produced there by either beam separately. Young gives no answer to this 'natural' question and the only conclusion that can be drawn is that the restriction was 'read off' the known refutations of his original, general interference principle.

Of course, an answer to the question was eventually given within the framework of the wave theory. Fresnel argued that the disturbance produced at any point of the ether would indeed be the vector sum of the disturbances from all the sources producing light. If, in particular, these disturbances happened to interfere destructively at that point then darkness would be produced there. We would expect this darkness to be fleeting, however, because the movements of matter creating the light seem likely to be subject to unpredictable fluctuations. This means that there would be no correlation between the fluctuations in different light sources, and so, in this case, the places at which destructive interference occurs would be constantly changing. Hence, since our visual apparatus is not fine-tuned enough to register these quick changes, the illusion of constant illumination is produced. It will only be when a change in the oscillations making up one 'portion' of light is matched automatically by changes in the other 'portions' that the interference pattern produced will be constant and so observable. And matching changes in oscillations will occur only when the various 'portions' left the same source at more or less the same time – matching changes will certainly *not* occur in the light from two different sources, like the two candles. Given Fresnel's theory, then, we would expect *in advance* observable interference fringes in the case of light from the same source and no observable fringes in the case of light from two different sources. It is surely reasonable, then, to regard Fresnel's theory as confirmed by the fact that both these expectations prove correct empirically – quite independently of the historical accident that both results were known before Fresnel formulated his theory. But in Young's case, and independently of any facts about his psyche, he failed to give the scientific community of his time any reason for the restriction of his principle to the case of 'portions' of light originating from the same source, beyond the fact that this was the way things had already been observed to be. So Young's principle is less well supported by these facts than Fresnel's even though both entail the facts. Most importantly, the factors necessary to decide whether or not there is support in the heuristic sense can be discovered in Young and Fresnel's works – no doubt a certain amount of interpretation is necessary to discern them, but no more, I would claim, than is required in discerning precisely what theory is being propounded by a scientist.

If the heuristic path which led to a theory can indeed be specified objectively, then the alleged subjectivism or relativism of my account of support disappears. As for the purported *reductio* (that the same theory may be arrived at in two different ways and hence receive different degrees of support on this account), this can be seen to be in fact a straightforward consequence of the three-term nature of the proposed support relation. If support is a relation not just between an empirical result and a theory, but between an empirical result, a theory, and the way in which the theory was arrived at, then any apparent contradition here disappears: *e* may support *T* relative to one heuristic path, while the same *e* does not support the same *T* relative to a different heuristic path. Indeed, far from this being a *reductio* of the approach, it is precisely what is needed to capture our intuitions: while we may not have historical cases of exactly the same theory being arrived at by different routes, we certainly do have such cases of theories with exactly the same set of observational consequences and yet which intuitively are differently supported by the facts.[19]

One final internal objection to the heuristic account of empirical support runs as follows. 'Admittedly in the case of Thomas Young, say, the historical order seems clear: first he produced a very general principle, then noticed that it was experimentally refuted, and then produced a modified principle by "incorporating the exceptions". But what if Young had not published until this final stage? Worse still, what if Young had simply *dreamed-up* the principle in its already modified form without having to take any notice of empirical results?' It would seem that my account penalizes a careful scientific thinker like the real-life Young who first formulates a general theory and then takes observed exceptions into consideration, and that it rewards a lucky imaginary Young who just happened to dream up the modified, and empirically correct theory.[20]

My answer to this powerful-sounding objection can, I think, only be this: science just isn't like that. Scientific theorizing is *never* a question of throwing out conjectures, more or less at random, most of which are then refuted. On the contrary, as I have already argued, scientific theories are *designed* to meet certain requirements – one of the foremost of which is that they capture in some way or another their predecessors' empirical success. It is massively improbable that an empirically successful theory from one of the more developed sciences will simply have been dreamed-up fully fledged; and scientists in fact seem to take this improbability as an impossibility.

Consider, for example, some successful theory containing parameters fixed at certain values – say the wave theory of light with its precise values for the wavelengths of monochromatic light from particular parts of the solar spectrum. It is, of course, logically possible that some early wave-theorist one night in his sleep received a message from beyond that the wavelength of some monochromatic light from the extreme violet end of the spectrum is some precise number (around 4×10^{5} cm) – this is possible but massively improbable, and apparently never condoned as a practical possibility in science. The scientist will want a *reason* for this precise value. Sometimes the answer will be that the value was dictated by basic and general theoretical considerations; more often, as in this particular case, the answer will be that it could be 'read off' one of several possible experimental results. If so, then this needs to be taken into account in determining empirical support.

(b) *The Question of Justification. Why Should it Matter How a Theory was Arrived At?*

Finally, I want to see what can be done to meet a different type of critic than the ones considered so far. This new critic is ready to allow that the ideas behind the heuristic criterion can be made sufficiently precise, and even perhaps might allow that it is this criterion which scientists intuitively apply, but he questions the *justification* for the criterion. Perhaps scientists do, as a matter of fact, act as if it mattered how a theory was constructed, but why on earth *should* they act in this way?

John Stuart Mill raised something like this point against William Whewell. (Mill 1843, p. 328) As I already suggested, the basic idea behind the heuristic criterion of confirmation is undoubtedly to be found in Whewel's writings. Unfortunately in presenting his ideas on what he called the 'consilience of inductions',[1] Whewell sometimes conflated the heuristic view with the 'purely temporal' account. This gives some plausibility to Mill's complaint that, although successful predictions are well calculated to impress the 'ignorant vulgar' and while they might even, *as a matter of fact*, be given special weight by those of 'scientific attainments', this fact is nonetheless 'strange' – that is, unjustified. As I already said, I agree with Mill that there is no reason on earth why the time-order of theory and evidence should in itself have any relevance for the question of how well a theory objectively stands up to the evidence.

However, once the heuristic account is clearly differentiated from the temporal, and once it is made clear that it is the heuristic account only which is being defended, then the situation surely changes. The justification for the *negative* aspect of the heuristic account seems almost self-evident. If the theory was adjusted so as to yield a certain result, then its yielding that result tells us something only about the ingenuity of man; it tells us nothing about the likelihood that the theory reflects some part of the blueprint of the universe, or even about its 'rational acceptability'. This is especially clear in the most blatant form of adjustment to known results: namely, 'tacking'.

Take some theory T (perhaps a wild one with all sorts of implausible, but untested consequences – though the point is independent of this supposition). Now 'tack' onto T any independent empirical result e which is already known to hold. If someone were to point to the fact that the resulting theory entails *e* as a reason for 'accepting' the overall theory and residing some degree of 'rational belief' in the theory's other consequences, then he would rightly be laughed out of court.

The real (and of course ancient) problem is to develop some contrast with the negative case by giving some reason why we *should* regard correct, genuine predictions as lending positive support to the theory which makes them. (Again assuming that such a judgment of positive support has some consequences – no matter how heavily hedged – for the rationality of working on, or applying, the theory concerned.)

In a recent paper Laudan (1981) accuses Whewell of doing nothing more than 'handwaving' in response to this problem. But Whewell *did* attempt an argument. He claimed that, underlying the attribution of special weight to predictions, whether temporally novel or not, is the assumption that the theory cannot have made the correct prediction *by chance*. He wrote:

> Men cannot help believing that the laws laid down by discoverers must be in a great measure identical with the real laws of nature, when the discoverers thus determine effects beforehand in the same manner in which nature herself determines them when the occasion occurs. Those who can do this must, to a considerable extent, have detected nature's secret ... Such a coincidence of untried facts with speculative assertions cannot be the work of chance, but implies some large portion of truth in the principles on which the reasoning is founded. (Whewell 1837, II, p. 64)

And again:

> No accident could have given rise to such an extraordinary coincidence. No false supposition could, after being adjusted to one class of phenomena, exactly represent a different class, where the agreement was unforeseen and uncontemplated. (Whewell 1837, II, p. 68)

What are we to make of this argument? Does it indeed amount to anything more than 'handwaving'?

I am sure that Whewell is right that when confronted with a successful prediction (in the more general sense which includes predictions of already known results) we intuitively dismiss the possibility that this success is due to chance. The problem – as Mill already essentially pointed out – is that the dismissal of this possibility is certainly not based on pure logic: all false theories have true consequences, a theory which has been strongly supported on the heuristic criterion *may* nonetheless be radically false and we just happened to have tested 'unrepresentative' consequences. As Mill pointed out, the fact that the wave theory of light, for example, has made successful empirical predictions does not prove it to be correct. Indeed:

> Though twenty such coincidences should occur, they would not prove the reality of the undulatory ether; it would not follow that the phenomena of light were the results of the laws of elastic fluids, but at most that they are governed by laws partially identical with these... (Mill 1843, Ch. XIV, pp. 326-331)

And Mill's point has, of course, been driven home by the subsequent development of science. Science now sees the classical wave theory of light as, strictly speaking, false. It follows that Whewell's claim that predictive success was never enjoyed by a theory which subsequently proved to be false is itself false.

In defense of Whewell, it should be remarked that a rather liberal notion of truth is implicit in his view. He continues to speak, for example, of Kepler's laws being true while explicitly acknowledging that Newton's theory "corrected" them; and in the first passage quoted (p. 28) he speaks of successful predictions *not* as showing that the theory is true, but rather that it 'must be *in a great measure* identical with the real laws of nature', or that it contains 'some large portion of truth' So his claim that successful predictive evidence has 'never yet been produced in favour of a falsehood' is not as easily dismissed as might at first appear. Whewell's claim should, perhaps, be interpreted as asserting that no theory was ever positively supported on the heuristic criterion but subsequently found to be "totally false" or found *not* to contain "important truth".

This saves Whewell's claim from falsity only by rendering it vague. While it is clear what it means for a theory to be true or false in the strict sense, a great deal of effort (especially of late) has so far failed to provide any sort of precise acceptable sense for the notions of 'approximate truth' or 'non-radical falsehood'. Indeed I remain sceptical that any precise notion of 'approximate truth' will ever be produced which will allow that once accepted but now superseded theories may be 'approximately true'. There is undoubtedly an important element of continuity in science *at the empirical level*, but this continuity does *not* extend to the theoretical level (at any rate to the level of what is sometimes called 'interpreted theory'). The ether, and therefore waves in it, are totally rejected by modern science, and so it is hard to see how the wave theory of light is, on present standards, anything other than outright false. Of course, many of its empirical consequences are correct – so we can certainly continue to say that light behaves in certain respects *like* a wave in a medium. But latest theory tells us that light *is not* such a wave. Unless we espouse some positivistic reduction of theories, we are, I think, stuck with sharp discontinuities at the fully-fledged theoretical level in science.[22]

The question of whether Whewell's claim can be modified so as to rationalize the weight given to successful prediction in science is one that I can only raise here and must leave open. It involves difficult problems about induction and about scientific realism. But, while leaving the question of its justification open, I still claim that the heuristic criterion is the one which is, as a matter of fact, applied intuitively by scientists. Whether or not convincing general arguments can be produced for the practice, we do, both in science and in everyday life, intuitively dismiss the chance explanation of predictive success.

The London School of Economics

NOTES

*At various points in this paper (and especially in section 2(c)) I am indebted to Elie Zahar. I completed the paper during a most enjoyable and stimulating stay as visiting fellow at the Center for Philosophy of Science, University of Pittsburgh. I should like to thank Larry Laudan, Nicholas Rescher and especially Adolf Grunbaum for making that stay possible. I received interesting comments on an earlier version from my fellow visiting fellows, especially Ron Giere and Ron Laymon.

1 I should say immediately that this thesis is far from new. As will be seen, it is to be found, though in a rather impure form, in Whewell's writings. It was revived by Elie Zahar (1973) and I tried to develop Zahar's arguments and respond to some early criticisms (Worrall 1978).

2 One recent philosopher who has insisted that the tacking paradox must be given due attention is Clark Glymour (1980). While very much agreeing with Glymour about the importance of the problems he raises, I believe he exaggerates the novelty (and adequacy) of the solutions he advocates (Worrall 1982).

3 For more details of the corpuscular optics program see (Worrall 1976).

4 The *real* situation was rather different – see especially (Zahar 1973).

5 Popper quotes this passage from Schlick in his (1979) - written in the 1930s but only recently published. I was directed to the passage by Elie Zahar.

6 See especially (Musgrave 1974).

7 This objection was raised by Alan Musgrave (1974).

8 J. S. Mill (1843), p. 328. In later editions 'ignorant vulgar' was replaced by some more tepid phrase.

9 See, for example, French (1971, pp. 5-6).

10 The restriction to near-parallel beams was based on Young's theory that the optical disturbance is *longitudinal*. For details see (Worrall 1976).

11 This was Fresnel's first theory too – it is of course incorrect. It was Fresnel and Fresnel alone who eventually discovered the "true" wave-theoretic account of diffraction in which the diffracting object plays no role except that of absorbing the light which falls on it.

12 Fresnel's theory of interference was indeed such a theory – although it was also independently supported.

13 I learned to phrase the thesis in this way during discussions with Ron Giere. Giere also holds a version of the heuristic view, but one which differs from mine in some important respects. See especially (Giere 1983).

14 I have, throughout this paper, simplified the presentation by assuming that the important question is always whether or not a given, *single* result was used in the construction of a theory. But, quite often, the situation is that there is a whole class of results, any of several subsets of which could tie down the free parameters, hence producing a theory which predicts the other results in the class. To take a simple example, say that T implies the equation $y = ax + b$ and that the known relevant empirical results are e: $y = 5$ when $x = 1$, e': $y = 8$ when $x = 2$, and e'': $y = 11$ when $x = 3$. Any two of the three results could be used to fix parameters and the resulting relation $y = 3x + 2$ used to *predict* the remaining result. Or, to take a real example, the wave theory of light makes several predictions of fringe-spacings as one-to-one functions of the wavelengths of the light involved. Any of the corresponding experiments can be used to fix the wavelength and the results of the rest then become genuine, explicit predictions of the theory.

Generally, therefore, the question is not does e support T? But rather how much support does (e , e , ...,e) lend to T? My claim is then that the size of the smallest subset of this evidence set needed to tie down initially free parameters in T must be taken into account in answering this question. For clearly if T entails every element of (e , e ,......,e) without having used any of them, while T' needed n-1 of these results to fix parameters and then predicts the nth, then T is considerably better supported by (e , e ,....,e) than is T'. (For detailed consideration of an example see (Elie Zahar 1978).)

15 None of this is said in an attempt to belittle Newton's achievement – indeed in my opinion it enhances it. It is meant only to show that scientific creativity is rather different from artistic, and much closer to *mathematical*, creativity. The framework within which creativity is possible is much more tightly constrained than might be imagined – see (Duhem 1904).

16 The above account of Newton's inference to universal gravitation is of course oversimplified – for a fuller, but

still somewhat simplified account see (Zahar 1983). I am heavily indebted to Zahar's account. For a better worked-out treatment of a rather simpler case of "inference" to a general theory, see (Worrall 1983).

[17] Indeed the way that a scientist really arrived at the theory and the way the theory was constructed out of available materials may be quite different. For more details see (Worrall 1983).

[18] This is shorthand for 'whether or not we regard the empirical results which supported Kepler's laws as in turn supporting Newton's theory'.

[19] One example of such a pair of theories is the Poincare-Lorentz modified classical physics and the special theory of relativity.

[20] Peter Urbach raised this objection in response to an earlier version of my account.

[21] Whewell's ideas on 'consilience' are more complicated than I suggest here – but underlying them is the thesis that there is a great difference between genuine predictions and results which have been 'written into' a theory.

[22] For a more detailed defense of the position outlined in this paragraph see Worrall (1982a).

REFERENCES

Duhem, P. (1904) *The Aim and Structure of Physical Theory*.

French, A. (1971) *Newtonian Mechanics*. Cambridge: MIT Press.

Gardner, M. (1982) "Predicting Novel Facts", *British Journal for the Philosophy of science, 33*.

Giere, R. (1983) "Testing Theoretical Hypotheses", *Testing Scientific Theories*. ed. J. Earman. Minnesota Studies in the Philosophy Science, vol. X. Minneapolis: University of Minnesota Press.

Glymour, C. (1980) *Theory and Evidence*. *Princeton*: *Princeton* University Press.

Keynes, J. M. (1921) *A Treatise on Probability*. Cambridge: Cambridge University Press.

Lakatos, I. (1976) *Proofs and Refutations*. Cambridge: Cambridge University Press.

_________ (1978) "Falsification and the Methodology of Scientific Research Programmes", *The Methodology of Scientific Research Programmes*: *Philosophical Papers*. *Volume I*. Cambridge: Cambridge University Press.

Laudan, L. (1980) "Why was the Logic of Discovery Abandoned", *Scientific Discovery, Logic and Rationality*. ed. T. Nickles. Dordrecht: D. Reidel.

_________ (1981) "A Study of Some Philosophical Controversies about Ether", *Conceptions of Ether*. eds. Cantor and Hodge. Cambridge: Cambridge University Press.

Mill, J. (1843) *A System of Logic* 1st ed.

Musgrave, A. (1974) "Logical versus Historical Theories of Confirmation", *British Journal for the Philosophy of Science, 25*.

_________ (1978) "Evidential Support, Falsification, Heuristics and Anarchism", *Progress and Rationality in Science*. eds. Radnitzky and Andersson. Dordrecht: D. Reidel.

Popper, K. (1979) *Die beiden Grundproblem der Erkenntnistheorie*. *Mohr-Siebeck*

_________ (1973) "The Aim of Science", *Objective Knowledge*. Oxford: Oxford University Press.

Whewell, W. (1837) *History of the Inductive Sciences*. 3 vols., 1st ed.

_________ (1858) *Philosophy of Discovery*.

Worrall, J. (1976) "Thomas Young and the 'Refutation' of Newtionian Optics", *Method and Appraisal in the Physical Sciences*. ed. Howson. Cambridge: Cambridge University Press.

_________ (1978) "The Ways in Which the Methodology of Scientific Research Programmes Improves on Popper's Methodology", *Progress and Rationality in Science*. eds. Radnitzky and Anderson. Dordrecht: D. Reidel.

_________ (1982) "Broken Bootstraps", *Erkenntnis, 18*.

_________ (1982a) "Scientific Realism and Scientific Change", *Philosophical Quarterly*, 32.

_________ (1983) "Hypotheses and Mr. Newton", *Boston Studies*. Forthcoming.

Zahar, E. (1973) "Why did Einstein's Programme Supersede Lorentz's?", *British Journal for the Philosophy of Science, 24*.

__________ (1978) "'Crucial' Experiments: A Case Study", *Progress and Rationality in Science* eds. Radnitzky and Anderson. Dordrecht: D. Reidel.

__________ (1983) "Logic of Discovery or Psychology of Invention", *British Journal for the Philosophy of Science,* forthcoming.

DIDERIK BATENS

MEANING, ACCEPTANCE AND DIALECTICS*

1. INTRODUCTION

It is the aim of this paper to present a general framework for meaning, acceptance, and their dynamics. The central adequacy condition for this framework is that it should enable us to understand science as a rational enterprise; and by 'science' I obviously mean the scientific enterprise as it is depicted by contemporary philosophy of science. I do indeed believe that the available frameworks prove inadequate if they are compared with the results arrived at by the philosophy of science. Apart from this, the framework presented in this paper should also enable us to understand meaning and rational belief as they occur outside the realm of science properly. The view that rationality is not specific to science is rapidly gaining consent, and rightly so. Nevertheless, science stays a privileged source for an inquiry into rationality because of its specific organizational aspects, the systematic way in which its results are organized, and the availability of detailed historical studies.

The final section of this paper deals with dynamic dialectical logics. My aim in adding this section is to show how, with the help of those logics, the outlook defended here enables us to solve several extremely important problems in a completely natural way.

My approach to meaning and acceptance (certainty, rational belief, etc.) is contextual in a specific sense. There is of course nothing very original about taking a contextual outlook to science and to knowledge in general. I should warn the reader, however, that the contextual outlook I shall defend is rather extreme.

The motivation for the approach lies in the fact that science is a dynamic, collective enterprise, and that its progressive character depends essentially on the communication between adherents of competing theories and competing research traditions. It follows that we should develop a theory about scientific communication and about its impact on scientific change.

J. C. Pitt (ed.), Change and Progress in Modern Science, 333–360.
© 1985 by D. Reidel Publishing Company.

What I mean by 'a context' is precisely a communication situation. A context is characterized by (i) a set of participants, (ii) the problem that one tries to solve, (iii) the set of statements that are regarded as certain and in this sense define the set of possible answers to the problem, (iv) the set of aspects that are considered relevant to the problem, and (v) the set of methodological do's and don'ts that are judged appropriate with respect to the problem. It would of course be a question-begging oversimplification if I were to suppose that the context, defined in this way, is identical for all participants. For this reason, I state from the outset that the context may be different for each participant, and that it may change – in the sense that one of the elements changes – for one participant, while remaining constant for the others. In other words, one participant should be singled out in order to describe a context.

In general, contexts may change very rapidly. If one of the participants judges that he is misunderstood by one of the others and starts explaining the sense in which he uses some term, the context changes and, if there is minimal communication, changes for all participants. If, after this *interludium*, the participants return to the original problem, it is quite possible that they do not return to their original contexts because some of the statements previously regarded as certain by some participant, may have been given up by this participant. Or consider a scientist who performs an experiment in the presence of another scientist, because he believes that the experiment constitutes a problem for the theory which the other accepts. There is a context in which the outcome of the experiment is determined. There is another context in which it is decided whether the outcome of the experiment does constitute a problem for the theory of the second participant. And there is a third context in which it is decided whether this problem, together with other problems, requires the rejection of one theory in favor of the other.

I do of course realize that some scientific activities occur in situations in which there is only one participant. Nevertheless, I shall still employ the term 'context', on the one hand because, such situations may be seen as limit cases of communication situations, and on the other hand because, if it really concerns a scientific context, the participant behaves in principle as if other participants were present.

2. MEANING

It goes without saying that philosophers of science are not in any direct way interested in problems about meaning. Some connected problems belong to their direct concern, however, e.g. the relation between linguistic entities and some set of human activities, and the problem of communication among scientists. It has been argued that such problems should be solved with the help of a theory of meaning, but the trouble is that the theories of meaning that are usually invoked in this connection fail to do the job. For example, they fail to explain how it is possible that there is communication between scientists that work in different research traditions. Some fail because their alleged explanation works only at the expense of a drastic simplification of the relation between different conceptual frameworks; others fail because they are inadequate even with respect to the actual situation as depicted by their adherents.

Needless to say, this failure constitutes a serious drawback. Nobody denies the possibility of communication between scientists that work in different research traditions or paradigms. Given that science is a collective enterprise, the problem of communication is a central one. And it clearly belongs to the job of a theory of meaning to explain how communication is possible.

Most views on meaning that we encounter in philosophy of science texts contain or presuppose two myths. The first is the MYTH ABOUT COMMUNICATION (MC):

MC. Communication between X and Y is only possible if either all terms have the same meaning for X and Y (identity), or at least certain terms, viz. the relevant ones, have the same meaning for X and Y (overlapping), or the meanings of all or some terms resemble each other well enough (somewhat in the sense of Putnam's "charitable account of meaning", (Putnam 1978, p. 22) or all terms of some metalanguage have the same meaning for X and Y, or all terms of some specific part of the language have the same meaning for X and Y (common sense).

Feyerabend has argued against the myth about communication in numerous places, and it seems to me that his arguments are extremely convincing, whereas in their replies his opponents either oversimplify the actual differences between conceptual frameworks (e.g., Krajewski 1977), or else avoid differences in meaning by redefining 'meaning' in an extremely vague way (e.g., Putnam 1978).

The second myth is the MYTH ABOUT MEANING (MM):

> MM. For any term of a given language, there is *the* meaning of this term with respect to this language.

Incidentally, I am convinced that one of the reasons why this myth is widely accepted, is that it is believed to be a necessary condition for the possibility of communication. My objection to the myth about meaning is not that meanings may change over time or that there are different aspects to meaning, for all this is compatible with the myth. My objection is that meanings vary constantly from one context to the other. Even in periods in which the conceptual framework is said to remain constant, there is no such thing as *the* (stable) meaning of a term, not even for one and the same scientist.

I shall now try to sketch a view on meaning in which both myths are given up. I need not argue for the following two theses on meaning:

> M1. The meanings of terms (in a given context) determine and are determined in part (and only in part) by the criteria one applies (within this context) to (provisionally) accept statements on the basis of observation.

> M2. The meanings of terms (in a given context) determine and are determined in part (and only in part) by the situations that are considered logically possible (within this context). (I shall from now on use the phrase 'possible worlds' for such situations).

The set of possible worlds is restricted by conscious decisions, which give rise to conventions, and by prejudices, which make some people believe in "intuitively obvious logical truths".

All of us agree that there is not actually some single set of all possible worlds; scientists working in different paradigms or research traditions do not actually share the same set of possible worlds. Most of us will even agree that there is not some single "correct" set of all possible worlds. Idealists of different brands will agree to this, as will certain realists, e.g., Ilkka Niiniluoto, who holds that THE WORLD "is not carved up into 'pieces' and 'facts' without human conceptualizing" (Niiniluoto 1980, p.441). Even if there were some single "correct" set of all possible worlds, it is obvious that what some person means by a term does not depend on this "correct" set, but depends on *the person's* set of possible worlds.

This brings me to a third thesis, which will require some more argumentation.

> M3. Sets of possible worlds, and hence meanings, vary with individuals, not with groups.

I do not want to say, of course, that there is no resemblance at all between the meanings of terms for two scientists, X and Y, that belong to the same research tradition. What I do claim is that in some cases it may be demonstrated that X and Y use terms with different meanings, and have even different sets of possible worlds; in general there is no good reason to believe that X and Y use all terms with the same meaning, or even that they share the same set of possible worlds. I mention three arguments.

The first argument is rather vague. If we take meaning seriously, if we take intensions and connotations into account, if we realize that an individual does not receive but, rather constructs meanings while interacting with this social environment, then we can conclude only that meanings vary from one individual to the other.

For the second argument I start by considering an objection. Suppose one describes Newtonian mechanics at some stage of its development, i.e., one describes the calculus, the intended domain of application, the set of exemplars, the instruments, the measuring techniques, etc.

One might claim that each term has some single meaning, which is fully determined by the enumerated entities. This claim is false for reasons that become clear later on. But more importantly, the claim is irrelevant to my present point. The theory under consideration is a fossil (and incidentally, a reconstructed one); philosophy of science, even where it takes a synchronic viewpoint, should care about real, living theories. The working, creative scientist does not deal with his theories in the way we may deal with fossils. For him the aforementioned aspects or elements of theories are not "given" but are worked upon. He has personal views, personal prejudices, personal misunderstandings, personal ideas, and he may be able to express some of these in language. So, he has *his* set of possible worlds and *his* meanings.

Now the third argument. The idea that meanings are common to all members of some group may seem to be acceptable from the point of view of Kuhn's paradigms. If you look closer, there are doubts as to whether this is possible. And if you look at Laudan's characterization of research traditions, then you can only reject the idea. Laudan points out the methodological importance of the relations between a scientific theory and theories from other domains, including metaphysical and other "non-scientific" theories. He shows that theories may travel from one research tradition to the other. And, most importantly, Laudan argues that a scientist may belong to different research traditions, some of which may even belong to another domain than the one the scientist is working in. In view of all this, no one can reasonably hold that meanings, or sets of possible worlds are shared by all scientists that work in the same research tradition.

In defending M3 I have been attacking the MYTH ABOUT MEANING. Yet, my attempt at destroying this MYTH is still incomplete, as may be seen from my fourth thesis about meaning;

> M4. Meanings are local, not global; they vary (for the same person) from one context to the other.

Expressed differently, a person may consider A to be logically possible in one context and not consider A a logical possibility in another context. In arguing for M4 I shall proceed in two steps.

First, the set of statements that are regarded as certain, varies from one context to the other. If an astronomer is trying to determine the position of a star, he accepts as certain, e.g., optical laws according to which his telescope has been constructed, a statement about the speed of light, and so on, and of course he accepts as certain that he is looking at the stars and not at a movie picture, that he is not having a dream, that the color of his eyes and the presence of a pot of flowers in the room are both irrelevant for the reliability of his result, etc. But it is obvious that each of these statements may be questioned in some other context. Examples of this kind may be repeated *ad nauseam*. We all know that lots of presuppositions are involved in any observation, in any actual explanation, in any prediction, in any judgment about the acceptability or rejectability of a hypothesis or theory. And it is obvious that the set of accepted statements may differ drastically from one context to the other. Incidentally, some people, e.g. Carnap (1968, pp.146-150) and Bar-Hillel (1968, pp.150-161), have argued that not a single general statement may ever be accepted, but I shall explain in Part Three why their arguments do not hold water.

Second, the variety of sets of statements that are regarded as contextually certain, results in a variety of contextual *logical* truths, and hence in a variety of contextual *meanings*. Let us use the term 'absolutists' for those who believe in context-invariant meanings. It should be pointed out that my disagreement with the absolutists is not a verbal one. The disagreement is not about how we use the phrase 'the meaning of a term'; it is about what people mean when using some term in some context.

With respect to each context there are statements that are regarded as certain, and there are other statements that are relevant to the problem that one tries to solve within the context. Some of the latter may be or become accepted in the context, but this is immaterial to my disagreement with the absolutists.

The absolutist will say that there are, within a given context C1, two kinds of statements that are regaded as certain, viz. logical truths and factual truths. Clearly, he can only make this distinction by referring to one or more other, related contexts C2, C3..., in which all "logical" truths of the first context C1 are kept, whereas no "factual" statement that is regarded as certain in C1, is regarded as certain in C2. I will not discuss the very complex implications of the absolutist's distinction in detail but merely point out one of them: the absolutist cannot accept that there is a context C3 (which is "more general" than C2) in which some logical truth of C1 is not a logical truth. If he were to accept this, he would admit that his choice of C2 to distinguish between logical truths and factual truths in C1, is plainly arbitrary; hence his criterion would break down. It follows that the absolutist must believe that there is some context, in which all and only logical truths are regarded as certain, and that there is no context in which some logical truth is questioned. Let us call this *the most general context*.

Even the absolutist will agree that any conceptual system contains prejudices, that the meaning of any term contains factual presuppositions. Even if we do not know which are the prejudices or presuppositions, we know that there are such. Hence, the most general context cannot be most general *in principle*, and the absolutist may at best refer to the most general context that some scientist may *actually* imagine.

Consider a scientist X who accepts some theory T. He meets another scientist Y who adheres to a rival theory T', that was unknown to X, and they discuss their respective theories. X continues to accept T, but, as a consequence of the communication, (typically) discovers some prejudice in the conceptual framework of T. In other words, the most general context that X could actually imagine, is replaced by a more general one. The absolutist is then driven to the following absurd conclusion: although X continues to accept T, this is *not the same* theory T, for the meanings of its terms have change *in all contexts*.

The absolutist might try to reply that X may introduce a meaning postulate, and in this way bring the terms back to their original meanings. However, in using this argument the absolutist would contradict himself. Indeed, the meaning postulate is false in the most general context X may imagine, and consequently the meaning postulate is a *factual* truth by the very criterion the absolutist introduced in the first place.

According to my view, however, X may introduce such meaning postulates in some contexts, and, consequently, may keep believing the same theory as before. It is precisely by defining meaning as context-dependent that we may avoid the conclusion that our views about the world change whenever we realize that some term may have another meaning than the one in which we used it in some context.

Apart from this theoretical argument there is the very simple fact that we may all easily imagine a situation in which a chemist *means* something about molecules and atoms when he uses the term 'water'. Such examples however, do not convince absolutists, because they will claim that there is some way to analyse what is meant by the chemist in terms of a combination of meanings on the one hand and of accepted factual truths on the other. By offering the theoretical argument I hoped to show that such complications are superfluous and useless.

Incidentally, some will think that the term 'logical truth' is too strong in connection with contextual meaning, but it is not. The meaning of logical terms is just as context dependent as the meaning of referring terms, and I shall show in Part Four that the meaning of negation changes if we move to a larger set of possible worlds than the one considered by two-valued logicians.

Up to now I have mainly considered contexts in which there is only one participant. I did so in order to avoid unnecessary complications, but now it is time to mention some contexts in which there is more than one participant and to explain under which circumstances there is communication between them.

Let us consider a very simple example of an adherent of classical mechanics, X, and an adherent of relativistic mechanics Y who together want to measure with a given precision the length of some normal seized object. In order to describe the context for each of them, we have to state both for X and for Y which aspects are considered relevant to the problem and which is the set of statements regarded as certain. It is obvious that the context for X is different from the context for Y: their sets of logical truths are different; their worlds, although restricted to the relevant aspects, are populated by different objects; and the meaning of the term 'length' is different. Nevertheless, there will be communication in this pair of contexts. *In general* there is communication between X and Y in a given pair of contexts if both X and Y believe truly that there is a one-one relationship R between the possible worlds of X and the possible worlds of Y such that, if RwXwY, then X will decide in this context that wX obtains under the circumstances under which Y decides in his context that wY obtains, and *vice versa*.

Several points are in need of clarification. I require that X and Y believe this condition *truly* because, if they merely believe it, there might be no communication, even if they believed that there is. I also require that X and Y *believe* that the condition is fulfilled, for if they do not believe so, there can be no communication. Next, both X and Y may describe in their own language the circumstances under which they decide to wX and wY respectively. But this is not a hindrance for the communication: X believes that Y will decide to wY under the circumstances he describes by means of his own language; and analogously for Y. Finally, X and Y need not believe that they in general decide under the same circumstances to wX and wY respectively; all that is required is that they truly believe that they do in *this* pair of contexts.

The aforementioned criterion may easily be adapted to define communication between more than two persons, partial communication, "one-sided" communication, pseudo-communication with an imaginery partner like one's former self, or an actually absent person, etc.

On first sight it might seem that the fact that meanings are contextual is a hindrance for communication. It is true that communication would be easier and less partial if all humans shared the meanings of all terms. However, the very fact that this is not so, makes the contextual character of meanings a necessary condition for communication. If the set of possible worlds were not restricted to the aspects that are relevant to the problem of the context, the one-one relation mentioned in the criterion could never obtain. Moreover, if the absolutist's account were correct, and meanings would not be determined within some context by *all* statements regarded as certain, then again a one-one relation would never obtain between the respective sets of worlds of two scientists belonging to different research traditions.

It is important to point out that the kind of communication we arrived at is a very powerful one. It leads to a form of communication which allows for collective action, the exchange of observational reports, etc. Moreover, this form of communication is sufficient for the existence of scientific communities, for the possibility of cooperation in research. Within a scientific community, communication will be guaranteed with respect to a variety of problems: problems about observation, problems about the theories, problems about methodology. This communication is furthered by the fact that the members of a scientific community make the same decisions in a large number of contexts. For this reason we may consider scientific communities themselves as actors and as decision makers in a large number of contexts.

3. CONTEXTUAL CERTAINTY AND ITS DYNAMICS

There are not many left, today, who believe in the viability of Carnap's original program in inductive logic. Nevertheless, I think that it is instructive to take a quick look at part of the evolution of Carnapian inductive logic.

When Rudolf Carnap started developing his inductive logic in the early fourties, his aim was to find an explicatum for "the methods of inductive reasoning which are at present applied in science and statistics" (Carnap 1950, p. vii), and he meant inductive logic "in the wide sense of nondeductive or nondemonstrative reasoning" (Carnap 1950, p. vii). Carnap soon discovered that the most obvious "confirmation function" was inadequate with respect to his aim, and in 1952 presented an infinite continuum of confirmation functions that are defined with the help of a parameter which may be interpreted as expressing the degree of caution with respect to predictions based on inductive reasoning (Carnap 1952).

From 1965 on, Jaakko Hintikka presented a two-dimensional continuum in which a second parameter, α, was introduced.[1] It measures the degree of caution with respect to inductive generalization (I remember that, in infinite languages, the degree of confirmation of any contingent general sentence with respect to finite evidence is always zero for almost all confirmation functions of Carnap's continuum).

Even before Hintikka's results were published, Carnap and Kemeny had tried to avoid an inadequacy by introducing yet a further parameter: Carnap's and Hintikka's. The inadequacy was that confirmation functions allow for induction with respect to predictions that are identical to elements of the factual evidence, but do not allow for induction with respect to predictions that are only partly analogous with respect to the elements of the evidence.[2] This means that, if I write my name on my pencil, I cannot predict what will happen when I drop it, for I have never observed a pencil which had my name on it and was dropped.

As Carnap and Kemeny's solution applied to a special case only, I tried, more than ten years ago to find a general solution. I soon discovered that the problem of analogy and partial analogy is far more complex than it seems at first sight. I found a solution, which was published in 1975 and which I still take to be right from the viewpoint of Carnap's program (Batens 1975).

However, I am now convinced that this solution shows that Carnap's program is *completely misguided*. I mean that a probability measure, defined for all sentences of some language, and which agrees at least with our most simple "intuitions" on inductive support, does not exist. The reason is that any such probability measure should solve the problem of partial analogy, and that the simplest such measure is too complex. I do not mean that it is too complex from a technical point of view. The fact that an explicatum for some set of intuitions is mathematically complex need not be an objection. What I mean is that the measure is too complex from a methodological point of view. In using arguments by analogy, we do not and cannot take into account all considerations that have to be taken into account according to the probability measure. And, incidentally, this measure is still far from complex enough to be applied to such generalizations as Galileo's law of free fall.[3]

There is a wide range of non-Carnapian approaches to induction, rational credence, etc. Instead of discussing some of these, I shall try to explain the outlook to which I came, partly in consequence of a long and intense involvement with Carnapian inductive logic. This outlook is a natural consequence of the contextual view on meaning I sketched in Part Two.

I shall defend a contextual view of rational belief, certainty, and induction. Let me point out from the beginning that making degrees of rational belief contextual is not in itself much of an advantage. After all, there are very general contexts, and we need degrees of rational belief in these too. As I see it, the main advantage of the approach I shall defend, is that it enables us to describe and understand as more or less rational the dynamics of our beliefs, and this includes the dynamics of our methodological beliefs. For quite some time now, Isaac Levi has been defending a contextual view on credal probability – I refer especially to his 1980 book – and many take his theory to be extreme; I shall argue that it is not extreme enough.

For a start, I claim that the statements that are regarded as certain within some context, and in this sense determine the certainty of possible worlds, are completely certain within the context: they are logically true, they receive probability 1, there is no room for doubt about them *within* the context. Of course, a scientist who accepts some singular or general statement within some context, may (under certain conditions on which I shall comment later) move to another context in which this statement is questioned; but this does not prevent the statement from being considered completely certain within the first context.

Some people will agree, but will add that such statements are only certain "in some sense". I claim that there is no meaningful and relevant sense in which those statements are not contextually certain. Maybe some statements are regarded as certain in more contexts than others. I doubt whether it is possible in general to numerically compare sets of possible contexts, but, even if it were, the number of contexts in which some statement is regarded as certain does not add anything to its certainty within a given context. And if the claim were that those statements are not, or at least should not be considered absolutely certain, then I readily agree, but only because no statement is justifiably considered absolutely certain, in that no statement is justifiably considered certain in all contexts.

In order to consider a statement as logically true and hence as completely certain within some context, it is not required that the certainty of the statement has been demonstrated or may in principle be demonstrated within some other context. Consider a theory that is regarded as logically true and completely certain by an engineer in a given context C1. It is quite possible and of course desirable that it has been demonstrated in some context C2 that the theory has to be preferred over its rivals and that it is reliable with respect to the domain to which the engineer applies it. But it is not and cannot be required that it has been demonstrated in C2 that the theory is true or that we may be absolutely certain about it. Or consider a context in which some measuring instrument is considered reliable. It obviously cannot be required that it has been demonstrated in some other context that we may be completely certain about the reliability of this instrument. No one could ever demonstrate this.

The fact that some statement is regaraded as logically true and hence gets probability 1, does not entail that we should keep believing forever that the statement is true. Of course, if A is logically true, its relative probability will always be 1. But this holds only within some context. We can move to another context, in which the statement is not logically true, in which, if it is an empirical generalization, it may be tested, or in which its acceptability is examined on the basis of some set of (contextually contingent) statements.

Before I proceed about contextual certainty, I shall spell out the way in which different contexts interact in general, and the way in which this interaction leads to changes to the set of beliefs, i.e. changes to the sets of contextual logical truths.

Any subject, whether an individual or a scientific community, has certain explicit or implicit beliefs about how to approach certain problems. For some problems, the set of beliefs will determine the selection of a set of logical truths, a set of relevant aspects and a set of methodological instructions. For kinds of problems the subject is not familiar with, the subject will first move to another context, in which the problem is to select the statements that have to be regarded as certain, the relevant aspects and the methodological rules that characterize the context in which the original problem has to be solved. Needless to say, any rational subject will sometimes declare itself incompetent to solve a problem.

Philosophers of science should not worry about how one arrives at one's first set of beliefs, but about the justificatory problems connected with given sets of beliefs. In this connection I am on the side of those who, like Isaac Levi, hold that a belief need not be questioned unless there is a good reason to do so. But, if there is a good reason, any belief should be open to investigation, even if it is methodological or logical in nature. One of the good reasons to question a belief might be the very fact that some subject sets itself the aim to question some set of beliefs. It would be wrong, however, to characterize science as an institutionalized form of questioning beliefs. The aim of science is to explore the world, to extend the available set of beliefs, and it is only by way of a fortunate side effect that the resulting process leads to good reasons to question beliefs.

Good reasons to question beliefs are always internal. A subject would not question some belief which conflicts with empirical evidence, if he had not among his beliefs some methodological rules which sanction the empirical evidence as reliable. It is extremely important that we consider methodological rules as beliefs, and not as unquestionable statements of some higher type. In general, any context-independent hierarchy of statements should be rejected. Methodological statements are taken for granted in contexts in which the problem is to decide about the merits of some empirical generalization, but the latter are regarded as completely certain in contexts in which the problem is to decide about the merits of methodological rules.

Although good reasons to question beliefs are always internal in the sense explained, their origin may be either internal or external. It is internal if the attempt to solve some problem shows that the set of beliefs is either inconsistent or incomplete; it is external if a problem originates either from observation or from communication. The progressive character of science is essentially connected to the fact that some good reasons to question beliefs have an external origin, and especially that some find their origin in communication.

The good reasons of an internal origin may be of two types. Those of the first type arise when a context proves inadequate in that the problem under consideration is solved in a wrong way or in a way that appears later to be wrong. Expressed in Laudan's terminology, this corresponds to falsification in the case of empirical problems, to inconsistency in the case of internal conceptual problems, or to implausibility or inconsistency in the case of external conceptual problems. The other possibilities of unsolved empirical or conceptual problems lead also to an inadequacy of the context, vix. that the problem cannot be solved because the context is incomplete. Under such conditions one also moves to another context, viz. in order to complete the set of relevant beliefs, but no statement of the original context need to be questioned.

Scientists and scientific communities may spontaneously generate problems and this is the second internal source of good reasons to question beliefs. Theories may be subjected systematically to critical examination, or some empirical domain may be explored. In the absence of good reasons of an external origin, this spontaneous problem generation does not occur very frequently and is not very fruitful. The impediment is that, in the absence of good reasons of an external origin, it is not clear why, e.g., one accepted theory should be questioned rather than another. Moreover in a context in which one theory is questioned, several other theories will be regarded as logically true; if the questioned theory is rejected, this rejection will be contingent on the order in which theories are questioned. But as it is not clear why one theory should be questioned rather than another under the considered circumstances, the rejection will be arbitrary. If spontaneous problem generation leads to the exploration of some empirical domain, the impediment is that, in the absence of good reasons of an external origin, the results obtained may be completely uninteresting.

Some good reasons to question staements regarded as certain in some context, have an external origin. They are of two kinds. Those of the first kind derive from the inadequacy of the context to solve a problem which is not itself generated by the theory or research tradition. What I mean is this: it is a problem for a research tradition that its latest theories do not enable one to predict, say, whether A or not A will be the case; but whether or not A will be the case is itself a problem, and is not in any way generated by the research tradition. It seems to me, incidentally, that, in his *Progress and its Problems*, Laudan's formulation is sometimes confusing in this respect. Most of the time he discusses problems of the first kind, and such problems are indeed central for the comparative assessment of theories and of research traditions. But the problems which theories have to solve are those of the first kind. Scientific theories have to solve problems that have an external origin, whereas problems that have an internal origin are solved by the rejection or acceptance of existing theories, or by projecting new theories.

But let me return to good reasons to question contextually certain statements. Some good reasons of an external origin are of a second type; these are related to the communication between research traditions or, more generally, to the communication between subjects that hold conflicting views. In this connection I refer to the detailed and well argued analysis offered in Laudan's *Progress and its Problems*. This analysis enables us to see how, as a result of communication between rival research traditions, each of these gets confronted with new problems and hence with new contexts. It also enables us to see how, if a context proves inadequate, one may, essentially as a result of this communication between research traditions, have good reasons to question some accepted statements rather than others.

According to the view I am defending, it is excluded that a unique acceptance criterion would take care of the acceptance of all statements. The acceptability of some statement is essentially related to the question whether the contexts in which the statement is regarded as logically true or is accepted as a methodological instruction, are adequate with respect to solution of their problems. A unique acceptance criterion will not do because, first, there is a variety of contexts, in which a variety of problems has to be solved, and, second, the statements that may be accepted as certain within some context are quite different in nature. The justification of the acceptability of some such statements will be that they are acceptable as nomologically true; for other statements the justification will be that they are acceptable as (merely) actually true; for still other statements the justification will be that they are true in most (but not all) cases. Such distinctions are discussed in my 1975 book (pp. 280-298), and although my general outlook on induction has changed, the proposals I offered there may easily be adapted to fit to my present view. In general, acceptance criteria should depend on the contexts in which the accepted statements have to serve.

I hope that I have made the relations between contexts sufficiently clear by now, and return to the notions of contextual certainty and contextual probability.

The choice of a probability distribution for some context is a methodological choice. Once we take the view that methods are subject to discussion and change, there is no reason for not treating methodological statements on a par with other accepted statements. And there is overwhelming evidence that methods are subject to change, that such changes may be justified (Laudan 1981) and that non-methodological beliefs are relevant to the justification. Only an extreme contextual view can account for this.

The justification of the choice of an inductive method for some context is extremely different from and extremely less complicated than the choice of a general inductive method. The problem of partial analogy, to take one example, is almost completely eliminated in view of the presence of statements about the world and of accepted statements about relevant aspects that are regarded as certain. The latter statements will enable us to disregard all kinds of information and, if the problem under consideration is the choice of a prediction about a partly known object (or fact or process), they will save us the complications that derive from the fact that the object is partly analogous to other objects with respect to properties that are considered irrelevant. Also (other) accepted statements about the world will save us lots of complications. If a statement expressing the melting point of a crystalline chemical compound is regarded as certain, the prediction of the temperature at which a sample of this compound will melt, will not require any reference to partial analogy considerations. And if it is given that all crystalline chemical compounds have a melting point, then, even if nothing is specified about the melting point of some such compound, one statement about the temperature at which a sample of this compound melted will enable us to predict, without taking into accouont *any* partial analogy considerations, the point at which another sample of this compound will melt. Under the same circumstances we may derive a general statement about the melting point of this compound. In all such cases there is no need to choose some value of a parameter; we do not even need an inductive method. Several methodological choices required by Carnap's model are thus taken care of in the contextual model by accepted statements *about the world*. This argument for local induction is well known.

In other cases we shall have to make methodological statements within the context. The (well known) advantage of contextual models over global models in this connection is that in the former methodological choices are much more restricted and may be different for different contexts. Moreover, even for the same context, the methodological choices need not be identical with respect to all properties. So, for example, we may take analogy in some respects as relevant to some prediction, while taking analogy in other respects as irrelevant. In other words, the overall choice of the value of parameters in the Carnapian model may be replaced by specific statements about the inductive behavior of certain properties. And of course each of the specific statements may be questioned and investigated by moving to an appropriate other context. Incidentally, it follows that a number of the symmetry considerations which Carnap introduced, should be rejected on the contextual view.

Apart from these arguments for a contextual approach and against global inductive methods, I also refer to Isaac Levi's arguments in favor of the thesis that only "serious possibilities" (Levi 1980) should be considered. My present position is rather close to Levi's position. One of the differences, however, is that Levi distinguishes logical possibilities from serious possibilities, and that he considers *the* set of logical, set theoretical and mathematical truths as the minimal set of accepted statements. The arguments I gave in Part One may be employed in defense of my position against Levi's. In this connection it is worth considering an objection by Henry Kyburg (1976, pp.191-215) against an earlier version of Levi's position. Kyburg's argument may be paraphrased as follows: It does not help much to restrict probability measures to contexts because we need more embracing probability measures as soon as a statement which was accepted at some point is questioned; eventually we need a global probability measure. I hope that it is clear that this objection does not apply to my contextual approach. If a statement that is regarded as certain in a context C1 is questioned, and we investigate its acceptability in context C2, the probability measure employed in C2 need not be an extension of the probability measure employed in C1. For one thing, most statements that are contingent and relevant in C1 will be either irrelevant or even non-contingent in C2.

There is another very important difference between the contextual view I am defending and Levi's position, and I need to comment on it because it concerns an essential aspect of my view, viz. that the set of statements regarded as certain may be different from one context to the other. According to Levi's model there is, at any time, a corpus of accepted statements. This corpus remains the same for all contexts, i.e. is not dependent on changing aims or problems. Under certain conditions statements may be added to the corpus (expansion); under other conditions statements may be removed from the corpus (contraction). It is essential for Levi as well as for me that such changes occur only if there is a good reason for them; as long as there is no reason to change the corpus, one should stick to it without having to justify its elements.

I do agree with Levi that it is possible in some cases to define such a corpus with respect to some set of contexts. It is perhaps even possible to define one corpus of accepted statements with respect to all contexts which concern the set of scientific problems as defined at some point in time with reference to some research tradition. The set of statements regarded as certain within some context will then consist of those statements from the corpus which are relevant to the problem investigated in the context.

Even if it is possible to define a corpus in this sense with respect to some research tradition at some point in time, the contextual view I am defending has some clear advantages. Consider a scientist X from this tradition who discusses with a scientist Y from a rival tradition about the merits of their respective theories. For the sake of the discussion both X and Y will have to move to a context in which some of their previously accepted statements are not regarded as certain. But suppose that the discussion does not convince X to give up his theory, and that X has not been able either to demonstrate that Y's theory is to be rejected. Under these circumstances X will certainly not give up his theory but neither will he decide that both his and Y's theory are equally acceptable and that he has to take both of them into account for all further purposes. It is extremely important that X sticks to his theory, that he keeps developing it, that he goes on using it to look at the world, and that, looking at the world from the viewpoint of his theory, he tries to find arguments against Y's theory. On Levi's model, however, this is impossible. X may stick to his theory without questioning it as long as there is no good reason to do so. But as the discussion compelled him to remove the theory from his corpus, he cannot, according to Levi's view, reintroduce it into his corpus unless he justifiably may do so. But the example is precisely a case in which X cannot justifiably choose between his and Y's theory.

A second argument against a context-independent corpus of accepted statements is that even a scientist may have his philosophical moments and question the truth-status of his theories in general. In such context he would be begging the question if he would stick to his background knowledge. Moreover, some form of Cartesian doubt, or some arguments for scepticism, or some arguments about the frequency of conceptual change in the history of science, may convince him that there is no good reason to believe that his present theories are true, or may even convince him that his theories are almost certainly false. But this will not, or by all means should not, prevent him from regarding those theories as certain within other types of contexts. If he became a sceptic even in the contexts in which he is engaged in solving problems as a scientist, he would become quite a failure as a scientist. To express it in a sloganlike way: a rational person may at the same time believe that some theory is completely certain and that it is almost certainly false, provided these beliefs relate to different contexts.

A third argument against a context-independent corpus of accepted statements is that the set of statements that are accepted by the members of a research tradition, may be inconsistent. The trouble is sometimes that the statements regarded as certain within some single context are inconsistent, but more often that the union of the sets of statements that are regarded as certain in different contexts, is inconsistent. Such a situation will constitute a problem, which should be solved within an appropriate context. But as long as the problem has not been solved, and perhaps in order to solve it, the scientists working in the research tradition will have to continue solving other problems in other contexts. According to my view, the set of statements regarded as certain may very well be consistent for most if not all of these contexts. But, according to Levi's position, the corpus is the same for all contexts. This argument may easily be generalized. Although the theories that are accepted within some research tradition at some time are, in the optimal case, those that deserve most to be accepted on the (methodological and other) premisses of this research tradition, those theories are far from unproblematic most of the time. This truism of present-day philosophy of science entails that all accepted statements are not and should not be regarded as certain in all contexts.

Although I have argued against Levi's idea of a corpus of accepted statements, my kind of contextual approach requires that something very close to Levi's corpus is to be sought. In a sense it is an optimal situation for a research tradition that there be a corpus of statements that are accepted in all contexts, except for those in which the investigated problem presupposes that they are not accepted. There are three reasons why this is an optimal situation: (i) the need to understand reality as a coherent system, (ii) the fact that, in the presence of such a corpus, the organization of the beliefs is drastically simpler and more economic, and (iii) the fact that the presence of such a corpus will permit the generation of an appropriate context for dealing with new, unexpected problems. Levi's idea of a corpus, if modified in the aforementioned way, may be considered as an ideal for scientific research traditions. But even if so modified, it is not an appropriate means to deal with those numerous cases in which, unfortunately, scientific research traditions do not meet this ideal.

4. DYNAMIC DIALECTICAL LOGICS

One of the typical properties of the contextual approach is that even formal logical systems may differ from one context to the other. In order to show the extremely important advantages of this situation, I now describe some logical systems which differ drastically from the usual ones, which are required for the solution of a family of central methodological problems, and the application of which should clearly be restricted to some contexts only.

It has been argued for quite some time now that inconsistencies frequently occur in connection with actual scientific theories. Some theories are inconsistent all by themselves. In other cases there is an inconsistency between two accepted theories. In still other cases there is an inconsistency between one or more theories and a set of accepted observational statements. In all such cases we face two different problems: (i) how to continue scientific research in the presence of an inconsistency, and (ii) how to get rid of the inconsistency. It appears that scientists solve such problems frequently. But they do not know exactly how they do it, and philosophers of science will fail to understand what is going on, as long as they try to approach those cases in terms of classical logic.

There is some similarity between the aforementioned problems and Hegel's notion of a dialectical process, and for this reason I use the expression 'dialectical logics'; I devised these logics precisely in order to deal with theories which, although they were intended to be consistent and were believed to be consistent for some time, turned out to be inconsistent. The addition of the term 'dynamic' is clearly redundant from Hegel's viewpoint, but it happens that Richard Routley and Robert K. Meyer use the term 'dialectical logic' for systems which they consider as a first step in the formalization of dialectics, but which are *static*, as Routley and Meyer explicitly admit.

The dynamical character of the logics lies not with the fact that one theory is replaced by another, but with the inferential process itself. Whenever an inconsistency has actually been arrived at in a derivation from a set of premisses, the rules of inference change. I do not mean that they are changed or should be changed by us; they change all by themselves. Moreover, this change is local; it affects only some statements, viz. those that proved to be inconsistent. As a result, some statements that were derivable before the inconsistency has been arrived at, may not be derivable afterwards. This applies even to inconsistencies. One or both "halves" of an inconsistency that has been derived at some point, may not be derivable any more at a later point as a consequence of the derivation of another inconsistency.

I realize that all this sounds quite extravagant, and for that reason want to point out that I am not talking about some vaguely formulated obscure system, but about a completely decent logic, which has an exactly formulated syntax and semantics (completeness and soundness proofs are available), a logic which has an exactly formulated derivability relation, an exact definition of the consequence set, etc. The consequence set is defined for any set of premisses, i.e., for any set of premisses. it is determined from the outset to which set of consequences the dynamic inferential process will lead if, as is also required for the more common logics, it is carried out in a suitable way. All such technical details may be found in my "Dynamical Dialectical Logics" (Batens 1983); I shall merely present two applicataions of dynamic dialectical logics here. Before doing so, however, I have to make two further short introductory remarks.

The first is that my only technical paper on the matter deals with the propositional level only, whereas I need predicate logic for the applications. I can, however, assure the reader that there are no special troubles on the predicative level, except for some undecidability results.

The second remark is that both applications concern the first of the two problems I mentioned at the outset, viz. how to continue scientific research in the presence of an inconsistency. The removal of the inconsistency should lead to an "enrichment" of the theory. This problem is typically a heuristic one, and although some possible enrichments may very well be arrived at by algorithmic means, no algorithm could possibly lead to the or a "optimal enrichment".

As a first application of dynamic dialectical logics, consider a theory which was meant and believed to be consistent, and for this reason was formulated with classical logic as its basis. Suppose, however, that we derive an inconsistency from the theory as it stands. We might of course simply reject the theory because it is trivial and false, but if no rival theory is available, this decision will have disastrous consequences. Furthermore, we know that scientists do not in general reject a theory merely because it is inconsistent (in itself or with another theory, or with observational reports), even if competitors are available. And several philosophers of science have argued that, under certain circumstances, scientists are justified in not rejecting an inconsistent theory.

Another alternative is that we change the theory drastically in that we replace its logic by a logic which does not, in the presence of an inconsistency, lead to triviality. One such logic, or rather logical machinary, has been presented at least as early as 1964 by Nicholas Rescher. This machinary, however, leads to results which are essentially dependent on the way in which the theory is axiomatized, and in some cases this is a disadvantage. A further possibility is to replace the logic of the theory by one of the numerous paraconsistent logics which have been presented from 1963 on.[4] If we do so, however, then the theory becomes extremely poor. In order to avoid that any statement be derivable from an inconsistent set of premisses, paraconsistent logics drastically cut down the rules of inference. As a consequence, they do not only avoid that an inconsistency leads to triviality, but also eliminate a large number of consequences that are in no way whatsoever related to the aforementioned inconsistency. Of course, this is not what scientists want or should want. What they want, and what they actually use, is the theory *in its full richness, except for the pernicious consequences of its inconsistency*. This is *exactly* what some dynamic dialectical logics lead to.

The idea is that we start from an extensional paraconsistent logic, i.e. a logic which is exactly like classical logic, except in that the meaning of negation has been weakened in order to make inconsistencies logically possible (Batens 1980, pp.196-234). For all inferences that are correct according to classical logic but are incorrect according to the paraconsistent logic, we have the following result within the paraconsistent logic: there is a set of statements such that either each of these statements is true together with its negation, or else the inference is correct.

I now define a dynamic dialectical logic by stating under which conditions a statement is derivable at some point in a demonstration. There are two cases.

(i) If, at some point in a demonstration, A is derivable from previous steps according to the paraconsistent logic, then A is derivable at this point and from those steps according to the dynamic dialectical logic.

(ii) If, at some point in a demonstration, A is derivable from previous steps according to the paraconsistent logic *on the condition* that the statements $B_i, \ldots, B_n$ behave consistently, at this point in the demonstration, then A is derivable at this point and from those steps according to the dynamic dialectical logic.

On later points in the demonstration A may not be derivable any more because one of the B may have been shown to behave inconsistently. (I leave out one small complication which would require too much time but is perfectly natural).

It is possible to define the notion of a *final consequence* of a set of premisses in terms of the set of all demonstrations that fulfill certain regularity conditions. The set of final consequences of a set of premisses is provably independent of the way in which a specific demonstration proceeds. This is very important, because we do not want the notion of derivability to be contingent on the accidental way in which the proof proceeds. Semantically the set of all final consequences of some set of premisses is exactly the set of all statements that are true in all models in which all premisses are true and which are precisely as inconsistent as is required to make the premisses true. In other words, the set of final consequences gives us the theory in its full strength, except for the pernicious consequences of its inconsistency.

I now quickly consider another application. Consider a scientist who has to make predictions on the basis of a falsified theory together with some set of observational statements. If there is only one falsifying instance, the theory will obviously not be considered as falsified; we would rather reject the falsifying instance. So, if the theory is falsified, we have at least an accepted empirical generalization which contradicts the theory. Suppose that 'All A are B' is derivable from the theory, whereas 'All A that are C, are not B' is the accepted empirical generalization (to take a simple example). Under such circumstances the scientist will predict that an object does not have property B if he knows that it has properties A and C, whereas he will predict that an object has property B if he knows that it has property A and not property C. If he does not know about C, he will refrain from making a prediction about B.

Incidentally, the scientist may have good reasons to make predictions in another way, but this is irrelevant to the point I am making. All I need is that the scientist will have a more or less demarcated opinion about the interplay between the theory and the accepted empirical generalization for his decision to predict B or not-B or neither. Whichever the criterion, the scientist may apply the aforementioned dynamic dialectical logic to make derivations from the theory together with the empirical generalization and observational statements, and then apply this criterion in order to eliminate, whenever possible, one half or both halves of the inconsistency (and to eliminate also all other statements for the derivation of which the eliminated half of the inconsistency has been employed). If one half of the inconsistency has been eliminated, he will refrain from making any prediction on the matter. This result corresponds exactly with what he wanted to do intuitively.

By way of a final remark, I want to stress that it would be completely wrong to conclude from all this that dynamic dialectical logic is *the* correct logic because it leads to correct results with respect to both consistent and inconsistent theories. What I did want to show is precisely the opposite, viz. that if some context proves inadequate to solve the problem under consideration, we have to move to another context to decide what is the right remedy; and in some cases the right remedy is to change the logic, and to change it only in some context, because only this context is affected by this inadequacy.

University of Guelph

NOTES

* I am indebted to Jon Dorling, Ilkka Niiniluoto, Nicholas Rescher, and several other participants in the Fourth International Conference on the History and Philosophy of Science (Blacksburg, Va., November 2-6, 1982) for helpful comments. But my gratitude goes in the first place to Leo Apostel for the stimulating discussions that accompanied my writing of the first draft.

For references see, e.g., Niiniluoto and Tuomela (1973).

[2] Expressed more precisely, some confirmation functions lead to the desired results with respect to predictions that share all their known properties with other known objects; but no confirmation function of those continua lead to the desired results with respect to predictions about objects that do not share all their known properties with some other known object.

[3] Carnap distinguished clearly between inductive logic and the methodology of induction; cf., e.g., his (1950 p. 202, ff). According to the spirit of Carnap's position, all analogy problems mentioned should be solved by inductive logic in that some set of "confirmation functions" should lead to the desired results; the task of the methodology of induction is merely to guide us in the application of inductive logic.

For an instructive overview, see A. Arruda (1983.)

REFERENCES

Arruda, A. (1983), "Aspects of the Historical Development of Paraconsistent Logic", *Paraconsistent Logic*. eds. R. Routley and G. Priest. (in press)

Bar-Hillel, Y. (1968) "The Acceptance-Syndrome", *The Problem of Inductive Logic*. Amsterdam: North-Holland.

Batens, D. (1975) *Studies in the Logic of Induction and in the Logic of Explanation*. Brugge: De Tempel.

Bar-Hillel, Y. (1968) "The Acceptance-Syndrome", *The Problem of Inductive Logic*. Amsterdam: North-Holland.

Batens, D. (1975) *Studies in the Logic of Induction and in the Logic of Explanation*. Brugge: De Tempel.

________ (1980) "Paraconsistent Extensional Propositional Logics", *Logique et Analyse*.

________ (1983) "Dynamical Dialectical Logics", *Paraconsistent Logic*. eds. R. Routley and G. Priest. (in press)

Carnap, R. (1950) *Logical Foundations of Probability*. Chicago: University of Chicago Press.

________ (1952) *The Continuum of Inductive Methods*. Chicago: University of Chicago Press.

________ (1959) *Induktive Logik und Wahrscheinlichkeit*. Wien: Springer.

________ (1968) "On Rules of Acceptance", *The Problem of Inductive Logic*. Amsterdam: North Holland.

Krajewski, W. (1977) *Correspondence Principle and the Growth of Science*. Dordrecht: Reidel.

Kyburg, H. (1976) "Local and Global Induction", *Local Induction*. Dordrecht: Reidel.

Laudan, L. (1977) *Progress and its Problems*. Berkeley: University of California Press.

________ (1981) *Science and Hypothesis*. Dordrecht: Reidel.

Levi, I. (1980) *The Enterprise of Knowledge*. Cambridge, Mass.: MIT Press.

Niiniluoto, I. and R. Tuomela (1973) *Theoretical Concepts and Hypothetico-Inductive Inference*. Dordrecht: Reidel.

___________ (1980) "Scientific Progress", *Synthese*, 45.

Putnam, H. (1978) *Meaning and the Moral Sciences*. London: Routledge & Kegan Paul.

Rescher, N. (1964) *Hypothetical Reasoning*. Amsterdam: North-Holland.

NICHOLAS RESCHER

EXTRATERRESTRIAL SCIENCE

1. COULD SCIENCE IN ANOTHER SETTING OVERCOME THE LIMITATIONS OF OUR HUMAN SCIENCE?

Our science as we humans cultivate it here on earth – is limited and imperfect, and is bound to remain so. It thus becomes tempting to wonder whether another, astronomically remote civilization might be scientifically more advanced than ourselves. Is it not plausible to suppose that an alien civilization might overcome the limitations of our science and manage to surpass us in the furtherance of this enterprise?

On first thought, the question seems very clear-cut, for, as one recent discussion puts it:

> Any serious speculations concerning the capabilities of intelligent biological life and automata must take into account technical societies that may be millions or even billions of years more advanced than our own. (MacGowan and Ordway 1966, p. 248.)

However, this seemingly straightforward issue is actually one of great complexity. And this complexity relates not only to the actual or possible facts of the matter, but also – and crucially – to somewhat abstruse questions about the very ideas or concepts that are relevant here.

To begin with, there is the question of just what it is for there to be another science-possessing civilization. Note that this is a question that *we* are putting – a question posed in terms of the applicability of *our* term "science." It pivots on the issue of whether *we* would be prepared to call certain of *their* activities – once we understood them – as engaging in scientific inquiry, and whether *we* would be prepared to recognize the product of these activities as constituting a (state of a branch of) science. At the very least this requires that we be prepared to recognize what those aliens are doing as a matter of forming beliefs (theories) as to how things work in the world and to

J. C. Pitt (ed.), Change and Progress in Modern Science, 361–392.
© 1985 by D. Reidel Publishing Company.

acknowledge that they are involved in testing these beliefs observationally or experimentally and in applying them in practical (technological) contexts. We must, to begin with, be prepared to accept them as (non-human) *persons*, duly equipped with intellect and will, and must then enter upon a complex series of claims with respect to their beliefs and their purposes.

A scientific civilization is not merely one that possesses intelligence and social organization, but one that puts this intelligence and organization to work in a very particular sort of way. This opens up a rather subtle issue of priority in regard to process versus product, namely whether what counts for a civilization's "having a science" is primarily a matter of the substantive *content* of their doctrines (their belief structures and theory complexes) or is primarily a matter of the *aims and purposes* with which their doctrines are formed.

The matter of content turns on the issue of how similar their scientific beliefs are to ours. And this is clearly something on which we would be ill-advised to put much emphasis at the very outset. After all, the speculations of the nature-theorists of pre-Socratic Greece, our ultimate ancestors in the scientific enterprise, bear little resemblance to our present-day sciences, nor does contemporary physics bear all that much doctrinal resemblance to that of Newton with respect to its substantive claims. We would do better to give prime emphasis to matters of process and purpose.

Accordingly, the matter of these aliens "having a science" is to be regarded as turning not on the extent to which their *findings* resemble ours, but on the extent to which their *project* resembles ours – of deciding whether we are engaged in the same sort of rational inquiry in terms of the sorts of issues being addressed and the ways in which they are going about addressing them. The issue thus is at bottom not one of the *substantive similarity* of their "science" to ours, but one of the *functional equivalency* of the projects at issue in terms of the characterizing goals of the scientific enterprise: description, explanation, prediction, and control over nature.

2. THE POTENTIAL DIVERSITY OF "SCIENCE"

Such a perspective enjoins the pivotal question: To what extent would the *functional equivalent* of natural science built up by the inquiring intelligences of an astronomically remote civilization be bound to resemble our science? In considering this, one soon comes to realize that there is an enormous potential for diversity here.

To begin with, the *machinery of formulation* used in expressing their science might be altogether different. Specifically, their mathematics might be very unlike ours. Their "arithmetic" could be anumerical – purely comparative, for example, rather than quantitative. Especially if their environment is not amply endowed with solid objects or stable structures – if, for example, they were jellyfish-like creatures swimming about in a soupy sea – their "geometry" could be something rather strange, largely topological, say, and geared to flexible structures rather than sizes or shapes. Digital thinking in all its forms might be undeveloped, while certain sorts of analogue reasoning are highly refined. Or again, an alien civilization might, like the ancient Greeks, have "Euclidean" geometry without analysis. In any event, given that the mathematical mechanisms at their disposal could be very different from ours, it is clear that their description of nature in mathematical terms could also be very different. (And not necessarily truer or falser, but just different.)

Secondly, the *orientation* of the science of an alien civilization might be very different. All their efforts might conceivably be devoted to the social sciences – to developing highly sophisticated analysis of psychology and sociology, for example. Then too their natural science might deploy mechanisms very different from ours. Communicating by some sort of "telepathy" based upon variable odors or otherwise "exotic" signals, they might devise a complex theory of empathetic thought-wave transmittal through an ideaferous aether. Electromagnetic phenomena might lie altogether outside the ken of alien life-forms; if their environment does not afford them loadstones and electrical storms, etc., the occasion to develop electromagnetic theory might never arise. The course of scientific development tends to flow in the channel of practical interests. A society of porpoises might lack crystallography but develop a very sophisticated hydrodynamics; one comprised of mole-like creatures might never dream of developing optics. One's language and thought-processes are bound to be closely geared to the world as one experiences it. As the difficulties we experience in bringing the language of everyday experience to bear on subatomic phenomena illustrates, our concepts are ill-attuned to facets of nature different in scale or structure from our own. We can hardly expect a "science" that reflects our parochial preoccupations to be a universal fixture.

The science of a different civilization would inevitably be closely geared to the particular pattern of their interaction with nature as funnelled through the particular course of their evolutionary adjustment to their specific environment. After all, those aliens might scan nature very differently. The "forms of sensibility" of radically different life-forms (to invoke Kant's useful idea) are likely to be radically different from ours. The direct chemical analysis of environmental materials might prove highly useful, and bioanalytic techniques akin to our senses of taste and smell could be highly developed providing them with environmentally oriented "experiences" of a very different sort. Or think of the beings who would populate Strawson's purely auditory world (Strawson 1959, Chapter 2). But then again, acoustics might mean very little to them, while other sorts of pressure phenomena – say the theory of turbulence in gasses – might be the subject of intense and exhaustive investigation. Rather than sending radiowaves or heat signals, they might propel gravity-waves through space.

Philosophers have often said things to the effect that people whose experience of the world is substantially different from our own are bound to conceive of it in very different terms. Sociologists, anthropologists, and linguists say much the same sort of things, and philosophers of science have recently also come to talk in this way. According to Thomas Kuhn, for example, scientists who work within different scientific traditions – and thus operate with different descriptive and explanatory "paradigms" – actually "live in different worlds." (Kuhn 1962, chap. X.)

Supporting considerations for this position have been advanced from very different points of view. One example is a *Gedankenexperiment* suggested by Georg Simmel in the last century – that of envisaging an entirely different sort of cognitive being, intelligent and actively inquirying creatures (animals, say, or beings from outer space) whose experiential modes are quite different from our own. (Simmel 1895, pp. 40-41.) Their senses respond rather differently to physical parameters – relatively insensitive, say, to heat and light, but substantially sensitized to various electromagnetic phenomena. Such intelligent creatures, Simmel held, could plausibly be supposed to operate within a largely different framework of empirical concepts and categories – the events and objects of the world of their experience might be very different from those of our own – their phenomenological predicates, for example, might have altogether variant descriptive domains. In a similar vein William James wrote:

> Were we lobsters, or bees, it might be that our organization would have led to our using quite different modes from these [actual ones] of apprehending our experiences. It *might* be too (we cannot dogmatically deny this) that such categories unimaginable by us to-day, would have proved on the whole as serviceable for handling our experiences mentally as those we actually use. (James 1907, p. 171)

Different cultures and different intellectual traditions, to say nothing of different sorts of creatures, are bound to describe and explain their experience – their world as they conceive it – in terms of concepts and categories of understanding substantially different from ours. They would diverge radically with respect to what the Germans call their *Denkmittel* – the conceptual instruments they employ in thought about the facts (or purported facts) of the world. They could, accordingly, be said to operate with different conceptual schemes, with different conceptual tools used to "make sense" of experience – to characterize, describe, and explain the items that figure in the world as they view it. The taxonomic and explanatory mechanisms by means of which their cognitive business is transacted might differ so radically from ours that intellectual contact with them would be difficult or impossible.

A comparison of the "science" of different civilizations here on earth suggests that it is not an outlandish hypothesis to suppose that the very *topics* of their science might differ dramatically from those of ours. In our own case, for example, the fact that we live on the surface (unlike whales or porpoises), the fact that we have eyes (unlike worms or moles) and thus can *see* the heavens, the fact that we are so situated that the seasonal positions of heavenly bodies is intricately connected with agriculture, are clearly connected with development of astronomy. It is clear that the mode of emplacement in nature of alien inquirers might be so different as to focus their attention on entirely different aspects or constituents of the cosmos.

If the world is sufficiently complex they might concentrate upon aspects of their environment that mean little or nothing to us.

Accordingly, the constitution of the alien inquirers – physical, biological, and social – emerges as a crucial consideration. It would be bound to condition the agenda of questions and the instrumentalities for their resolution – to fix what is seen as interesting, important, relevant, and significant. In determining what is seen as an appropriate question and what is judged as an admissible solution, the cognitive posture of the inquirers must be expected to play a crucial role in shaping and determining the course of scientific inquiry itself.

In addition, the *conceptualization* of an alien science might be very different. We must reckon with the theoretical possibility that a remote civilization might operate with drastically different system of concepts in its cognitive dealings with nature. It is (or should be) clear that there is no single, unique, ideally adequate concept-framework for "describing the world." The botanist, horticulturist, landscape gardener, farmer, and painter will operate from diverse cognitive "points of view" to describe one selfsame vegetable garden in very different terms. It is mere mythology to think that the "phenomena of nature" can lend themselves to only one correct style of descriptive and explanatory conceptualization. There is surely no "ideal scientific language" that has a privileged status for the characterization of reality. Different sorts of creatures are bound to make use of different conceptual schemes for the representation of their experiences. To insist on the ultimate uniqueness of science is to succumb to "the myth of the God's eye view." Different cognitive perspectives are possible – no one of them more adequate or more correct than any other independently of the aims and purposes of their users.

To motivate this idea of a conceptually different science, it helps to cast the issue in temporal rather than spatial terms. The descriptive characterization of *alien* science is a project rather akin in its difficulty to that of describing our own *future* science. It is a key fact of life in this domain that ongoing progress in the science/technology area is a process of *ideational* innovation that always places certain developments outside the intellectual horizons of earlier workers. The very concepts we think in terms of become available only in the course of scientific discovery itself. Like the science of the remote future, the science of the remotely distant must be presumed to be of such a nature that we really could not achieve intellectual access to it on the basis of our own position in the cognitive scheme of things. Just as the technology of a more advanced civilization would be bound to strike us as magic, so its science would be bound to strike us as incomprehensible gibberish – until we had learned it "from the ground up." They might (just barely) be able to *teach* it to us, but they could not *explain* it to us by transposing it into our terms.

The most characteristic and significant sort of difference between one conceptual scheme and another arises when the one scheme is committed to something the other does not envisage at all – something that lies outside the conceptual range of the other. The typical case is that of the stance of Cicero's thought-world with regard to questions of quantum electrodynamics. The Romans of classical antiquity did not hold *different* views on these issues; they held no views at all regarding them, because they lay outside their conceptual reach. This whole set of relevant considerations lay outside their conceptual repertoire. The diversified history of *our* terrestrial science gives one some miniscule inkling of the vast range of possibilities along these lines.

The "science" of different civilizations, may well, like Galenic and Pasteurian medicine, simply *change the subject* in key respects so as no longer "to talk about the same things," but treat things (e.g., humors and bacteria, respectively) of which the other takes no cognizance at all. The difference as to "conceptual scheme" between modern and Galenic medicine is not that the modern physician has a different theory of the operation of the four humors from his Galenic counterpart, but that modern medicine has *abandoned* the four humors, and not that the Galenic physician says different things about bacteria and viruses, but that he has *nothing* to say about them. If, for example, certain intelligent aliens were a diffuse assemblage mass of units comprising wholes in ways that allow of overlaps (compare Ehrensvard 1965, pp. 146-148), then the role of social concepts might become so paramount for them that nature as a whole would be viewed in social terms. The result would be something very difficult for us to grasp, seeing that the concepts involved are based on a mode of "shared experience" with which we ourselves have no contact.

It is thus only reasonable to presume that the conceptual character of the "science" of an alien civilization would depart so radically from ours as to orient their thought about "the nature of things" in altogether different directions. Even in strictly functional terms, their "functional equivalent" of our science might be very unlike ours. Their explanatory mechanisms, their predictive concerns, and their modes of control over nature might all differ from ours in at least some fundamental ways. It is only natural to expect that the interests of creatures shaped under the remorseless pressure of evolutionary adaptations to be very different – and endlessly variable – environmental conditions should be oriented in directions very different from ours.

As long as the fundamental categories of thought – the modes of spatiality and temporality, of structural description and explanatory rationalization – are not seen as necessary features of intelligence as such, but as evolved cognitive adaptations to particular contingently constituted ways of emplacement in and interaction with nature, there will be no basis for claims to uniformity. Sociologists of knowledge tell us that even here on earth, our science is but one of many competing world-pictures – one not particularly privileged in comparison to such other systemic schemes of explanatory orientation in the environment as the well-known magical beliefs of the Azande. When one turns outwards towards space at large, the prospects of diversity become literally endless. It is a highly problematic contention even that beings constituted as we are and emplaced in an environment such as ours must inevitably describe and explain natural phenomena in our terms.

Natural science – broadly construed as inquiry into the ways of nature – is something that is in principle endlessly plastic. Its development will trace out an historical course closely geared to the specific capacities, interests, environment, and opportunities of the creatures that develop it. We are deeply mistaken if we think of it as a process that must follow a route roughly parallel to ours and issue in a comparable product. It would be grossly unimaginative to think that either the journey or the destination must be the same – or even substantially similar.

The more one reflects on the matter, the more firmly one is led to the realization that our particular human conception of the issues of science is something parochial, because we are physically, perceptually, and cognitively limited and conditioned by our specific mode of emplacement within nature. Given that we might be dealing with intelligent beings of a physical and cognitive nature profoundly different from ours, one clearly cannot assert with confidence what the natural science of such a creature would be like.

3. THE ONE-WORLD, ONE-SCIENCE ARGUMENT

One recent writer raises the question "What can we talk about with our remote friends?", and answers with the remark that: "We have a lot in common. We have mathematics in common, and physics, and astronomy." (E. Purcell in Cameron 1963, p. 142.) Another writer maintains that "we may fail to enjoy their music, understand their poetry, or approve their ideals; but we can talk about matters of practical and scientific concern." (Anderson 1963, p. 130.) This line of thought begs some very big questions.

Our alien colleagues scan nature for regularities using (at any rate to begin with) the sensors provided them by their evolutionary heritage. They note, record, and transmit those regularities which they found to be intellectually interesting or pragmatically useful – and then develop their inquiries by theoretical triangulation from this basis. Now this is clearly going to make for a course of development that closely gears their science to their particular situation – their biological endowment, ("their sensors"), their cultural heritage ("what is interesting"), their environmental niche ("what is pragmatically useful"). Where these key parameters differ, there too we can expect that the course of scientific development will differ as well.

Admittedly there is only one universe, and its laws and materials are, as best we can tell, the same everywhere. We share this common universe with all life-forms. However radically we differ in other respects (in particular those relating to environment, to forms of life, and its mode of civilization) we have a common background of cosmic evolution and a common heritage of natural laws. And so if intelligent aliens investigate nature at all, they will investigate the same nature we ourselves do. But the sameness of the object of contemplation does nothing to guarantee the sameness of ideas about it. It is all too familiar a fact that even where human (and thus *homogeneous*) observers are at issue, very different constructions are often placed upon "the same" occurrences. Primitive peoples thought the sun a god and the most sophisticated among the ancients thought it a large mass of fire. We think of it as a large thermonuclear reactor, and heaven only knows how our successors will think of it in the year 3,000. As the course of human history clearly shows, there need be little uniformity in the conceptions held about one selfsame object by different groups of thinkers.

It is surely naive to think that because one single object is in question, its description must issue in one uniform result. This view ignores the crucial impact of the describer's intellectual orientation. A piece of driftwood is viewed very differently indeed by the botanist, the painter, the interior decorator, the chemist, the woodcarver, etc. It's a matter of what "aspects" of the thing are focused upon as important or interesting. Minds with different concerns and interests and different backgrounds can deal with the same items in ways that yield wholly disjoint and disparate results because altogether different features of the thing are being addressed. The *things* are the same, but their *significance* is altogether different. Different cognitive vantage points, different "perspectives of consideration" are at issue.

The universality anad intersubjectivity of our science, its finder-independence, still leaves matters at the level of *human* science. As C. S. Peirce was wont to insist, the aim of scientific inquiry is to allay *our* doubts – to resolve the sorts of questions we ourselves deem worth posing. Other beings might well ask very different sorts of questions.

Different conceptual frameworks take us into literally different spheres of thought. They go their separate ways in very much the same sort of way as different subject-matter specialties dealing with altogether different sorts of issues even when (as per the soldier, the farmer, and the geologist) they address themselves to something which appears as "the same thing" to a detached bystander (the same piece of terrain). It is notorious that observers approach their observations via the predispositions of their cognitive preparation and interests seeing only what their preestablished cognitive resources enable them to see.

No one who has observed how very differently the declarations of a single text (the Bible, say, or the dialogues of Plato) have been interpreted and understood over the centuries – even by people of a common cultural heritage – can be hopeful that the study of a common object by different civilizations must lead to a uniform result. Yet such textual analogies are oversimple and misleading, because with scientific inquiry there is no one predetermined original text. It would be a grave mistake to think that there is just one fixed basic text – the "book of nature writ large" – which different civilizations can decipher in different degrees. The scientific study of nature is not a matter of decoding a preexisting text. Nature's discourse lies in the ears of the beholder; its "messages" are ones whose substance we ourselves influence For science is always the result of *inquiry* into nature and this is inevitably a matter of a *transaction or interaction* in which nature is but one party and the inquiring beings another. The result of such an interaction depends crucially on the contribution from both sides – from nature and from the intelligences that interact with it. A kind of "chemistry" is at issue where nature provides only one input and the inquirers themselves provide another – one that can massively and dramatically affect the outcome in such a way that we cannot disentangle the respective contributions of nature and the inquirer. Things cannot of themselves dictate the significance an active intelligence can attach to them.

After all, throughout the earlier stages of man's intellectual history, different human civilizations developed their "natural sciences" in a substantially different way. And the shift to an extraterrestrial perspective is bound to amplify this diversity. The "science" of an alien civilization may be far more remote from ours than the "language" of our cousin, the dolphin, is remote from our language. Perhaps reluctantly, we must face the fact that on a cosmic scale the "hard" physical sciences have something of the same cultural relativity that one encounters with the materials of the "softer" social sciences on a terrestrial basis.

We must accept with respect to natural science a rough functional equivalent of the "historicist' position urged by Wilhelm Dilthey and others with respect to history. Science is a cultural artifact whose status is akin to that of other such artifacts in *point* of change and transience. We can attain no Hegelian "position of reason itself" outside of and contradistinguished from the fallible and imperfect positions that constitute the changing sequence of historical stages. Such progress as the development of science affords – and substantial it is – is not a matter of discernible evolution toward some history-external, transcendental, position, but rather a matter of exchanging for one imperfect and ephemeral position yet another position which, while indeed having some significant point of advantage over its predecessor in points of its substantiating warrant, is for all that itself no more ultimate or definitive. And what holds here with respect to *historical* changes may be presumed to hold with respect to the situational changes of mutually remote civilizations as well.

Each inquiring civilization must be expected to produce its own, perhaps ever-changing cognitive products – all more or less adequate in their own ways, but with little if any actual overlap in conceptual content. Human organisms are essentially similar, but there is not much similarity between the medicine of the ancient Hindus and that of the ancient Greeks. There is every reason to think that the natural science of different astronomically remote civilizations should be highly diversified. Even as different creatures can have a vast variety of lifestyles for adjustment within one selfsame physical environment like this earth, so they can have a vast variety of thought-styles for cognitive adjustment within one selfsame world.

With respect to his hypothetical Planetarians, the ingenious Christiaan Huygens wrote three centuries ago:

> Well, but allowing these Planetarians some sort of reason, must it needs be the same with ours? Why truly I think 'tis, and must be so; whether we consider it as applied to Justice and Morality, or exercised in the Principles and Foundations of Science. . . . For the aim and design of the Creator is every where the preservation and safety of his Creatures. Now when such a reason as we are mastors of, is necessary for the preservation of Life, and promoting of Society (a thing that they be not without, as we shall show) would it not be strange that the Planetarians should have such a perverse sort of Reason given them, as would necessarily destroy and confound what it was design'd to maintain and defend? But allowing Morality and Passions with those Gentlemen to be somewhat different from ours . . . yet still there would be no doubt, but that in the search after Truth, in judging of the Consequences of things, in reasoning, particularly in that fort which belongs to Magnitude or Quan-

> tity, about which their Geometry (if they have such a thing) is employ'd, there would be no doubt I say but that their Reason here must be exactly the same, and go the same way to work with ours, and that what's true in one part will hold true over the whole Universe; so that all the difference must lie in the degrees of Knowledge, which will be proportional to the Genius and Capacity of the Inhabitants. (Huygens 1698, pp. 41-43.)

With the exception of a shift from a theological to a natural-selectionist rationale, this analysis could still be advanced today.

Perhaps it seems plausible to argue thus: "Common problems constrain common solutions. Intelligent alien civilizations have in common with us the problem of cognitive accommodation to a shared world. Natural science as we know it is *our* solution of this problem. Ergo it is likely to be *theirs* as well." But this tempting argument founders on its second premiss. Their problem is not common with ours because their situation must be presumed substantially different, since they live in a significantly different environment and come equipped with significantly different resources. To presuppose a common problem is in fact to beg the question.

In philosophical perspective, the "one world, one science" doctrine reflects a Hegelian monism. Its basis is the idea that there is one single "position of reason alone." For – so it is maintained – while the process or course of scientific development is something variable and contingent, the ultimate product that will eventually issue from these diversified strivings is fixed in preordained uniformity, with all different cognitive routes leading inexorably and inevitably to the same destination. This thesis clearly represents a problematic and implausible bit of metaphysics.

Our science reflects not only our interests but also our capacities. It addresses a range of issues that are correlative with our specific modes of physical interaction with nature, the specific ways in which we monitor its processes. It is selective – the science of a being that gets its most crucial information through sight, monitoring developments along the spectrum of electro-magnetic radiation, rather than, say monitoring variations of pressure of temperature. It is certainly not a phenomenalistic science geared to the feel of things or the taste of things. The science we have developed depends on our capacities and needs, our evolutionary heritage as a being inserted into the orbit of natural phenomena in a particular way.

The fact that all intelligent beings inhabit the same world does not countervail against the no less momentous fact that we inhabit very different ecological niches within it, which bring about different sorts of *modus operandi* – physically and cognitively as well, since we cannot separate *theoria* from *praxis*. The chain that links cognition to power, power to a power base, and a power base to a mode of enmeshment in nature is wrought of unbreakable links.

Moreover, we should never lose sight of the fact that science is an evolving process. It is the product of a historical course of development along certain particular lines – a development producing a product very different 100 years ago from its condition today, and one that will no doubt look very different 100 years hence. The history of our own science illustrates the radical differences in the nature of the "science" arrived at even by cultures that are actually in communicative touch with one another. The position we have reached today was certainly not inevitable. Nor is it final. Our conceptions and our technical capabilities – our theories and our instruments – will surely continue to change massively in the future. And given this variety and variability even with respect to ourselves, there is little alternative to supposing that the natural science developed by beings radically different from ourselves will only have remote points of tangency with our own.

All such factors as capacities, requirements, interests, and course of development affect the shape and substance of the science and technology of any particular place and time. Unless we narrow our intellectual horizons in a parochially anthropomorphic way, we must be prepared to recognize the great likelihood that the "science" and "technology" of a remote civilization will be something *very* different from science and technology as we know it. We are led to view that our human sort of natural science may be *sui generis*, adjusted to and coordinate with a being of our physical constitution, inserted into the orbit of the world's processes and history in our sort of way. It seems that in science as in other areas of human endeavor we are prisoners of the thought-world that our biological and social and intellectual heritage affords us.

The one-world, one-science doctrine begs some pretty big questions – to begin with the question if indeed there is only one world. The one reality argument shatters against the fact that different *thought-worlds* can be at issue in the elaboration of a "science." For alien intelligences may simply fail to inhabit "our world" in the sense needed for the "one world, one science" argument to achieve its intended aim.

There is no quarrel here with "the principle of the uniformity of nature." But this principle merely tells us that when exactly the same question is put to nature, exactly the same answer will be forthcoming. However, the development of a science hinges crucially on this matter of questions – to the sorts of issues that are addressed and the sequential order in which they are posed. And here the prospect of variation arises; for we must expect alien beings to question nature in ways very different from our own. There is no reason to think that the sciences of different civilizations will exhibit anything more than the roughest sorts of family resemblance.

4. COMPARABILITY AND JUDGMENTS OF RELATIVE ADVANCEMENT

We have insisted from the outset that "science" must be understood in terms not of the substantive content of assertions, but of a rather generic sort of *functional* equivalency of enterprises. The question, however, that concerns us at present is not whether a remote civilization has a "science" of some sort, but whether it is *scientifically more advanced* than ourselves. Now if *advancement* is to be at issue, and the question is to be one of *relative* sophistication, then an actual substantive comparison must be provided for. We are now dealing not with the essentially quantitative aspects of intelligence as such – with data processing in volumetric terms – but with the quality of the orientation of intelligence to substantive issues. If the natural science of an alien civilization is to represent an *advance* over ours, we must clearly construe it as *our sort* of science in much more particularized and substantive terms. Even a superficial scrutiny of the terrestrial situation suffices to show that the development of natural science is certainly not inevitable for cognitive beings, but is closely geared to their internal constitution and their cultural thought-orientation. And given the *immense* diversity to be expected among the various modes of "science" and "technology," the number of extraterrestrial civilizations posessing a science and technology that is duly consonant and contiguous with ours – and in particular, heavily geared towards the mathematical laws of the electromagnetic spectrum – must be judged to be very small.

It must be recognized that sciences can vary (1) in their formal mechanisms of *formulation* – their "mathematacs," (2) in their *conceptualization*, that is, in the kinds of explanatory and descriptive concepts they bring to bear, and (3) in their *orientation* towards the manifold pressures of nature, reflecting the varying "interest"-directions of their developers. While "science" as such is clearly not anthropocentric, science *as we have it* – the only "science" that we ourselves know – is a specifically human artifact that must be expected to reflect in significant degree the particular characteristics of its makers. Consequently, the prospect that an alien "science"-possessing civilization has a *science* that we would acknowledge (if sufficiently informed) as representing the same general line of inquiry as that in which we ourselves are engaged seems extremely implausible. The possibility that *their* science-and-technology is "sufficiently similar" in orientation and character to be substantively proximate to *ours* must be viewed as extremely remote. We clearly cannot estimate this as representing something other than a very long shot.

Just such comparability with "our sort of science" is, however, the indispensable precondition for judgments of relative advancement or backwardness *vis-a-vis* ourselves. The idea of their being scientifically "more advanced" is predicated on the uniformity of the enterprises – doing better and more effectively the kinds of things that *we* want science and technology to do. Any talk of advancement and progress is predicated on the sameness of the direction of movement: only if others are traveling along the same route as we can they be said to be ahead or behind us. The issue of relative advancement is linked inseparably to the idea of doing *the same sort of thing* better or more fully. And this falls apart when "this sort of thing" is not substantially the same. One can say that a child's expository writing is more primitive than an adult's, or that the novice's performance at arithmetic or piano-playing is less developed than that of the expert. But we can scarcely say that Chinese cookery is more or less advanced than Roman, or Greek pottery than Renaissance glassware. The salient point for present purposes is simply that where the enterprises are sufficiently diverse, the ideas of relative advancement and progress are inapplicable for lack of a *sine qua non* condition. (Of course, even "doing the same sort of thing" at a very remote level of generality may not suffice for the comparability judgments of doing it better or worse that are needed to underwrite judgments of relative progressiveness. The condition at issue is merely necessary and not sufficient. Butterflies and sparrows both fly, but, given the differences in their *modus operandi*, it makes little sense to say that one does so better than the other.)

Our reasoning here has occasionally used the analogy of spatial and temporal remoteness. But in this context of advancement or progressiveness there now arises a crucial disanalogy between *alien* science and *future* science. Future science is inevitably comparable with ours. For it is developmentally connectible with it: it is the successor-state of a successor-state of a successor-state. And advantageousness is transmitted along this chain of succession. If a state of science were not significantly more advanced than one already in place it would (presumably) not establish itself as a successor. The very conditions of historic change in our terrestrial science assure its progressiveness in point of prediction and control, and thus presumptively in point of explanatory adequacy as well. No such in-principle assurance of relative comparability is available in the case of an alien "science."

The matter of entitlement to claim scientific superiority does not automatically emerge from their capacity to make many splendidly successful predictions. For this could be the result of precognition or empathetic attunement to nature or such-like. What is wanted is not just a matter of *correct*, but of cognitively underwritten and thus *science-guided* predictions. Yet that's just exactly what is to be proved.

Again, "technological wonders" do not necessarily reflect *scientific* accomplishment. After all, bees can do all sorts of things we would like to do but cannot. The technology in question must surely be the product of intelligent contrivance rather than evolutionary trial and error. To betoken advancement, their performatory wonders must actually issue from (superior) theoretical knowledge – i.e., from superior science. And this too will beg the question.

Nor would the matter of scientific superiority be settled by the consideration that an extraterrestrial species might be more "intelligent" than we in having a greater capacity for the efficient and comprehensive monitoring and processing of information. After all, whales or porpoises, with their larger brains, may (for all we know) have to manipulate relatively larger quantities of sheer data than ourselves to maintain effective adaptation within a highly changeable environment. What clearly counts for scientific knowledge is not the *quantity* of intelligence in sheer volumetric terms but its *quality* in substantive, issue-oriented terms. Information manipulation does not assure scientific development. Libraries of information (or misinformation) can be generated about trivia – or dedicated to matters very different from science as we know it.

It is perhaps too tempting for humans to reckon cognitive superiority by the law of the jungle – judging as superior those who do or would come out on top in outright conflict. But surely the Mongols were not possessors of a civilization scientifically superior to that of the Near Eastern cultures they decimated in medieval times. We earthlings might easily be eliminated by not very knowledgeable creatures able to emit at will – perhaps *a la* H. G. Wells' *War of the Worlds* – a naturally secreted biological or chemical agent capable of killing us off. Their *physical* superiority may have nothing to do with *scientific* superiority.

The key point then, is that if they are to effect an *advance* on our science they must both (1) do roughly our sort of thing in roughly our sort of way, and (2) do it significantly better. In speaking of the "science" of another civilization as "more advanced" than our own we contemplate the prospect that they have developed *science* (*our* sort of science – "science" as we know it) further than we have ourselves. And this is implausible. Even assuming that "they" develop a "science" at all – that is, a *functional equivalent* of our science – it seems unduly parochial to suppose that they are at work constructing *our* sort of science in a *substantively* oriented sense of this time.

Diverse life-modes have diverse requirements; diverse requirements engender diverse technologies; diverse technologies make for diverse modes of science. For if a civilization of intelligent aliens develops a science at all, it seems plausible to expect that they will develop it in another direction altogether and produce something that we, if we could come to understand it at all, would regard as simply detached in content-orientation – though perhaps not in intent – from science as we ourselves cultivate it.

There just is no unique itinerary of scientific/technological development that different civilizations travel in common with mere differences in speed or in staying power (not withstanding the penchant of astrophysicists for the neat plotting of numerical "degrees of development" against time in the evolution of planetary civilizations (cf. Ball 1980, p. 658). In cognition and even in "scientific" evolution we are not dealing with a single-track railway, but with a complex network leading to many mutually remote destinations. Even as cosmic evolution involves a spatial red shift that carries different star systems ever farther from each other, so cognitive evolution may well involve an intellectual red shift that carries different civilizations into thought-worlds ever more remote from each other.

The literature created by extraterrestrial-intelligence enthusiasts is fraught with the haunting worry: "Where is everybody?" "Why haven't we heard from them?" Are they simply too distant – or perhaps too cautious (see Nozick 1972) or too detached (see Ball 1973)? Our present discussion offers yet another line of response: they are simply too busy doing their own thing. Radiocommunication is ours, theirs might be something very, *very* different. If alien civilizations inhabit alien thought-worlds, then this probable lack of intellectual communion might well explain the lack of physical communication.

The prospect that an alien civilization is going about the job of doing *our* science – a "science" that reflects the sorts of interests and involvement that *we* have in nature – better than we do ourselves must accordingly be adjudged as extremely far-fetched. Specifically two conditions would have to be met for the science of an intelligent civilizatioin to be in a position to count as comparable to ours.

(1) That, *given* that they have a "science" and a developing "technology," they have managed to couple the two and have proceeded to develop (unlike the ancient Greeks and Chinese and Byzantines) a *science-guided* technology.

(2) That their science-guided technology is sufficiently oriented towards issues regarding natural processes sufficiently close to those at which our science-guided

technology is oriented that a comparison can reasonably be made between them.

To judge by terrestrial experience, it seems rather optimistic to estimate these probabilities as anything but *very* small. If those alien scientists are differently constructed (if they are silicon-based creatures, for example), or if their natural environment is very different, their practical interests and its accordant technology will be oriented in very different directions from ours. Even we ourselves might well have aimed our technological development not at "mastery over external nature" but at the advancement of "control over oneself" in the manner of some oriental civilizations. Analogously, their technology might be wholly independent of "hardware," oriented not towards physical machinery, but towards the software of mind-state manipulation, telepathy, hypnotism, autosuggestion, or the like. (Ray Bradbury's Martians destroy an expedition from Earth armed with atomic weapons by thought-control.) We must not keep our imagination on a short leash in this regard. Given the diversity of different modes of "science" and the enormous spectrum of possible issues and purposes in principle available to extraterrestrial aliens, the prospect must be recognized that the direction of their science-guided technology might be vastly different from our own.

If "being there" in scientific regards means having *our* sort of scientifically guided technology and our sort of technologically channeled science, then it does not seem all that far-fetched to suppose that *we might be there alone* – even in a universe teeming with other intelligent civilizations. The prospect that remote aliens might be doing "our sort of thing" in the scientific sphere better than we can do it ourselves seems very remote.

There are, no doubt, various ways in which the prospect of alien civilizations might cause us legitimate concern. But their being scientifically more advanced than ourselves is not one of them. For it only makes sense to speak of them being "more advanced" or "more backward" when the parties are engaged in a common journey. But this is hardly likely to be so in the present case.

5. FIRST PRINCIPLES

People frequently seem inclined to reason as follows:

> There are, after all, an immense number of planetary objects running about in the heavens. And proper humility requires us to recognize that there is nothing all that special about the Earth. If it can evolve life and intelligence and civilization and science, then so can other planets. And given that there are so many other runners in the race we must assume that – even though we cannot see them in the cosmic darkness – some of them have got ahead of us in the race.

As one recent writer formulates this familiar argumentation, "Since man's existence on the earth occupies but an instant in cosmic time, surely intelligent life has progressed far beyond our level on some of these 100,000,000 (habitable) planets (in our galaxy)." (Calvin in Cameron 1963, p. 75.) But such plausible-sounding arguments overlook the numerical complexities. Even though there are an immense number of solar systems, and thus a staggering number of planets (some 10^{22} by current estimates), nevertheless, a substantial number of conditions must be met for "science" (as we understand it) to arise. The astrophysical, physical, chemical, biological, psychological, sociological, and epistemological parameters must all be in proper adjustment. There must be habitability, and life, and intelligence, and culture, and technology, and a "science" coupled to technology, and an appropriate subject-matter orientation of this intellectual product, etc. On this reckoning, the number of civilizations that possess a technologized science as we comprehend it, is clearly not going to be very substantial – it is, in fact, going to be strikingly close to 1.

A great many turnings must go aright enroute to science of a quality comparable to ours. Each step along the way is one of finite (and often smallish) probability. And to reach the final destination, all these probabilities must be multiplied together, yielding a quantity that might be very small indeed. Even if there were only twelve turning points along this developmental route, each involving a chance of successful eventuation that is, on average no worse than a one-in-a-hundred, the chance of an overall success would be diminutively small, corresponding to an aggregate success-probability of merely 10^{24}.

It is tempting to say "The Universe is a big place; surely we must expect that what happens in one locality will be repeated someplace else." But this overlooks the issue of probability. Admittedly cosmic locales are very numerous. But probabilities can get to be very small: No matter how massive *N*, there is that diminutive 1/*N* that can countervail against it.

In the course of evolutionary development things can eventuate very differently at many junctures. George G. Simpson has very rightly stressed the many chancy twists and turns that lie along the route, insisting that

> the fossil record shows very clearly that there is no central line leading steadily, in a goal-directed way, from a protozoan to man. Instead there has been continual and extremely intricate branching, and whatever course we follow through the branches there are repeated changes both in the rate and in the direction of evolution. Man is the end of one ultimate twig.... Even slight changes in earlier parts of the history would have profound cumulative effects on all descendent organisms through the succeeding millions of generations .... The existing species would surely have been different if the start had been different, and if any stage of the histories of organisms and their environments had been different. Thus the existence of our present species depends on a very precise sequence of causative events through some two billion years or more. Man cannot be an exception to this rule. If the causal chain had been different, *homo sapiens* would not exist. (Simpson 1964, p. 773.)

The working of evolution – be it of life or intelligence or culture or technology or science – is always the product of a great number of individually unlikely events. The unfolding of developments involves putting to nature a series of questions whose successive resolution produces a process reminiscent of the game "Twenty Questions," sweeping over a possibility-spectrum of awesomely large proportions. The result eventually reached lies along a route that traces out one particular contingent path within a space of alternatives that provides for an ever divergent fanning out of alternatives as each step opens up yet further possibilities. An evolutionary process is a very iffy proposition – a complex labyrinth where a great many twists and turns in the road must be taken aright for matters to end up as they do.

Of course, it all looks simple with the wisdom of hindsight. If things had not turned out appropriately at every stage, we would not be here to tell the tale. The many contingencies on the long route of cosmic, galactic, solar-systemic, biochemical, biological, social, cultural, and cognitive evolution have all turned out aright – the innumerable obstacles have all been surmounted. In retrospect it all looks easy and inevitable. The innumerable possibilities of variation along the way are easily kept out of sight and out of mind. The Whig interpretation of history beckons comfortably. It is so easy, so tempting, to say that a planet on which there is life will of course evolve a species with the technical capacity for interstellar communication. (Cf. Cameron in Cameron (1963, p. 312), who fixes this conditional probability as 1.) It is tempting, but it is also nonsense. There are simply too many critical turnings along the road of cosmic and biological evolution. The fact is that many junctures along the way are such that, had things gone only a little differently, we would not be here at all (Cf. Road and Trefil (1981), who develop this line of thought in cogent and plausible detail.)

The ancient Greek atomists' theory of possiblity affords an interesting object-lesson in this connection. Adopting a Euclideanly infinitistic view of space, they held to a theory of innumerable worlds:

> There are innumerable worlds, which differ in size. In some worlds there is no sun and moon, in others they are larger than in our world, and in others more numerous. The intervals between the worlds are unequal; in some parts there are more worlds, in others fewer; some are increasing, some at their height, some decreasing; in some parts they are arising, in others failing. They are destroyed by collision one with another. There are some worlds devoid of living creatures or plants or any moisture. (Diels-Kranz 68 A 40 [for Leu-

cippus and Democritus]; in Kirk and Raven 1957, p. 411)

On this basis the atomists taught that every (suitably general) possibility is realized in fact someplace or other. Confronting the question of "Why do dogs not have horns: just why is the theoretical possibility that dogs be horned not actually realized?" the atomists replied that it indeed is realized but just elsewhere – *in another region of space*. Somewhere within infinite space there is another world just like ours in every respect save one, that its dogs have horns. For the circumstance that dogs lack horns is simply a parochial idiosyncracy of the particular local world in which we interlocutors happen to find ourselves. Reality accommodates all possibilities of worlds alternative to this through spatial distribution: as the atomists saw it, *all* alternative possibilities are in fact actualized in the various subwords embraced within one spatially infinite superworld.

This theory of virtually open-ended possibilities was shut off by the closed cosmos of the Aristotelian world picture, which dominated European cosmological thought for almost two milennia. The break-up of the Aristotelian model in the Renaissance and its replacement by the "Newtonian" model is one of the great turning points of the intellectual tradition of the West – elegantly portrayed in Alexandre Koyre's splendidly entitled book, *From the Closed World to the Infinite Universe* (New York, 1957). One may recall Giordano Bruno's near-demonic delight with the explosion of the closed Aristotelian world through one opening into an infinite universe spread throughout endless spaces. Others were not delighted but appalled – John Donne spoke of "all cohearence lost," and Pascal was frightened by "the eternal silence of infinite spaces" of which he spoke so movingly in the *Pensee* (pp. 205-206). But no one doubted that the onset of the "Newtonian" world-picture represented a cataclysmic event in the development of Western thought. Strangely enough, the refinitization of the universe effected by Einstein's general relativity produced scarcely a ripple in philosophical or theological circles, despite the immense stir caused by other aspects of the Einsteinean revolution. (Einsteinian space-time is after all, even more radically finitistic than the Aristotelian world-picture, which left open at any rate the prospect of an infinite future with respect to time.)

To be sure, it might well seem that the finitude in question is not terribly significant because the distances and times involved in modern cosmology are so enormous. But this view is rather naive. The difference between the finite and the infinite is as big as differences can get to be. And it represents a difference that is – in this present context – of the most far-reaching significance. For this means that we have no alternative to supposing that a highly improbable set of eventuations is not going to be realized in very many places, and that something sufficiently improbable may well not be realized at all. The decisive *philosophical* importance of cosmic finitude lies in the fact that in a finite universe only a finite range of alternatives can be realized. A finite universe must "make up its mind" about its contents in a far more radical way than an infinite one. And this is particularly manifest in the context of low-probability possibilities. In a finite world – unlike an infinite one – we cannot avoid supposing that a prospect that is sufficiently unlikely is simply not going to be realized at all, that in piling improbability on improbability we eventually outrun the reach of the actual. It is, accordingly, quite conceivable that our science represents a solution of the problem of cognitive accomodation that is terrestrially locale-specific.

6. THE IMPLAUSIBILITY OF BEING OUTDISTANCED

Our science is bound to be limited in certain respects by the very fact of its being *our* science. A tiny creature living within a leaf for a life-span of weeks could never recognize that the leaf is deciduous – itself part of a cyclic process. The processes of this world of ours (even unto its utter disappearance) could make no cognitive impact upon a being in whose body our entire universe is but a single atom. No doubt the laws of our world are (part of) the laws of its world as well, but this circumstance is wholly without practical effect. Where causal processes do not move across the boundaries between worlds – where the levels of operative law are so remote that nothing happening at the one level makes any substantial impact upon the other – there can be little if any overlap in "science." Science is limited to the confines of discernibility: as Kant maintained, the limits of our science are the limits of our experience.

A deep question arises. Is the mission of intelligence in the cosmos uniform or diversified? Two fundamentally opposed philosophical positions are possible with respect to cognitive evolution in its cosmic perspective. The one is a uniformitarian *monism* which sees the universal mission of intelligence in terms of a certain shared destination, the attainment of a common cosmic "position of reason as such." The other is a particularistic *pluralism* which sees each intelligent cosmic civilization as forging its own characteristic cognitive destiny. It takes the mission of intelligence as such to span a wide spectrum of alternatives and to realize a vastly diversified variety of possibilities, with each thought-form achieving its own peculiar destiny in separation from all the rest. The conflict between these doctrines must in the final analysis be settled not by armchair speculation from general principles but by rational triangulation from the empirical data. This said, it must, however, be avowed that the whole tendency of these present deliberations is towards the pluralistic side. It seems altogether plausible to see cognition as an evolutionary product that is bound to attune its practitioners to the local peculiarities of their particular niche in the world order.

The developmental path from intelligence to science is a road strewn with substantial obstacles. Here matters must be propitious not just as regards the physics, chemistry, biochemistry, evolutionary biology, and cognitive psychology of the situation; the social-science requisites for the evolution of science as a cultural artifact must also be met. Economic conditions, social organization, and cultural orientation must all be properly adjusted before the move from intelligence to science can be accomplished. For scientific inquiry to evolve and flourish there must, in the first place, be cultural institutions whose development requires economic conditions and a favorable social organization. And terrestrial experience suggests that such conditions for the social evolution of a developed culture are by no means always present where intelligence is. At this point we do well to recall that of the thousands of human civilizations evolved here on earth only one, the Mediterranean/European, managed to develop natural science as we understand it. The successful transit from intelligence to science is certainly not a sure thing.

For many there is, no doubt, a certain charm to the idea of companionship. It would be comforting to think that however estranged we are in other ways, those alien minds and ourselves share *science* at any rate – that we are fellow travellers on a common journey of inquiry. Our yearning for companionship and contact runs deep. It would be pleasant to think ourselves not only colleagues but junior collaborators whom other, wiser minds might be able to help along the way. Even as many in 16th century Europe looked to those strange pure men of the Indies (East or West) who might serve as moral exemplars for sinful European man, so we are tempted to look to alien inquirers who surpass us in scientific wisdom and might assist us in overcoming our cognitive deficiencies. The idea is appealing, but it is also, alas, very unrealistic.

In the late 1600's Christiaan Huygens wrote:

> For 'tis a very ridiculous opinion that the common people have got among them, that it is impossible a rational Soul should dwell in any other shape than ours.... This can proceed from nothing but the Weakness, Ignorance, and Prejudice of Men; as well as the humane Figure being the handsomest and most excellent of all others, when indeed it's nothing but a being accustomed to that figure that makes me think so, and a conceit ... that no shape or colour can be so good as our own. (Huygens 1698, pp. 76-77)

What is here said about people's tendency to emplace all rational minds into a physical structure akin to their own familiar one is paralleled by a tendency to emplace all rational knowledge into a cognitive structure akin to their own familiar one.

It is clear that life on other worlds might be very different from the life we know. It might be based on a multivalent element other than carbon and be geared to a medium other than water – perhaps even one that is solid or gaseous rather than liquid. In his splendid book entitled *The Immense Journey* (New York, 1957) Loren Eiseley wrote:

> Life, even cellular life, may exist out yonder in the dark. But high or low in nature, it will not wear the shape of man. That shape is the evolutionary product of a strange, long wandering through the attics of the forest roof,

> and so great are the chances of failure, that nothing precisely and identically human is likely ever to come that way again. (Pp. 160-61)

What is here written regarding the material configuration of the human shape would seem no less applicable to the cognitive configuration of human thought – scientific thought nowise excluded. It is plausible to think that alien creatures will solve the problems of *intellectual* adjustment to their environment in ways as radically different from ours as those by which they solve the problems of physical adjustment. The physics of an alien civilization need resemble ours no more than does their physical therapy. We must be every bit as leary of *cognitive* anthropomorphism as of *structural* anthropomorphism – they are certainly not writ larger in the inevitable scheme of things. (Fred Hoyle's *The Black Cloud* is thought-provoking in this regard. The cloud tells a scientist what it knows about the world. The result is schizophrenia and untimely death for the scientist: the cloud's information is too divergent and compelling.)

With respect to biological evolution it seems perfectly sensible to reason as follows:

> What can we say about the forms of life evolving on these other worlds?...[I]t is clear that subsequent evolution by natural selection would lead to an immense variety of organisms; compared to them, all organisms on Earth, from molds to men, are very close relations. (Shklovskii and Sagan 1966, p. 350)

It is plausible that much the same situation should obtain with respect to cognitive evolution: that the "sciences" produced by different civilizations here on earth – the ancient Chinese, Indians, and Greeks for example – should exhibit an infinitely greater similarity than obtains between our present-day science and anything devised by astronomically remote alien civilizations. The idea of a comparison in terms of "advance" or "backwardness" would simply be inapplicable.

The upshot of these deliberations is thus relatively straightforward. The prospect that some astronomically remote civilization is "scientifically more advanced" than ourselves – that somebody else is doing "our sort of science" *better* than we ourselves – requires in the first instance that they be doing our sort of science at all. And this deeply anthropomorphic supposition is extremely unlikely. To endorse the idea of a scientifically superior extraterrestrial civilization is to step beyond the realm of scientifically plausible fact. The quest for superhuman science should bring home to us the uniqueness of that artifact which is our human science. It should be stressed, however, that this consideration that "our sort of natural science" may well be unique is not so much a celebration of our intelligence as a recognition of our peculiarity.

University of Pittsburgh

REFERENCES

This listing is confined to those relevant materials which I have found particularly interesting or useful. It does not aspire to comprehensiveness. A much fuller bibliography is given in MacGowan and Ordway (1966).

Allen, T. (1965) *The Quest: A Report on Extraterrestrial Life*. Philadelphia: Chilton Books. (An imaginative survey of the issues.)

Anderson, P. (1963) *Is There Life on Other Worlds?* New York and London: Collier-Macmillan. (Chapter 8 "On the Nature and Origin of Science" affords many perceptive observations.)

Ball, J. (1973) "The Zoo Hypothesis,' *Icarus*, 19, pp. 347-349. (Aliens are absent because the Intergalactic Council has designated Earth a nature reserve.)

_______ (1980) "Extraterrestrial Intelligence: Where Is Everybody?", *American Scientist*, 68, pp. 565-663.

Beck, L. (1971) "Extraterrestrial Intelligent Life," *Proceedings and Addresses of the American Philosophical Association*, 45, pp. 5-21. (A thoughtful and very learned discussion.)

Beadle, G. (1960) "The Place of Genetics in Modern Biology," *Engineering and Science*, 23, no. 6, pp. 11-17. (An insightful analysis of the evolutionary process.)

Berrill, N. (1964) *Worlds Without End*. London: Macmillan. (A popular treatment.)

Bracewell, R. (1975) *The Galactic Club: Intelligent Life in Outer Space*. San Francisco: W. H. Freeman. (A lively and enthusiastic survey of the issues.)

Cameron, A. ed. (1963) *Interstellar Communication: A Collection of Reprints and Original Contributions*. New York and Amsterdam: W. A. Benjamin. (A now somewhat dated but still indispensable collection.)

Dick, S. (1982) *Plurality of Worlds: The Origins of the Extraterrestial Life Debate from Democritus to Kant*. Cambridge: Cambridge University Press. (A lively and painstaking survey of the historical background.)

Dole, S. (1964) *Habitable Planets for Man*. New York: Blaisdell; 2nd ed., New York: American Elsevier, 1970. (A painstaking and sophisticated discussion.) A more popular version is S. H. Dole and Isaac Asimov, *Planets for Man* New York: Random House, 1964.

Drake, F. D. (1962) *Intelligent Life in Space* New York and London: Macmillan, 1962). (A clearly written, popular account.)

Eddington, A. (1928) *The Nature of the Physical World*. Cambridge: Cambridge University Press. (Chapter VIII on "Man's Place in the Universe" is a classic.)

Ehrensvard, G. (1965) *Man on Another World*. Chicago and London: University of Chicago Press. (See especially Chapter X on "Advanced Consiousness.")

Firsoff, V. (1963) *Life Beyond the Earth: A Study in Exobiology*. New York: Basic Books. (A detailed study of the biochemical possibilities for extraterrestrial life.)

Hart, M. (1975) "An Explanation for the Absence of Extraterrestrials on Earth," *Quarterly Journal of the Royal Astronomical Society*, 16, pp. 128-135. (A perceptive survey of this question.)

Herrmann, J. (1963) *Leben auf anderen Sternen*. Gutersloh: Bertelsmann Verlag. (A thoughtful and comprehensive survey with special focus on the astronomical issues.)

Hoerner, S. (1973) "Astronomical Aspects of Interstellar Communication," *Astronautica Acta*, 18, pp. 421-429. (A useful overview of key issues.)

Hoylke, F. (1965) *The Black Cloud*. New York. (A provocative science-fiction tale.)

Hoyle, F. (1966) *Of Men and Galaxies*. Seattle: University of Washington Press. (Speculations by one of the leading astrophysicists of the day.)

Huang, S. (1960) "Life Outside the Solar System," *Scientific American*, 202, no. 4, pp. 55-63. (A useful discussion of some of the astrophysical issues.)

Huygens, C. (1698) *Cosmotheoros: The Celestial Worlds Discovered – New Conjectures Concerning the Planetary Worlds, Their Inhabitants and Productions*. (London, 1698; reprinted by F. Cass & Co., Ltd., London, 1968). (A classic from another age.) Cf. (Dick 1982).

James, W. (1907) *Pragmatism*. New York.

Jeans, J. (1948) "Is There Life in Other Worlds?", A 1941 Royal Institution lecture reprinted in H. Shapley et. al. eds., *Readings in the Physical Sciences*. New York: Appleton-Century-Crafts, pp. 112-117. (A stimulating analysis.)

Kaplan, S. ed. (1971) *Extraterrestrial Civilization: Problems of Interstellar Communication*. Jerusalem: Israel Program for Scientific Translations. (A collection of Russian scientific papers that present interesting theoretical work.)

Kirk, G. S. and Raven, J. E. (1957) *The Presocratic Philosophers*. Cambridge: Cambridge University Press.

Kuhn, T. (1962),*The Structure of Scientific Revolutions*. Chicago: University of Chicago Press.

Lem, S. *Summa Technologiae* (1964) Krakow: Wyd. Lt.. (To judge from the ample account given in Kaplan, 1971, this book contains an extremely perceptive treatment of theoretical issues regarding extraterrestrial civilizations. I have not, however, been able to consult the book itself.)

MacGowan, R. and Ordway, F. III, (1966) *Intelligence in the Universe*. Englewood Cliffs: Prentice Hall. (A careful and informative survey of a wide range of relevant issues.)

MacGowan, R. (1962) "On the Possibilities of the Existence of Extraterrestrial Intelligence," in F. I. Ordway ed., *Advances in Space Science and Technology*. New York and London: Academic Press, 4, pp. 39-111.

Nozick, R. (1972) "R.S.V.P. – A Story," *Commentary*, 53, pp. 66-68.

Pucetti, R. (1969) *Persons*. New York: Herder and Herder. (A most stimulating philosophical treatment.)

Rood, R. and Trefil, J. (1981) *Are We Alone: The Possibility of Extraterrestrial Civilization*. New York: Scribners.

Sagan, C. (1973) *The Cosmic Connection*. New York: Doubleday. (A well written, popularly oriented account.)

________ (1980) *Cosmos*. New York: Random House. (A modern classic.)

Shapley, H. (1958) *Of Stars and Men*. Boston: Beacon Press. (See especially the chapter entitled "An Inquiry Concerning Other Worlds.")

Shklovskii, I. and Sagan, C. (1966) *Intelligent Life in the Universe*. San Francisco, London, Amsterdam: Holden-Day, Inc.. (A well-informed and provocative survey of the issues.)

Simmel, G. (1895) "Ueber eine Beziehung der Selektionslehre zur Erkenntnistheorie," *Archiv fur systematische Philosophie und Soziologie*, 1, pp. 34-45.

Simpson, G. (1964) "The Nonprevalence of Humanoids," *Science*. 143, pp. 769-775 = Chapter 13 of *This View of Life: The World of an Evolutionist*. New York: Harcourt Brace. (An insightful account of the contingencies of evolutionary development by a master of the subject.)

Strawson, P. F. (1959) *Individuals*. London: Meuthen.

Sullivan, W. (1964) *We Are Not Alone*. New York: McGraw Hill, revised ed. 1965. (A very well written survey of the historical background and of the scientific issues.)

Wilkinson, D. (1977) *The Quarks and Captain Ahab or: The Universe as Artifact*. Stanford, Schiff Memorial Lecture. (Ruminations on first principles by a master scientist.)

INDEX

L

M

N

THE UNIVERSITY OF WESTERN ONTARIO
SERIES IN PHILOSOPHY OF SCIENCE

A Series of Books in Philosophy of Science, Methodology, Epistemology, Logic, History of Science, and Related Fields

Managing Editor:

ROBERT E. BUTTS

Editorial Board:

J. BUB, L. J. COHEN, W. DEMOPOULOS, W. HARPER, J. HINTIKKA, C. A. HOOKER, H. E. KYBURG, Jr., A. MARRAS, J. MITTELSTRASS, J. M. NICHOLAS, G. A. PEARCE, B. C. VAN FRAASSEN

1. J. Leach, R. Butts, and G. Pearce (eds.), *Science, Decision and Value.* Proceedings of the Fifth University of Western Ontario Philosophy Colloquium, 1969. 1973, vii+213 pp.
2. C. A. Hooker (ed.), *Contemporary Research in the Foundations and Philosophy of Quantum Theory.* Proceedings of a Conference held at the University of Western Ontario, London, Canada, 1973, xx+385 pp.
3. J. Bub, *The Interpretation of Quantum Mechanics.* 1974, ix+155 pp.
4. D. Hockney, W. Harper, and B. Freed (eds.), *Contemporary Research in Philosophical Logic and Linguistic Semantics.* Proceedings of a Conference held at the University of Western Ontario, London, Canada. 1975, vii+332 pp.
5. C. A. Hooker (ed.), *The Logico-Algebraic Approach to Quantum Mechanics.* 1975, xv+607 pp.
6. W. L. Harper and C. A. Hooker (eds.), *Foundations of Probability Theory, Statistical Inference, and Statistical Theories of Science*, 3 Volumes. Vol. I: *Foundations and Philosophy of Epistemic Applications of Probability Theory.* 1976, xi+308 pp. Vol. II: *Foundations and Philosophy of Statistical Inference.* 1976, xi+455 pp. Vol. III: *Foundations and Philosophy of Statistical Theories in the Physical Sciences.* 1976, xii+241 pp.
7. C. A. Hooker (ed.), *Physical Theory as Logico-Operational Structure*. 1979, xvii+334 pp.

8. J. M. Nicholas (ed.), *Images, Perception, and Knowledge.* Papers deriving from and related to the Philosophy of Science Workshop at Ontario, Canada, May 1974. 1977, ix+309 pp.
9. R. E. Butts and J. Hintikka (eds.), *Logic, Foundations of Mathematics, and Computability Theory.* Part One of the Proceedings of the Fifth International Congress of Logic, Methodology and Philosophy of Science, London, Ontario, Canada, 1975. 1977, x+406 pp.
10. R. E. Butts and J. Hintikka (eds.), *Foundational Problems in the Special Sciences.* Part Two of the Proceedings of the Fifth International Congress of Logic, Methodology and Philosophy of Science, London, Ontario, Canada, 1975. 1977, x+427 pp.
11. R. E. Butts and J. Hintikka (eds.), *Basic Problems in Methodology and Linguistics.* Part Three of the Proceedings of the Fifth International Congress of Logic, Methodology and Philosophy of Science, London, Ontario, Canada, 1975. 1977, x+321 pp.
12. R. E. Butts and J. Hintikka (eds.), *Historical and Philosophical Dimensions of Logic, Methodology and Philosophy of Science.* Part Four of the Proceedings of the Fifth International Congress of Logic, Methodology and Philosophy of Science, London, Ontario, Canada, 1975. 1977, x+336 pp.
13. C. A. Hooker (ed.), *Foundations and Applications of Decision Theory,* 2 volumes. Vol. I: *Theoretical Foundations.* 1978, xxiii+442 pp. Vol. II: *Epistemic and Social Applications.* 1978, xxiii+206 pp.
14. R. E. Butts and J. C. Pitt (eds.), *New Perspectives on Galileo.* Papers deriving from and related to a workshop on Galileo held at Virginia Polutechnic Institute and State University, 1975. 1978, xvi+262 pp.
15. W. L. Harper, R. Stalnaker, and G. Pearce (eds.), *Ifs. Conditionals, Belief, Decision, Chance, and Time.* 1980, ix+345 pp.
16. J. C. Pitt (ed.), *Philosophy in Economics.* Papers deriving from and related to a workshop on Testability and Explanation in Economics held at Virginia Poly-Technic Institute and State University, 1979. 1981.
17. Michael Ruse, *Is Science Sexist?* 1981, xix+299 pp.
18. Nicholas Rescher, *Leibniz's Metaphysics of Nature.* 1981, xiv+126 pp.
19. Larry Laudan, *Science and Hypothesis.* 1981, x+258 pp.
20. William R. Shea, *Nature Mathematized.* Papers deriving from the third international Conference on the history and philosophy of science, Montreal, Canada, 1981. Vol. I, 1983.
21. Michael Ruse, *Nature Animated.* Papers deriving from the third international Conference on the history and philosophy of science, Montreal, Canada, 1981. Vol. II, 1983.
22. William R. Shea (ed.), *Otto Hahn and the Rise of Nuclear Physics,* 1983, x+252 pp.
23. H. F. Cohen, *Quantifying Music,* 1984, xvii+308 pp.
24. Robert E. Butts, *Kant and the Double Government Methodology,* 1984, xvi+339 pp.
25. James Robert Brown (ed.), *Scientific Rationality: The Sociological Turn.* 1984, viii+330 pp.
26. Fred Wilson, *Explanation, Causation and Deduction.* 1985, forthcoming.